MIND THE GAP

The Labyrinthine Story of Planetary Orbits,
Mathematics, and the Titius-Bode Rule

MIND THE GAP

The Labyrinthine Story of Planetary Orbits, Mathematics, and the Titius-Bode Rule

Martin Beech
University of Regina, Canada

World Scientific

NEW JERSEY • LONDON • SINGAPORE • BEIJING • SHANGHAI • HONG KONG • TAIPEI • CHENNAI • TOKYO

Published by

World Scientific Publishing Co. Pte. Ltd.
5 Toh Tuck Link, Singapore 596224
USA office: 27 Warren Street, Suite 401-402, Hackensack, NJ 07601
UK office: 57 Shelton Street, Covent Garden, London WC2H 9HE

Library of Congress Control Number: 2023932606

British Library Cataloguing-in-Publication Data
A catalogue record for this book is available from the British Library.

MIND THE GAP
The Labyrinthine Story of Planetary Orbits, Mathematics, and the Titius-Bode Rule

ISBN 978-981-127-397-1 (hardcover)
ISBN 978-981-127-640-8 (paperback)
ISBN 978-981-127-398-8 (ebook for institutions)
ISBN 978-981-127-399-5 (ebook for individuals)

For any available supplementary material, please visit
https://www.worldscientific.com/worldscibooks/10.1142/13343#t=suppl

Typeset by Stallion Press
Email: enquiries@stallionpress.com

Contents

Chapter 1

Introducing the Labyrinth

1.1. Introduction

While this is a book about the intertwined history of mathematics and astronomy, it is essentially a book about origins, ideas, expectations, pattern-searching, intellectual triumphs, impositions, and delusion. Here, indeed, is a maze to traverse — a multistranded interweaving of egos and geniuses, logic and guesswork. Here, also, is a selected history of human thought and insight, all mixed and spun together, in a complex story that has rung down the centuries to inform and support the modern-day edifice that is called science.

The gaps referred to in our title are not specifically gaps in our collective knowledge, although that is a large part of the narrative. Indeed, the gaps correspond to both physical and mental spaces, voids if you will, where inspiration can be found, and where the imagination, if allowed, can run riot. Some gaps are trivial and narrow, others are cavernous but tractable, and some gaps will never be bridged, intellectually or otherwise, ever. The astronomical gap that is of primary interest in this book is both physical and intellectual, and it concerns planetary spacing. That is, it concerns the gaps that exist between the planets in our solar system. Indeed, the story pertaining to planetary gaps has a very long history, and it is still a vibrant topic of research to this day. The physical gaps may have been bridged in the modern era by spacecraft and measured by geometry, but an understanding of why the specific gaps exist remains unsolved.

Figure 1.1. The two-headed, lion-goat chimera of ancient Greek mythology. Image shows a chimera on a red-figure Apulian plate, c. 350–340 BCE. Louvre, Paris.

Gaps imply discord and they indicate places where expectation and observation are in tension. Once identified, however, the quickened question concerning a gap is: how do we close or explain its existence? Importantly, the recognition that something might be missing, making for a noticed or pregnant gap, can be a great philosophical motivator, and such recognition, along with its associated intellectual action, has proved vital to the mensuration of planetary orbits, and the understanding of the structure behind the solar system. Concomitant to this, observing the planets and measuring the gaps between their orbits has proved vital to the history and development of mathematics. Gaps are important — they tell us where both our understanding begins and where it ends, and they indicate where new insights might be found, and their solution illuminates the genius of the human intellect.

Gaps generate knowledge, they spark investigation, and they require solutions. These solutions are sometimes hard-won, requiring multiple lifetimes to resolve, and sometimes they appear spontaneously, out of the depths, when least expected: in a dream, on a bus, in the bath, or out walking the dog. In the latter spontaneous cases, all of a sudden, there it is, the solution or key step required to resolve some problem[1] inspired by a knowledge gap. Indeed, French mathematician Henri Poincaré tells us that, disgusted with his inability to solve a problem, he took a holiday at the seaside, and that after several days, while on a clifftop walk, "the idea came to me, ... , quadratic forms were identical with those of non-Euclidian geometry". The solution to the problem, Poincaré further informs us, came with, "brevity, suddenness and immediate certainty". Here, at play, is the marvel of the human mind, with the brain and its synaptic gaps firing in glorious union to ponder, grasp, and solve.

Even when we are not consciously thinking about a problem, the mind seems to keep it in subconscious play: trawling the depths of experience, quietly probing, and turning over the possibilities. The subconscious is like the proverbial fisherman casting a net into a seemingly bottomless ocean of ideas and insight — hoping for a catch. Most often nothing very exciting is caught, but sometimes, as in a great epiphany, lo! out of the depths swims a beautiful idea. Sometimes the emerging idea is the quarry sought, its metaphorical scales gleaming with gold and silver; but also, equally as often, it is a monster that emerges — a monster from the abyssal depths, set upon spawning confusion. Some ideas, honestly caught (perhaps) in the beginning, have turned out to be disastrous for humanity, while others have been more benign. These latter monsters are perhaps more chimeras than demons — monsters dressed as friendly prophets (figure 1.1). Astronomy, as with the other mathematical sciences, abounds with many such chimeras — indeed, they are a vital part of learning and the (so-called) scientific method. One must push and pull at an idea, teasing out its possibilities and predictions: Can it

---

[1] Pausing to cross the road in London, England, has resulted in at least two extreme examples of this kind of epiphany. One is related by Nobel laureate Roger Penrose, who recounts how during a walk by Birkbeck College he (subconsciously) solved the problem of how an object might smoothly collapse to form a black hole. The other extreme case relates to Leo Szilard, who, on crossing the road next to the British Museum, realized how to initiate a neutron-induced chain reaction and thereby create an atomic bomb.

solve this, or can it solve that, or, perhaps more importantly, does it synthesize theory and observations, making for a coherent body of understanding and experience capable of successful forward-looking predictions? But some of the chimerical monsters that swim into our intellectual ken, are just that, chimeras — fantasy ideas that have no founding in physical reality. They delude us, and while they may initially seem intriguing and even profitable, they generally tend to lead us into moribund seas where sensible navigation is no longer possible.

Surrounded, as we are, by the everyday world of inane advertising blather, the unexpected, dynamical chaos, and stochastic mayhem, we (as human beings) have developed the entirely unreasonable expectation of finding order, stability, and understanding in the universe. Indeed, it has been apparent for many an age that there is order, at least of a sort, to be found in the heavens and the workings of nature. Some of the observed order is well explained by science, having long been put to the test of experimentation and the interpretation of mathematical and physical theories. Some of the observed order, however, has long-defied explanations, and this begs the following wanting question: "Are we attempting to find order and numeration where none actually exists?" — or, more to the theme of this book: Are we slavishly trying to shore-up creaky-buildings that have been constructed on the endlessly flexible foundations of mathematics? Since at least the time of Kepler, now some 400 years ago, it has been tacitly assumed that if some phenomenon is observed in the heavens, then its behavior can be caught within the machinations of mathematics and then explicitly written down on paper. Perhaps for the past 150 years, and predominantly so in modern times, the reverse situation has grown to predominance: if it can be described mathematically on paper, then it must exist astronomically in the cosmos. One could say that mathematics and astronomy are a match made in heaven, but the question remains: Is the match true? To this question, there is a spectrum of possible answers. At one end of the spectrum, in his complex, but highly insightful book, *The Philosophy of Physical Sciences* (published in 1939), Arthur Eddington [1] argued that while theoretical physics is highly mathematical, "the mathematics is not there till we put it there". It hardly goes without saying that Eddington's arch rival, James Jeans [2] held an opinion from the other end of the spectrum, writing in his *The Mysterious Universe* (published in 1930) that, "the Great Architect of the Universe [is] a pure mathematician". Moving even further away from Eddington, and even Jeans, in an infinity-magnifying, but highly insightful book, *Our Mathematical Universe* (published in 2014), Max Tegmark [3] asserts that mathematics is everything, or more specifically, everything is a mathematical structure. Neither Eddington on the one side of the spectrum, nor Jeans, and Tegmark on the other, are likely correct in their assertions, but it is probably safe to say that the truth lies somewhere between the two extremes — or, as is perhaps even more likely, the truth resides somewhere entirely different, in a reality, as yet, unexplored or experienced by the human mind.

All aspects of the order, both the understood and the entirely mysterious, reside within the tumbling planetary dance played out in the solar system. Indeed, with respect to the planets, a great deal is known and fully resolved concerning their structure: we know their number, their size, their surface features, and their compositions. Equally, however, the currently observed spacing between the orbits of the planets, and how they formed and ultimately how they attained their observed placings, remains, on multiple levels, a profound mystery. It is here, however, that we find archipelagos of apparent order emerging from the greater ocean of randomness. Here, for example, resides the island mystery that is the Titius–Bode law (the topic of Chapter 9), a law that is in fact a rule (there are important differences between the two[2]) which seeks to explain the spacing of the planets in terms of a pure number sequence. This is an incredible state of affairs, and are we really to believe that the shifting and heaving of an infinitude of random gravitational encounters, perturbations, collisions, and myriad knocks, clunks, and break-ups, along with the multifold mergers that have been responsible for the production of the planets, are, when all is said and done, described by the integer number sequence beginning 0, 3, 6, 12, 24, 48, 96,

[2] A law implies that the result is based upon some sound physical theory. A rule is essentially a counting exercise with no physical foundation.

192, ...? It seems too easy, it seems too simple, and yet this basic arithmetic sequence of numbers, doubling on every successive step, has been responsible, in one way or another, for inspiring several important discoveries with respect to the contents and structure of our solar system. It is also a chimera.

Perhaps we shouldn't be surprised at the seeming success of the Titius–Bode rule; after all, our daily lives, largely lived in an orderly fashion, are really governed by the tumbling of random encounters, and the manic machinations of society and social-media-run computer algorithms. But then, and here is the rub, as human beings we rebel against the very predictive power and dizzying speed of the computer's logic, and we fight with all our innermost zeal against being placed within the straightjacket of impersonalized numbers and algorithmic compliance. Why, might we then ask, should nature, the great non-conformer, be any different in practice than us? The universe might well be described by mathematics, but the universe is certainly not a mathematician, and nor is it a sequential (even quantum) computer. We live in this contradiction: rebelling against being numbered ourselves, but pushing headlong to constrain the universe in the inescapable noose of algebra. The human notion of *expecting* the heavens to fit the shackles of simple (even complicated) numbers is nearly countless years old, and in spite of all the objections that can be made about the situation, it has proved to be a remarkably fecund idea. While Eugene Wigner (in 1959) famously questioned the unreasonable ability of mathematics to describe the universe, what is really unreasonable is our assumption that the universe is constructed according to laws that are accessible to mathematical expression.

The Titius–Bode rule, in one form or another, has been in place since the early 18th century, and it is a standard component of almost every astronomy textbook that has been published on the solar system during the past century. Again, this is a remarkable circumstance given that it has no clear and/or direct physical interpretation. Recently Titius–Bode-like rules have been applied, by numerous authors, to the domain of exoplanetary systems, and it appears that our solar system is not entirely unique. Nature does indeed like to construct planets and arrange their orbits according to straightforward near numerical sequences — although, so far, it appears that the number sequences are system unique, and the use of the word *near* is also fundamental. There is no universal Titius–Bode rule that describes all planetary systems; rather, in the mold of some Platonic ideal, there are number sequences, with an unbounded range of possible input parameters and multiplying terms, which successfully ape the highly complex processes by which nature builds planets. This realization alone should make us cautious of placing any real value in Titius–Bode-like rules. Italian physicist Enrico Fermi is credited with once arguing that "Give me three or more parameters and I can fit an elephant", such *fitting*, least-squares or otherwise, may indeed produce the perfect profile of an elephant, but it certainly does not make for a living, walking, pooping, and trumpeting elephant. The simplicity of the Titius–Bode law is likewise a fabrication; it is a glittering façade that hides the ramshackle workhouse that is the planet-building process. Yes, the Titius–Bode rule is a sham, *argumentum ad ignorantiam*, but importantly, and here is the philosophical intrigue, it has been a highly useful and important sham. As with myriad, every day, rules-of-thumb, the Titius–Bode rule sets bounds upon our expectations and it provides an envelope of possibilities. It does not, however, tell us how nature makes planets and/or arranges planetary orbits — at best, it provides some form of an executive overview of the salient end result. This is where the apparently predictive power of Titius–Bode-like rules reside — they codify and compress (in some obscure fashion) the emergent order that results from the multitude of randomly acting, yet finite and physically constrained interactions which govern the construction of planetary systems. Many academic research papers have been written about the Titius–Bode rule, and at least one book. We shall encounter some of these works within the following pages, but it is the book by physicist Michael Nieto [4] that is worth trying to find if you can. Indeed, Nieto's book has long been out of print, and while we do cover some of the same material (and much more, I should point out), his book covers more physical theory than will be encountered here. There are several reasons for this departure, but the main one is that none of the theories explored by Nieto actually provide a proof of, or indeed a convincing explanation for, the existence of the

Titius–Bode rule. The reason for this, as will be revealed, is that there is no one physical theory behind the emergence of the Titius–Bode rule — rather, it is the result of numerous competing factors, random interactions, and the nimbleness of numbers. The Titius–Bode rule is like finding the seeming center of a labyrinth, only to realize, on further investigation, that the labyrinth has no center.

Order and stability are key beginning and end points in the progress of science. As humans, we crave order, and we like simple rules — they make for a happy life. Nature, however, operates via complex rules in complex ways, but it too, at least on occasion, *likes* to build ordered systems. At an anthropic level, if planets never formed in an inherently orderly fashion, having orbits that remain stable for very long intervals of time, in the right place, of the right size, about the right mass parent star, then we, as human beings, would not exist to argue the point. Order and stability, however, are never permanently achieved in any planetary system, but during those moments of dynamical stasis an observer, either internally or externally placed with respect to the system, will likely find an arrangement that suggests design and demonstrates internal numerical harmony. The appearance of the design, however, does not mean that an actual designer has been at play. The natural world around us, the ultimate example of apparent design and fitness, as Richard Dawkins has beautifully explained [5], is very much the product of a blind watchmaker — the artisan in this case being the subtle (but often brutal) machinations of Darwinian evolution. For all this, the discovery of near and approximate numerical harmonious within the makeup of our solar system has driven much of the history of astronomy, and the Titius–Bode rule is by no means the only fortuitous numerical sequence that has inspired successful prediction. Furthermore, it is by no means the only situation in which the urge to find some simple mathematical explanation for the workings of the universe has led astronomers astray. It is within the edifice of mathematics, solid, strong, regular, and resolute, that the modern scientist sits, and it is from this hallowed structure that attempts to grind out the equations that underpin the workings of the universe are made. The task is a hard one, and the view from the scriptorium is only partial. We are, for so it seems, forever left with some uncertainty, just as surrealist artist Remedios Varo intones in her painting *Celestial Pablum*, to ponder if the real workings of the cosmos have truly been captured, condensed, and described by the equations so carefully derived.

Figure 1.2. *Papilla Estelar* (*Celestial Pablum*) by Remedios Varo. Oil on board, 92 × 62 cm (1958). Image courtesy of Museo FEMSA, Ciudad de México.

Science and scientific fields grow and flourish over time — they shift and they evolve. According to the observational data that is available, ideas and theories meld and develop. Astronomers presently have a very detailed and well-nuanced grasp on how planetary systems form. The data provides a clear indication of the starting conditions (accretion disks around newly forming stars), and the end results (exoplanetary systems), but as always, the devil is in the details. It is not possible to follow, with any certainty and/or continuity, the tortuous path that leads to planetary system formation. It remains true that we do not know exactly, step by step, blow by blow how our solar system came about. We know the end result well, and astronomers can now describe the planets, their moons, and some aspects of their orbital evolution in exquisite detail, but we will never (ever) know the full-formation story — this story is indeed, a gap too far. Never is always a problematic word to use — especially so in the sciences: the lesson of Auguste Comte's claim, in 1835, that we will never know the composition of the stars, reminds us of this fact. Likewise, William Thomson's (Lord Kelvin) brash claim, in 1900, that all of physics was understood and that no new discoveries were likely to be made — a claim shattered by the emergence of quantum mechanics and relativity theory — reminds us that predicting the future is a game that even angels should fear to play. For all this, it is the case that we will never (ever) know in exact detail how the planets in the solar system grew and evolved onto their present orbits. Certainly, the basic concepts and the associated physical mechanisms are known, and these will no doubt improve over time, but an exact path, from beginning to end, will forever escape enumeration. There will forever be gaps in our knowledge about the solar system. These gaps will endure, not because of a lack of detailed observation knowledge, or because of a lack of computational power to solve the evermore complicated sets of equations, but because the process of formation is messy, complex, chaotic, and entirely random at heart, and this messiness is beyond the reach of even mathematics to tidy away. This is not to say that mathematics won't take us a very long way into the story of the solar system.

Along with the astronomy of planetary spacings, the twin theme to be developed in this book is concerning the role of mathematics in cosmogonical history. It remains a profound mystery (and a point of great contention) as to why mathematics does such a good job at describing, and additionally predicting, the behavior of the world around us. There is no obvious reason for the miraculous reach of mathematics, and (I am sorry to say) no explanation for this ability will be provided here — for indeed, no one knows why. This being said, the story that is the astronomy of the gaps brushes up against, and is invigorated by, aspects of the strangeness of number theory. Indeed, the world of numbers is a truly marvelous realm, but is it really a true analog or template for the *real* world — the real world in which we live and laugh? The world of numbers and number theory is full of surprises, amusements, and great excitement — it is an abundant, overflowing world, but is it anything more than a system of logic and symbols, cleverly conceived and cultivated by the human mind? The answer to this riddle is partly writ large in the sky, and it can be glimpsed amid the gaps of planetary orbits.

As a final comment in this introduction, I should make clear that this book is not concerned with the theological issue of the *God of the Gaps*. This phrase, coined by applied mathematician Charles Coulson (chemist, Methodist Church leader, and Rouse Ball Professor of Mathematics at Oxford University), was introduced in the mid-1950s to counteract the notion that the handiwork of God might only be found in those places where scientific explanations (apparently) fail [6]. The theological danger in adopting such a viewpoint, of course, is that the continued advancements in scientific research and understanding might eventually squeeze God in the gaps out of all existence. There will always be gaps in our knowledge, and God may, or may not, reside in some of them. Here, we make no invocations of a lurking deity and solely address our gaze toward those gaps in the solar system (both in a mental and physical sense) that can be closed, and those gaps that will never be bridged — ever — by the combined might of mathematics and human imagination.

References

[1] Eddington, A. *The Philosophy of Physical Science*. CUP, Cambridge, 1939.
[2] Jeans, J. *The Mysterious Universe*. CUP, Cambridge, 1930.
[3] Tegmark, M. *Our Mathematical Universe*. Knopf, New York, 2014.
[4] Nieto, M. M. *The Titius-Bode Law of Planetary Distances: Its History and Theory*. Pergamon Press, Oxford, 1972.
[5] Dawkins, R. *The Blind Watchmaker*. Longman Scientific and Technical, Harlow, 1986.
[6] Coulson, C. A. *Science and Christian Belief*. University of North Carolina Press, Chapel Hill, 1955.

Chapter 2

Dealing with Pythagoras

2.1. The Mystery of Numbers

Counting and astronomy go way back — indeed, it could be said that they grew up together; the one stretching and nurturing the growth of the other. The ancient Greek Philosophers made much play of numbers and astronomy. Indeed, so moved were they by numerological doctrine that the Pythagorean philosophers invented an entirely new planet, the counter Earth, in order that the number of heavenly bodies would add up to the *Dekad*, the sum of the four rows pertaining to the tetractys. The tetractys, or mystic triad, symbolized the four elements (earth, water, air, and fire) and the organization of the dimensionality of space (from the point, to the line, to the plane, and then to the solids). For the Pythagoreans, it was the power of numbers, mystic or otherwise, that provided access to knowledge, and indeed, they were prepared to meld, invent, and re-invent the heavens so that it conformed to the dictate of numbers.

For all these philosophical foundations, the Pythagorean universe was very simple in form and not, as such, enumerated in scale. The cosmos, as described by Philolaus (who lived circa 470 to 385 BC) was finite, and enclosed within a fixed shell of stars which was situated just beyond the region occupied by planet Saturn. No specific mention is made of what might reside outside of the shell of stars, if anything, but it was predicated upon a central fire (Figure 2.1). The central fire is not the Sun, which was

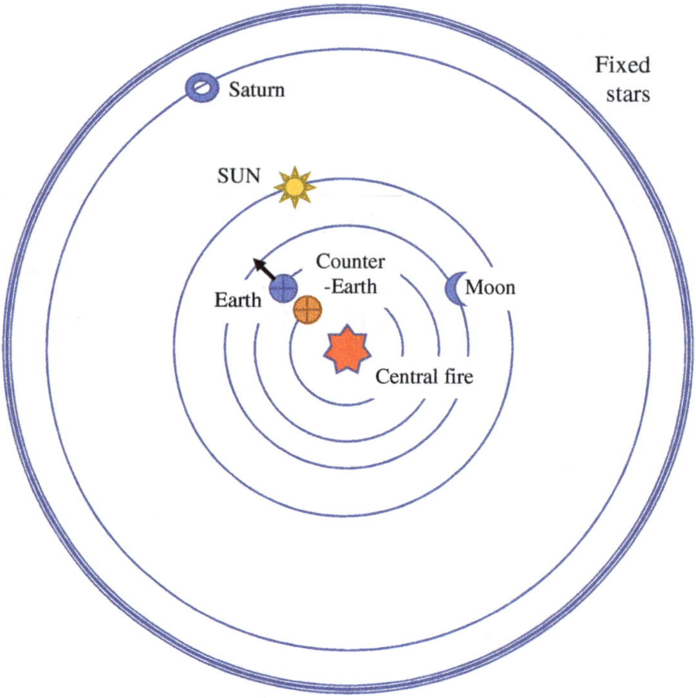

Figure 2.1. The Pythagorean cosmos. The planets orbit the central fire, and the Earth spins on its axis so that as it orbits the central fire (once per day) it retains a fixed radial orientation (as indicated by the arrow). The orbits of Mercury through to Jupiter (beyond the Sun and interior to Saturn) are not shown in the diagram for clarity.

considered to be a separate planet, beyond the orbit of the Moon, but it is rather the origin and essence of dynamics and motion in the cosmos — it is the central fire that drives the planets around in their orbit — as such it is not to be thought of as an illuminating fire, rather it is an animating and ordering spirit. In the cosmology of Philolaus, Earth is actually the second planet out from the center, and it orbits the central fire once a day with synchronous spin. In this latter manner, an observer on the Earth's surface retains a fixed orientation with respect to the central fire. Beyond the Earth are the orbits of the Moon and the Sun. The Sun takes one year to complete one orbit about the central fire, while the

Moon takes 29.5 days. Beyond the Sun are the planets Mercury through to Saturn.

The cosmos described by Philolaus is based upon the geometry of the sphere and the circle but contains many features that are purely ideology-based. The first such entity is the central fire, and the second is the introduction of a wholly new planet interior to the orbit of the Earth. This additional planet was needed under the guidance of the tetractys and *dekad*. With the introduction of a new planet, the so-called counter-Earth, the number of planets, from the counter-Earth, the Earth, the Moon, the Sun, Mercury through to Saturn, and the outer shell of stars, all added up to ten structures. The counter Earth was additionally introduced so as to restrict the view of the central fire from the Earth. It was also responsible for producing lunar and solar eclipses — although under the guise of the model these could only occur at either sunrise or sunset. The Pythagorean cosmological model did not, indeed, could not, account for solar or lunar eclipses in general, nor for the retrograde motion of the superior planets Mars through Saturn. For all this, it was structured in parallel with and according to the numerical relationships that describe the musical scale. Indeed, the Pythagoreans introduced the idea of *musica universalis*, in which the spacing of planetary realms was described in accordance with musical intervals. Here, the Pythagoreans projected the universality of numbers onto the heavens and ordered the system in terms of ratios as revealed by the harmonies of vibrating strings. Indeed, the Pythagoreans knew from experimentation that the pitch of a plucked string was related to the inverse of its length — the highest note is that sounded by the shortest string, the lowest note by the longest. Additionally, the Pythagoreans discovered three consonances, the octave, the 5th, and the 4th, with the string lengths associated with these being in the ratios 1:2, 2:3, and 3:4, respectively. These ratios struck a chord (bad pun intended) with the Pythagoreans because they were ratios constructed from the numbers 1, 2, 3, and 4, which sum to 10 (the *dekad*), and they are the numbers which form the four rows of the tetractys. In the Pythagorean cosmos, it was envisioned that each planet moved at a constant speed within its specific circular zone and emitted a fixed tone. The actual musical tones and ratios for the planetary spacings in the Philolaus cosmos are not actually known. The idea of using rational fractions to describe the spacings between the planets was later discussed, in a somewhat revised form by, Plato (see Chapter 3). Indeed, the idea of the music of the heavens was something that later scholars made much more play of than the ancient Greek philosophers.

2.2. The Burden of Proof

The Pythagoreans worked with integers and rational fractions, and it was taken that numbers were the expression of two essential phenomena of the real world — that is extension and measure. Within numbers, the Pythagoreans believed, lay a form of universal truth, and indeed, by investigating numbers universal truths could be discovered. In many ways, mathematicians, to this very day, operate under these same principles. Indeed, given the *treasures* that lie within the manipulation of numbers, modern mathematicians do something that most of us never concern ourselves with — they bother about all the nitty-picky details. Not just the small ones, but all the infinitely small and infinitely large ones. Accordingly, they are forced to deal with concepts, problems, and paradoxes that a greater number of human beings will never encounter in their entire lives. We do not need mathematics to survive in this world — it has zero evolutionary advantage, but it does, importantly, afford a very flexible and fecund language capable of explaining some aspects of the world around us. To the mathematician, every small detail is not just important, it is fundamental. To achieve mastery over all these infinite details, and to safely coral every logical possibility of every choice and every procedure, the modern mathematician toils long and hard, burning many midnight candles, to find some subtly crafted proof of a statement or theorem. Theorems are typically cast in the following general form: "the statements X_i, $i \in N^*$, always hold true under circumstances Y_j, $j \in N^*$, provided conditions Z_k,

$k \in N^*$ are in place".[1] The proof of the theorem then makes precise, non-arbitrary, statements about where, when, and how the various statements X_i, can be taken as indicating some specific and unchanging truth.[2] Writing in his insightful book *A Mathematician's Lament*, Paul Lockhart argues that, "a good proof should be an epiphany from the gods, not a coded message from the Pentagon". He is right, of course, but the ugliness of logical rigor more often than not gets in the way of divine revelation. In general However, the idea is that, in principle, anyone following through the steps in a proof will be fully convinced that it, and the theorem that it relates to, are correct.

Proofs of mathematical statements, or theorems, are not only remarkable for what they can establish, but they are remarkable for the fact that once established (provided the proof is correct) they are true for all eternity — no if, and, or buts. It is an absolute statement about the correctness of the theorem within what I shall call for the moment the mathematical universe. Given the unchanging, never-ending, certainty of mathematical proofs, however, and upon the basis of everyday life and experience, one is immediately prompted to ask, "What universe is it that the mathematicians and their proofs are actually working in?" because, assuredly, it is not the real world of human experience, where supposed absolutes and inalienable rights change all the time, and where there is no long-term certainty about almost anything that society or nature actually builds. The mathematician's universe is assuredly not the universe in which we

[1] As with all languages, mathematics uses symbols to describe specific objects and operations. A *set*, in mathematics, has a very specific meaning, and it is taken to be a well-defined collection of distinct objects, possibly an infinite number of objects, that can be considered an object in its own right. An element in a set is identified through the $\in$ symbol, so, for example, the statement $i \in N^*$ is taken to mean that i is an element of the set of positive integers (or natural numbers excluding zero) $N^* = \{1, 2, 3, \ldots\}$, where the curly brackets indicate a set off objects is being considered.

[2] While we all think we know the meaning of the word truth, it is, in fact, a very difficult and slippery word (quality) to define. Mathematicians take the idea of truth to be something provable, under the manipulation of symbolic logic, in some formal axiomatic system, although this is not the whole story.

reside, but it does provide a useful symbolic, simulacra-representation of our universe. Indeed, mathematics informs us about the universe in which we live, but it does not make our universe. Mathematics, for all its tortuous proofs, is a human invention that provides for a language that avoids the pitfalls of arbitrary interpretations and time-shifting meanings inherent to the spoken word. It is for this reason that the mathematician has to be so ultra-picky about every statement that they make, every axiom that they adopt, every action that they invoke, and every symbolic manipulation that they perform. For all this, mathematics and mathematicians do not work in a boundless universe — there are limits to what mathematics can describe, and perhaps more importantly for our theme, there are also consequences (some highly unwelcome) that follow from the mathematics that can be done. This was something that, for all their mathematical underpinning, the Pythagorean scholars found out to their great dismay.

The classic example of a mathematical theorem and its proof, which many readers will probably be familiar with, is that relating to Pythagoras's Theorem (**PT**). This theorem can be stated in a number of (equivalent) ways, but let us use the following.[3]

Statement of PT:

If a right-angled triangle has hypotenuse c and sides a and b, then $a^2 + b^2 = c^2$.

The proof of **PT** has been established many, many times over in many different ways — a result that begs the following question: Why? Surely one proof is enough? But for all this, Figure 2.2 provides the outline of a straightforward proof.

[3] There is always some unpacking to do in any set of mathematical statements. First, we have to appreciate what a triangle is and then know about a special subgroup of triangles in which one angle is exactly 90°. Second, we additionally need to know some terminology, and that in a right-angled triangle, the longest side of the triangle (subtended between the two sides that are at 90° to each other) is called the hypotenuse. Third, when it is said that a right-angle triangle "has hypotenuse a", this means that the length of the hypotenuse is a in some (usually unspecified) unit of measure.

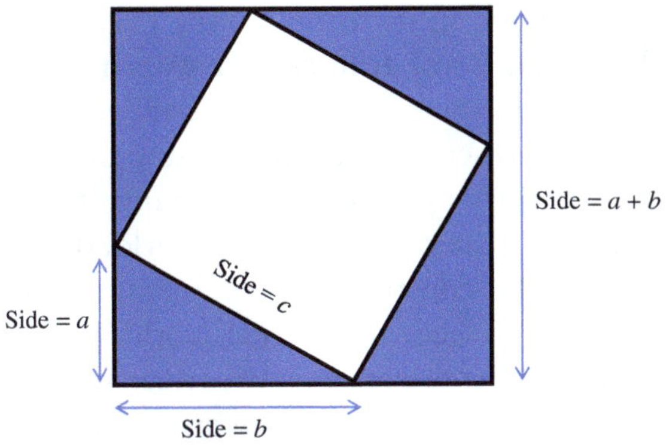

Figure 2.2. Proof outline for Pythagoras's Theorem. Here, the right-angled triangle T_1 has been drawn four times to make a square S_1 of side $(a + b)$ and area $(a + b)^2$. The central square S_2 has sides c and area c^2.

Proof of PT: Take the initial right-angled triangle (call it T_1) and with three identical triangles construct a square of side $(a + b)$ — call this S_1. Interior to this square is a second square of side c and area c^2 — call this square S_2. We can now proceed algebraically and look at the area relationships. Accordingly,

$$\text{Area}\ (S_1) = \text{Area}\ (S_2) + 4 \times \text{Area}\ (T_1) \tag{2.1}$$

The area of a triangle $T_1 = \frac{1}{2}ab$, so equation (2.1) can be re-cast as follows:

$$(a + b)^2 = c^2 + 4 \times \tfrac{1}{2}ab \tag{2.2}$$

Expanding the term on the left-hand side of equation (2.2) gives $(a + b)^2 = a^2 + 2ab + b^2$, and after the cancelation of the $2ab$ term from each side of the equal sign, we have our result and proof of **PT**: $a^2 + b^2 = c^2$.

The proof of **PT** just presented is far from fundamental, and it assumes many preconditions. First, a square can be constructed from four right-angle triangles. Second, it assumes knowledge of the formula for the area of a square and a triangle, and third it assumes that equation (2.1) is exactly correct — that is, there are no small gaps or excesses not incorporated in its formulation. This latter point highlights one of the fundamental differences between the mathematical universe and the real universe in which we live, where it is not possible to physically construct (that is manufacture or draw) four identical, true right-angled triangles and then arrange them in an exact square configuration. In the real world, our world, there is invariably some inaccuracy in the construction (or measurement of) a side length or some angle. Indeed, in the real world, there are no such physical entities as perfect triangles and perfect squares. That such inaccuracies, or unavoidable construction errors, do not enter into a mathematical proof is part and parcel of the power of mathematics — it is also the reason why the mathematical universe is different from the one we actually live in. The mathematician's universe is one of imagined ideals, a shadow world of perfect forms — not a real world that can be physically experienced, but a world constructed by the human mind.

Touted, and with good reason, as one of the most influential books ever written, Euclid's *Elements* (compiled circa 300 BC) does not make for light bedtime reading. The *Elements* is a vast mathematical treatise in 13 books, and it is a vast compilation of definitions, postulates, theorems (or propositions), and proofs. Euclid's aim in producing the *Elements* was to build up, step by step, a systematic description of geometry. His aim was to begin at the proverbial beginning, building from the simplest, undeniable statements (definitions and postulates) and then moving upwards to more and more complex results (the theorems). The point is that every theorem is then proved on the bases of the rigorously established results that preceded it. In essence, Euclid's motto, and that of every mathematician since is, from a firm and solid foundation, a mighty tower can be constructed. The **PT** is introduced at the end of Euclid's book 1, and it is described in proposition 47 (and proposition 48). Book 1 begins with a framework of 23 definitions, 5 postulates, and 5 common notions, and it is from these that Euclid builds up to a proof of **PT**. The definitions set the mathematical concepts that are going to be adopted without proof, and, for example, definition 1 states that, "a point is that which has no part". The other definitions go on to define what a line is, an area, and so on. The postulates begin to define some of the scaffolding of mathematics, and, for example, postulate 1 states, "It is possible to draw a straight line between any two points". The common notions are those observations that relate to arguments about entities, and, indeed, make common sense. Accordingly, Euclid's common notion 1 indicates that "things which equal the same thing also equal each other"[4] — that is, if $a = c$, and $b = c$, then $a = b$. With the various definitions, postulates, and common notion statements in place, Euclid launches into proposition 1, which concerns the construction of an equilateral triangle on a given finite line. From proposition 1, step by step, each proposition building upon the proof of its forerunners, Euclid prepares the way to prove **PT**, or proposition 47, "in right-angled triangles the square of the side opposite the right-angle [hypothenuse *c*] equals the sum of the squares on the sides [sides *a* and *b*] containing the right-angle". Euclid's proposition 48 then goes the extra mile and proves the converse of proposition 47, and shows that if one has a triangle for which the sides and hypothenuse satisfy the relation $a^2 + b^2 = c^2$, then it must be a right-angled triangle. With proposition 48 proved, Euclid's book 1 closes.

Having introduced and proved **PT**, we now have an *object* that has a history, and we have an *object* that leads directly to strangeness and revolution. Indeed, hidden within **PT** was a shock for the Pythagoreans — an unexpected, inexplicable infinity. To begin, however, the first piece of history concerning **PT** is that it is not a theorem due to the Ionian Greek philosopher Pythagoras (who lived circa 570–495 BC). Yes, we use his name, but the result was known and applied a thousand years and more before Pythagoras even lived. We have absolutely no idea who discovered **PT**, and, in fact, no primary texts by Pythagoras have survived the ravages of time to

[4] A version of this idea is found in the zeroth law of thermodynamics, where it is stated that if two thermodynamic systems are each in thermodynamic equilibrium with a third system, then they are in thermal equilibrium with each other.

Figure 2.3. Plimpton 322 showing a set of Pythagorean triples in cuneiform notation. The tablet was produced circa 1800 BC.

our era. The origins of **PT** most likely lie in its application to the construction of right-angled triangles (discussed below), and with respect to the surveying and laying out of fields and buildings. The earliest known records concerning **PT** are found on a Babylonian clay tablet (Plimpton 322 — see Figure 2.3) dated to circa 1800 BC. This tablet displays, in cuneiform notation, a set of Pythagorean triples (more is said about such numbers in Chapter 13). These numbers provide for special solutions to **PT**, namely they correspond to exact integer-sided right-angle triangles. In general, if one sets sides b and c to some random integer value, then hypothenuse a will not be an integer. For example, with $(a, b) = (1, 2)$, so $c = \sqrt{5} \approx 2.236067978\ldots$ (where the three dots indicate that the decimal expansion continues on forever — but more on this shortly). Probably, the best-known Pythagorean triplet is the (3, 4, 5) right-angled triangle. In this case, the sides are of length 4 and 3, and the hypothenuse is of length 5. Other such integer triplets are (5, 12, 13) and (8, 15, 17) — in fact there are an infinite number of such triplets, and a generalized formula can be established to find them. Indeed, the set of numbers (x, y, z such that $x^2 + y^2 = z^2$) will be a Pythagorean triplet if

$$x = a^2 - b^2, \quad y = 2\,ab, \quad \text{and} \quad z = a^2 + b^2 \tag{2.3}$$

when $a, b \in N^*$ and $a > b$ — this last condition says that a and b are any randomly chosen positive integers with the only restriction being that a is greater than b. So, for example, if we set $a = 54$, and $b = 40$, then (1316, 4320, 4516) is a Pythagorean triplet. By having a set of Pythagorean triplets available, one can design structures (buildings, fields, and so on) that will have integer properties — that is the sides and diagonals that can be paced out of measured exactly in some specific set of measuring units. This result was deemed to be of great practical value to many ancient civilizations, especially when field areas had to be determined for taxation purposes. There is even some evidence that the ancient Neolithic monuments and stone circles, found throughout Europe and dating back some 5000 years, were constructed according to the underlying properties of the (3, 4, 5) and (5, 12, 13) Pythagorean triplets.

As suggested earlier, not every right-angled triangle will satisfy the integer-sided condition, and it was through the study of the simplest possible right-angled triangle that the Pythagorean philosophers received their greatest shock. The simplest right-angled triangle has sides $(a, b) = (1, 1)$ with hypothenuse $c^2 = 2$. In modern times, we simply take our electronic calculator and evaluate $c = \sqrt{2} \approx 1.414213562...$ (with the three dots meaning the expression never repeats and goes on forever). But, wait, what is this number? Clearly, it is not an integer, that is, it is not taken from the set $N^* = \{1, 2, 3, 4, 5, ...\}$, but it is a number with an infinite tail of decimal places, truncated to a number composed of 10 digits, only because that's how many digits an electronic calculator can show. The fact that a number technically requires an infinite number of decimal places to complete its expression is very odd when we come to think about it. Where could such a number exist? Certainly not in the real world, where we cannot even write the number down.[5] And yet, for all this, we can make use of such numbers as the square root of 2 in the real physical world, by not bothering about the infinite tail of digits. Practical mathematics is brutal, and it ruthlessly cuts the infinite down to size, adopting just the first four, five, nine, or ninety decimal places: to nine decimal places, $\sqrt{2} = 1.414213562$ — note here, the truncation of the number (the missing three dots) and the use of the equal sign (=), rather than the approximately equal sign (≈). This truncation, tantamount to the brazen throwing away of an infinitely long string of digits, will work perfectly well for any practical calculation performed by any engineer or physicist, but it will assuredly not make for a happy mathematician. No life-or-death situation will ever depend upon the exact value of the 50th decimal place of the square root of 2. So should we really care about all those digits?

The fact that some numbers, like the square root of 2, have infinite decimal expression was not (at first) a specific philosophical problem for the Pythagoreans, who were quite happy to accommodate and corral infinite decimal expansions in terms of integer fractions. Accordingly, for example, 17/12 = 1.416666666..., or 577/408 = 1.414215686..., or 665,857/470,832 = 1.414213562... The integer ratios effectively tame and subdue the decimal infinites, and they can be written down in an exact fashion. It would seem, therefore, in a practical sense, that we have successfully navigated our way around the problem of infinite expansions — they are all expressible in some fraction form as a/b, where $a, b \in N^*$, and $b \neq 0$ (the latter qualification says that b cannot be equal to 0). Indeed, such numbers are called rational numbers.

Once again, there are algebraic methods by which square roots can be approximated in terms of integers. The so-called Babylon

[5] The square root of 2 has been evaluated (by electronic computer) to some 10 trillion decimal places, but for a number that is infinitely long in its actual expression, this is a mere nothing. The main reason for such calculations, however, is to see if the square root of 2 is normal — an odd expression, perhaps, but the idea behind a normal number is that each of the digits 0 through 9 appears equally as often in its expression. That is no digit appears more or less often than any other, so apparently, root 2 is normal.

method[6] is possibly the very one that is used in your electronic calculator when the square root button is pressed. The method proceeds by successive approximations, and if x is the number for which a square root is needed, then the successive approximations are given by

$$a_{n+1} = \frac{1}{2}\left(a_n + \frac{x}{a_n}\right) \tag{2.4}$$

In equation (2.4), the first term a_0 can be any integer — the simplest starting value is $a_0 = 1$, and n is simply a counting term label. The method relies on the fact that as the iterations proceed so the a_{n+1}th term provides a better and better approximation to $\sqrt{x}$. Taking $x = 2$, then the successive approximations to $\sqrt{2}$ are as follows:

$$a_0 = 1$$
$$a_1 = 1.5$$
$$a_2 = 1.416666\ldots$$
$$a_3 = 1.414215\ldots$$

and so on for $n = 4, 5, 6, 7, \ldots$

In fact, the approximations for $n = 2$ and 3 are just the integer ratios 17/12 and 577/408 given earlier. Equation (2.4) is a remarkably simple and easy-to-apply formula, and it provides the result that we need in the everyday world. In principle, it would seem, we can keep on going to larger and larger values of n, and approach a closer and closer approximation to the square root of 2 — indeed, getting as arbitrarily close as we like — the question, however, being can we actually attain the square root of 2 in a finite number of steps.

The Babylonians were certainly capable of some similar such calculation as given by equation (2.4) in finding square roots, and in

[6] There is absolutely no evidence that this method was actually known to Babylonian mathematicians — it is just a name that has (perhaps inadvisably) been given to the method.

Figure 2.4. Babylonian clay tablet YBC 7289, and the expression for the square root of 2 in sexagesimal cuneiform notation. The tablet was produced between 1800 and 1600 BC

the clay tablet known as YBC 7289 (thought to date from 1800 to 1600 BC — see Figure 2.4) the square root of 2 is given in cuneiform notation as follows:

$$1+\frac{24}{60}+\frac{51}{60^2}+\frac{10}{60^3}=\frac{30,547}{21,600}$$

The ratio 305,470/21,600 gives $\sqrt{2} \approx 1.414212963...$ This is not a bad (practical) representation for the square root of 2, accurate to 6 parts 10 million, but it leaves one wondering what practical reason the Babylonians had for evaluating it to such high accuracy — they certainly couldn't measure or construct anything similar to such precision. Sometimes, when it comes to number theory and the expression of a difficult calculation it is the challenge (the intellectual mind game) that is important, not the practical utility.

So, far, so good. But there is still a shock buried deep within the infinite decimal expansion of the square root of 2. The unexpected, and amazing detail hidden within $\sqrt{2}$ was found by Hippasus of Metapontum (to his supposed great detriment[7]), and it concerns the very nature of numbers as they were understood by the Pythagoreans. The idea of the ratios of numbers was of extreme interest to the Pythagoreans, and they developed a theory of the whole universe around the interplay of integers and integer ratios. Great numerical play and mysticism were made of the *dekad* and the tetractys, and it was incorporated into musical theory, as well as the description of the planetary system — but these are not our interests here. Rather, the shockwave unleashed by Hippasus was that the square root of 2 cannot, in fact, be represented by an integer ratio — it is not a rational number, but a number seemingly devoid of *logos*. The proof of this result is a classic, and worth seeing (at least once), and the type of proof used is an example of one of the most powerful tools available to the mathematician. The proof works by contradiction. Here, the idea is to assume the contrary statement, that is, that $\sqrt{2} = p/q$, where p and q are integers, and then show that this leads to a contradiction. In other words, the initial assumption must be wrong and accordingly $\sqrt{2} \neq p/q$.

Proof of the irrational nature of $\sqrt{2}$: Suppose that $\sqrt{2} = p/q$ where it is assumed that p and q have no common factors. Accordingly, squaring both sides and re-arranging we determine that $p^2 = 2q^2$. The factor of 2 tells us that p^2 must be an even number, and this can only be true if p itself is even. Given p is even then p^2 must be divisible by at least 4, and hence q^2 must be even and this means that q must be even. So, by this reasoning, both p and q are even, and this tells us that they must have a common factor of at least 2, which is in contradiction to our initial assumption that p and q have no common factors. Accordingly, the initial statement must be incorrect and $\sqrt{2} \neq p/q$.

---

[7] Some accounts suggest that the Pythagoreans were so upset by his result that they drowned Hippasus to keep his discovery from "going public". Other histories suggest that he was only banished from the Pythagorean sect.

This proof was shocking to the Pythagoreans, and it opened up the realization that there are such entities as irrational numbers. The square root of 2 is just one irrational number, other well-known examples of irrational numbers are $\pi = 3.141592654...$, the base of the natural logarithms $e = 2.718281828....$; and $\sqrt{3} = 1.732050808...$ Indeed, there are infinitely many irrational numbers. We now have the remarkable result that the spread between the infinities of the integers and the rational numbers is another infinity of irrational numbers. The additive combination of all the integers, all the rational, and all the irrational numbers is known as the set of real numbers R. The real numbers are often described according to the real, or number, line, where it is envisaged that the integers are equally spaced along an infinitely long line (stretching from — infinity to + infinity) with the rational and irrational numbers falling at appropriate points on the line between bounding integers. The real line (or continuum) is defined as a Euclidian space of dimension 1, which is ordered (that is the numbers on the line can always be distinguished according to the less than < sign), with no maximum or minimum element, and which is dense.[8]

From a mathematical perspective the appearance of numbers with infinite decimal representations, that cannot be described by the ratio of two integer numbers, is simply something that one has to accept — it is a product of the mathematical universe muscling into the real universe. For the physicist, using mathematical symbols and algebra to express a specific theory, however, this result holds a strange paradox — a paradox that we have to live with in order to use mathematics in the real world. The point is if numbers like π and e and $\sqrt{2}$ have infinite expressions that cannot be written down or directly evaluated in any finite amount of time, what is it that is being expressed when we include these terms in a physical formula? Clearly, the numbers are not *evaluated* in the real universe, their very nature says that they cannot be fully evaluated, so what is going on?

[8] This latter condition indicates that for any two points x and y on the real line, with $x < y$, there is another point z on the real line such that $x < z < y$.

2.3. Timeless Mathematics?

In this section, we leap forward some 2000 years in time, and our focus is directed toward the formative years of the 20th century. This is the moment in history when mathematicians finally begin to get to grips with the problem of numbers. The problem was not, of course, related to the use of numbers, or in doing mathematics, but rather it was concerned with the fundamental definitions of what numbers are and the establishment of axiomatic systems of knowledge. The generally accepted solution to these highly technical problems, however, resulted in schism — a split of ideologies that rages even to this very day. Not only at stake in this division is the manner in which mathematics is founded, but also the way in which mathematicians should proceed to construct their proofs.

What are numbers? This is a simple enough question, and everyone today knows the answer. But, of course, as with all such simple questions, the answer is both obvious and profoundly problematic. To the mathematician, the question is incredibly difficult to answer, because, recall, the mathematician has to worry about all those niggly details that the non-mathematician can happily ignore. To the Pythagoreans numbers were entities defined by their descriptive properties — there was a common property of oneness (unity), for example, about one sheep, one boat, one tree, and one rock. There was a similar common property about twoness and so on. Mathematicians, in the modern era, however, are no longer interested in defining numbers in terms of a property that can be attributed to physical objects — instead, they define numbers, specifically, the integers in terms of sets.

The natural numbers, since the 19th century, are defined in a set-theoretic fashion and can be expressed in the axiomatic system developed by German logician Ernst Zermelo, and Abraham Fraenkel (typically shortened to just ZFC[9]). Sets are just collections

[9]The C indicates the inclusion of the so-called Axiom of Choice — the full details of which need not concern us here, but basically, it says that given a collection of non-empty sets, it is possible to pick exactly one object from each set.

of objects, and the natural number system begins by accepting (that is, it is an axiom) that there exists a set Ø with no elements — this is appropriately called the empty set. With the empty set so given, each natural number can now be constructed entirely on the basis of the numbers that preceded it. The sequence is anchored by the empty set, with 0 = Ø. The next natural number 1, can now be defined as the set that contains the empty set, with 1 = {Ø}, and in a similar fashion, 2 = {0, 1} = {Ø, {Ø}}, 3 = {0, 1, 2} = {Ø, {Ø}, {Ø, {Ø}}}, and so on to infinity. So, now we can count in terms of positive integers, and the remainder of the ZF axioms (there are 9 altogether) eventually allow for the rigorous development of the mathematics of numbers. While this formalism satisfies the rigors of mathematics, it is important to note that numbers are defined in terms of the empty set — they are assemblages of absolutely nothing. Numbers are an illusion of something rather than actually being something. Numbers are, in this manner, intangible entities dragged into a form of crepuscular existents, *ex nihilo*, by the human mind.

For all its foundation on nothingness, the ZPC formulation affords us a systematic way of generating the integers, but what about those real numbers in the continuum with their infinitely long digital tails? Two fundamental approaches to this problem have been proposed. On the one side of the argument was German mathematician David Hilbert, and on the other was Dutch mathematician Luitzen Brouwer. Hilbert was specifically concerned with promoting a formalized approach to mathematics and insisted that real numbers (with all their infinite tails) were complete, fully formed, and distinct real objects. Brouwer, in complete contrast, argued that each point of the continuum was represented by a never-ending sequence that developed over time. The real numbers are accordingly provisional. The expansion of √2, for example, under Brouwer's hypothesis, is not a completed sequence of digits, and accordingly it carries only a finite amount of information — to a degree (presumably) that is consistent with the age of the universe.[10] The older the universe

[10] Here, we take the Universe to be of a finite age — some 13.8 billion years. In the 1920s when the Hilbert–Brouwer debate began, it was obviously clear that the universe was not infinitely old.

gets, so the greater the decimal determinacy of $\sqrt{2}$, and the more information that it will (number wise) carry. This is a profoundly different approach to that envisioned by Hilbert, and it effectively adds a time component to mathematics. To Hilbert, and his followers, the infinite digital tails to the real number were all, somehow and somewhere, in place,[11] and fully expressed. To Brouwer, and his followers, real numbers are continually under construction and always finite in expression. Well, history tells us that Hilbert essentially won the commonly accepted argument, and time has accordingly been expunged from the mathematics of numbers. The intuitionistic approach, as Brouwer's methodology has become known, has not gone away, however, and it may, in fact, provide some answers (or hints toward answers) relating to how we experience physical reality. Indeed, time is very much a part of our, that is human, experience of reality. We do have a past, a present, and a future. To envisage an event is to speak of it occurring at some moment in time — a specific "now". Modern physics, however, has been formulated on the basis that time is effectively an illusion, rather than a fundamental, absolute (flowing) property of the universe. Between Hilbert's full determinism and Brouwer's evolving enumeration, we essentially have a choice to make: either all real numbers are fully determined, from the very first moment that the universe came into existence (and earlier?), and physics is accordingly formulated in a timeless sense. Or, real numbers are still evolving their digital tails, and that physical theory really does contain an ever-changing series of continuous events that happen as time passes. Nicolas Gisin (University of Geneva) has summarized this latter situation along the lines, that, "a finite volume of space can't contain more than a finite amount of information", and that, "physically relevant numbers can't contain infinite information".

Intuitionistic mathematics also introduces a middle ground in logical reasoning. Specifically, in intuitionistic logic, a proposition could be neither true nor false. This seeming strangeness follows

[11] The somewhere mentioned here, in which the real numbers exist in all their infinite glory, is discussed in the following chapter.

since intuitionistic logic does not allow for the law of the excluded middle — that is, if you will, the continuum cannot be cleanly cut. The law of the excluded middle dates back to at least the times of Aristotle who argued that of two contradictory propositions, one must be true and the other must be false. In a sense, in intuitionistic logic, the law of the excluded middle can only be applied after the end of all time, when the real numbers are all fully expressed. This point was further emphasized by Leopold Kronecker who insisted that there could be no existence without some form of construction, and that to construct something mathematically required that it was composed of a finite number of digits. Hilbert would countenance none of this, however, and he is said to have commented that the intuitionistic dismissal of the excluded middle was tantamount to prohibiting a boxer from using his fists in a fight. For all this, the philosophical tensions between the two approaches, that of Hilbert and that of Brouwer, continues to rumble on to this very day — although, as with most such fundamental debates, one has to accept one viewpoint or the other, and then work within its particular domain. Either way, practical mathematical results still follow.

Incredibly, the innocent-looking Pythagorean theorem, born of the properties of simple right-angled triangles, not only brought down the Pythagorean ethos of numerical perfection, it revealed the existence of an infinitely large set of numbers (the irrational numbers) that have infinitely long decimal tails which cannot be expressed by any integer ratio, but which, according to Hilbert and his followers, exist fully formed "somewhere". For the mathematician, the outcome of all this investigation, this burden of proving something, is that, if you are going to have such things as mathematical proofs, then you must be prepared to live with strange and unexpected consequences — in the case of the Pythagorean theorem, this was the discovery of numbers that defied sensible representation. In a direct sense, the proof of the Pythagorean theorem brought out the existence of what might be called good and bad infinities. The good infinity was the timelessness of a correct proof — it will hold true for all time. The bad infinity, from the Pythagorean school perspective, is that the proof of the theorem brought into existence the irrational numbers with uncorralled infinite decimal expansions. The mathematical universe is full of strangeness, and for all its otherworldly separateness it can generate monsters in the real, everyday world in which we live.

2.4. Beyond Pythagoras

Not only do mathematicians niggle over the smallest of details, and worry about every twist and turn in a given problem, but they also like to extend problems. By extension here, the idea is to look at other forms of a specific (solved) problem in order to see what they might reveal. One reasonably obvious extension of the Pythagorean theorem is one that asks for solutions to the general equation $a^N + b^N = c^N$, for $N \geq 1$. There are solutions to this formula when $N = 1$, this is just ordinary addition. There are solutions to the formula when $N = 2$ — these are the Pythagorean triplets. What about $N > 2$, however? Are there integer solutions to the equation $a^3 + b^3 = c^3$, for example, or $a^{33} + b^{33} = c^{33}$ for that matter? Can, indeed, the Pythagorean theorem be extended to a whole range of numbers for N? The answer to this question is a remarkably, no. And the proof of this statement is far from easy to grasp, and it is wrapped within a historical mystery. Indeed, the generalized Pythagorean problem is typically discussed under the guise of Fermat's Last Theorem.

Pierre Fermat was a lawyer by profession, but a mathematician by inclination, and he was one of the early practitioners to study differential calculus. His association with the generalized Pythagorean problem, however, involves a marginal note that he made within his copy of *Arithmetica* — a book composed by Diophantus of Alexandria in the 3rd century AD. This particular text deals with the methods needed to solve a list of algebraic equations, and within its pages, the problem of splitting a given square number into two other squares (as in the Pythagorean equation) is considered. To this section, however, Fermat added the comment, *it is impossible to separate a cube into two cubes, or a fourth power into two fourth powers, or in general, any power higher than the second, into two like powers. I have discovered a truly marvelous proof of this, which this margin is too narrow to contain.* Presumably, Fermat thought that he had found a proof of these statements, but it is unlikely that it was a truly general one for all powers $N > 2$. This is the historical mystery surrounding Fermat's Last Theorem. Fermat made many unsubstantiated claims of proofs to various mathematical problems. The appellation of being his *last* theorem came about as it was the last of his various claims that defied verification by other mathematicians. Indeed, Fermat's Last Theorem stood unproven for over 350 years. A proof was eventually constructed by Andrew Wiles (using advanced mathematical techniques completely unavailable to Fermat) and described during a series of lectures at Cambridge University in June of 1993. Poignantly, the proof of Fermat's Last Theorem followed as a final corollary to the work that Wiles actually presented — a marginal note in a complicated series of lectures. The techniques applied by Wiles were far removed from any mainstream mathematical methods, and they were understood (it is said) by perhaps 20 to 25 specialists, but the announcement of the proof made newspaper headlines around the world. The story was an engaging one. Here was a problem easy to state, straightforward to understand, and yet fiendishly difficult to prove. Indeed, over the centuries, Fermat's Last Theorem has attracted a veritable army of devotees, all intent on proving the theorem in just a few lines of algebra. Wiles worked on his methodology, largely in isolation and in secrecy, for many years, and it transpired that there was a fatal flaw in his initial proof. This technical flaw, however, was, with the help of several colleagues, eventually overcome, with the final polished proof being published in 1995.

While Fermat's Last Theorem is now proven and put to bed, indicating that there are no solutions to $a^N + b^N = c^N$, for $N > 2$, other variants of the problem remain open. The generalized Fermat problem $a^N + b^M = c^Q$, for non-equal values of N, M, and Q still requires the development of a fully annotated list of solution conditions. Likewise, the solution conditions to the switched Fermat problem, in which $N^a + N^b = N^c$, also requires enumeration — this formula works for $N = 2$, $a = b = 1$, $c = 2$, but what about other values? The future is indeed rich with numerical problems and possibilities.

2.5. Beyond Space and Time

Let us leave the oddities of numerical problems aside for a moment, and turn, once again, to the theorem of Pythagoras. While we have seen that hidden within the exactitude of its proof there are unanticipated infinities and numerical conundrums, it is also the case that it is predicated on simple triangles and geometry — the latter being the graphical embodiment of finite things. Indeed, within the 2-dimensional plane, the inexpressible infinites of irrational square roots can be corralled. Remarkably, while the irrational nature of $\sqrt{2}$ dictates that it is a number that can never be fully evaluated and/or written down, it can be *visualized* by drawing a right-angled triangle with sides of unit lengths 1 and 1. Such a perfect triangle, of course, can never be physically drawn in the real world, where measuring and construction error must always exist — so we are still in a hypothetical, idealized realm when considering even this simplest of constructions. However, the geometrical approach allows us to visualize numbers in objective terms — that is, in the same way as the Pythagoreans would have. Indeed, to the Pythagoreans numbers were objects — the number 3, for example, being expressed as a physical straight line with a length corresponding to 3 units. The square root of 2 is, by this methodology, realized as the length of the hypotenuse of the unit-sided right-angle triangle — its infinitely extended decimal tail squeezed into a finite space. But, why stop at the square root of 2, what about other irrational numbers such as $\sqrt{3}$ and $\sqrt{5}$, and so on? Well, these inexpressible numbers can also be drawn and corralled geometrically. For these latter numbers, the graphical methodology used to visualize $\sqrt{2}$ can be continued indefinitely through the continued construction of right-angled triangles, with each new triangle being built upon the hypotenuse of its predecessor. This process will

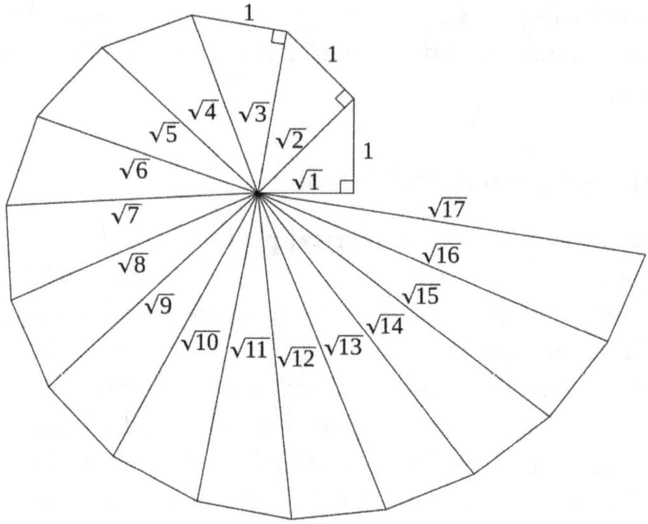

Figure 2.5. The Pythagorean spiral. Image courtesy of Pbroks13: https://commons.wikimedia.org/wiki/File:Spiral_of_Theodorus.svg.

generate the Pythagorean spiral[12] as illustrated in Figure 2.5. The sequence of numbers swept out corresponds to the hypotenuse sequence: 1, √2, √3, 2, √5, √6, √7, √8, 3, √10, √11, √12, √13, √14, √15, 4, √17, ... Coiling ever outward the Pythagorean spiral gathers together and embraces both the infinitely small and the infinitely large.

Long held as a potent symbol of nature and art, seen, for example, in the radiating florets of the sunflower,[13] the trailing arms of galaxies, the swirling of a hurricane, the double helix of DNA, as

[12] Also known as the spiral of Theodorus. There is no evidence that Theodorus actually described this particular spiral, although Plato tells us in his *Theaetetus* that Theodorus had demonstrated the irrationality of √2, √3, ... , √17. How Theodorus established this result is not known, but amateur mathematician J. H. Anderhub suggested in 1941 that he might have used the spiral generated by the sequential construction of right-angle triangles. It need hardly be said that the Pythagoreans had nothing to do with the Pythagorean spiral.

[13] In this case, the florets follow a Fibonacci spiral.

well as the works of surrealist painters,[14] the spiral is an entirely mesmerizing curve. The spiral pulls the eye to its center but carries its boundary to infinity — indeed, such curves would seem to spiral out of all control. This is not so, however, the infinities are contained, with each step in the construction being described by a specific formula[15] or a fixed sequence of actions. For the Pythagorean spiral, the set of construction actions is simply to keep building right-angled triangles on top of right-angled triangles. The length of the *n*th hypothenuse is $L_n = \sqrt{n}$, $n = 2, 3, 4, \ldots$, and this tends to infinity as n goes to infinity. Here, we have the infinitely large beginning to appear. The successive apex angles (situated at the center of the spiral) are given by $\varphi_n = \tan^{-1}(1/\sqrt{n})$, and this angle becomes smaller and smaller as n increases. Here, we have the infinitely small appearing. Overlaps of triangles, corresponding to an accumulation of more than 360° of rotation, occur at $\sqrt{18}$, $\sqrt{55}$, $\sqrt{112}$, $\sqrt{189}$, $\sqrt{286}$, and so on — these overlaps define spiral windings of index 1, 2, 3, 4, and 5, and they continue indefinitely. The spiral will continue to wind ever more times around itself as its radial opening stretches ever more outwards, as ever more triangles are added, one on top of the other, out to infinity. If we look at the spacing between successive windings of the spiral, a new (but well known, and perhaps not unexpected) constant appears. For the first winding, the approximate separation is $\sqrt{18} - \sqrt{1} = 3.24264$. For the fifth winding, the separation between spiral arms is $\sqrt{286} - \sqrt{189} = 3.16381$ — the trend is starting to become clear (although not fully justified here), but as the number of triangles grows larger and

[14] Perhaps one of the best such examples being the painting *Spiral Transit (1962)* by Remedios Varo.

[15] A simple formula for a spiral is $r = a\varphi^n$, where r is the radius, a is constant, and φ is the angle. According to the value of n, different spirals emerge, each with their characteristic appearance. For $n = -1$, we have the hyperbolic spiral; for $n = 1$, we obtain the spiral of Archimedes; for $n = ½$, the spiral of Fermat appears; for $n = -½$, the so-called Lituus of John Cotes is formed. Indeed, one can construct many spiral forms; some of which expand for ever, while others approach some asymptotic boundary.

larger, so the distance between successive windings of the spiral approaches π. Out of the rigidity of triangles emerges the quintessential number of the circle. Yet another infinity involving π lurks within the Pythagorean spiral, and this one was discovered by Erich Teuffel in 1958. Specifically, Teuffel showed that the manner in which the angles at the center of the spiral are swept-out dictates that no two hypotenuses will coincide — this is equivalent to saying that there are no positive integers n, m, and p for which the equation $\varphi_{n+m} - \varphi_n = p\pi$ is satisfied. As the Pythagorean spiral winds ever outwards, the hypothenuse of its constituent triangles completely blanket the 2-dimensional plane with no overlap. What a remarkable result.

The Pythagorean spiral (as illustrated in Figure 2.5) is constructed within, and confined to the 2-dimensional plain, but we can imagine it being constructed in another way, with each new triangular addition being accompanied by a twist into a new dimension. The spiral now takes on a multi-dimensional form taking it from the initial 2-dimensional plane to 3-dimensional space, to 4-dimensional space, to 5-dimensional space, and so on. After the construction of the second triangle, however, we lose sight of the remaining constructions since they reside in dimensions we cannot see. With the eyes of the mathematician, however, the multi-dimensional spiral can still be enumerated. What now of the Pythagorean theorem? The hypotenuse of the first triangle (in the 2-dimensional plane) is given by $(l_2)^2 = x^2 + y^2$, while the hypothenuse of the second triangle (in 3-dimensional space) is given by $(l_3)^2 = x^2 + y^2 + z^2$ and so on. Extending this result to m-dimensional space gives the hypotenuse of the final triangle as $(l_m)^2 = \Sigma(x_i)^2$, where the summation runs from $i = 1$ to m (in this notation $x_1 = x$, $x_2 = y$, $x_3 = z$ and so on through to x_m in the mth-dimension).[16] The theorem survives its multi-dimensional extension.

In principle the Pythagorean formula can be extended to as many orthogonal dimensions as one likes, and although we cannot draw or even imagine such structures beyond 3-dimensional space, the

[16] Technically, this is called the Euclidian distance or measure.

Pythagorean law will always hold true. This is an amazing result and one that goes against expectations — as human beings we have no expectation for there being anything beyond 3-dimensional space. We are 3-dimensional beings living in and experiencing a 3-dimensional world.[17] What then is the meaning behind the fact that the ancient Pythagorean Theorem extends to and remains true in more dimensions than we can physically experience? Of course, we should step back just a little here and draw attention to the fact that there is no specific reason why the universe should be constrained so that we, as human beings, can make sense of it, or, more to the point, experimentally measure it. This being said, the fact that the Pythagorean theorem holds true in more dimensions than we can experience begs the following question: Does this imply there is some shadow world in which dynamics and other physics can take place and yet be entirely inaccessible to study? Alternatively, we can also ask, because the universe of mathematical equations extends to more than 3-spatial dimensions, must this mean that physical reality extends to more than three dimensions? This is a problematic question that physicists have struggled with for many years. On the large scale, above that of the subatomic particles, there is no evidence (whatsoever) to suggest that there are more than 3-spatial dimensions to our universe — this is worth stating again: there is zero evidence that the universe is composed of more than 3-spatial dimensions. Not one iota of experimental data provides for the *prima face* argument that more than 3-spatial dimensions exist. This may change, of course, but as of writing this sentence, the total number of known spatial dimensions is three — end of story.

Let us consider the possibilities of what might be imagined (or realized) in universes composed of various (time, space) dimensions. In terms of the time dimension, the only ones that have any meaningful cross-over with our experience of the observable universe are

[17]Technically, Einstein reveled in his theory of general relativity that we live in 4-dimensional spacetime. That is, space is composed of 3-spatial dimensions and 1-time dimension, which is to be thought of as one integrated structure, rather than a 3-dimesnional space with time simply happening within it.

those of 0 and 1; for the spatial dimensions, we have 0, 1, 2, 3. The following (note, finite) list of arrangements can be considered:

(Time, space)	Comments
(0, 0)	No time, no space — a rather dull universe where absolutely nothing can happen ever
(1, 0)	Time, but no space for anything to happen in
(1, 1)	The universe on a line — no complex physics or chemistry
(1, 2)	The universe in the plane — more can happen than in the (1, 1) universe, but, again, no complex physics or chemistry
(1, 3)	Our universe — much complex physics and chemistry

The various combinations of (time, space) dimension single out the (1, 3) combination as having sufficient spatial dimensions for interesting (indeed, life-enabling) physics and chemistry to occur, and it dictates that causality is preserved. It is certainly the case that one can mathematically consider universes in which there are (>1, >3) dimensions, but none of these universes have characteristics that share any commonality with the observable universe. When there are more than 3-spatial dimensions, however, it can be shown that all dynamic systems must be unstable — there are no stable planetary orbits and no stable atoms, and everything is in a state of complete and utter chaos. For more than 1-time dimension, nothing is predictable, and it is the very existence of predictability, and the fact that cause always drives and precedes effect, that makes the observable universe special and indeed, what it is observed to be.

To sum up, the entire content of the observable universe is encapsulated within a domain that provides no more freedom of motion than that of 3-dimensional space: moving up-down, right-left, and backwards-forwards (all in a relative sense, of course). This is the observed reality in which we live. There is no implied necessity for the nature of the observed universe to be the same and behave in a manner identical to that which is allowable and possible in the mathematical universe. The two universes are distinct, and the possibilities and actions in one do not require that similar possibilities and actions exist in the other.

2.6. The Pythagorean Genie

Dealing with the Pythagoreans has brought us a long way through time — from the ancient past to the present day. Not only did the Pythagoreans begin to open up the story that has become the narrative of modern astronomy and physics, they also pioneered the use of mathematics as a tool to investigate the universe. The Pythagoreans set about using numbers and geometry to explain phenomena, and this has proved to be a very fertile philosophical practice. For all this, however, the Pythagoreans also unleashed un-looked for mathematical demons — numbers that can never, even if one were sentenced to an infinite servitude of toil, time, ink, and paper, be fully written down. By imposing mathematical rules and insisting upon the exactness of mathematical proofs, the Pythagoreans began to ensnare the real universe in a framework of geometry and numbers. The Pythagoreans unleashed a mathematical Genie into the world. Released from its deep, seemingly boundless, and dark bottle, the mathematical Genie began to cast a deep magic upon the world, but as with all magic, the seemingly miraculous is, more often than not, nothing more than a sleight of hand, and a grand illusion — it is not reality, but a simulacrum of reality that mathematics describes.

The motto that hung outside the entrance of the Pythagorean school was simply "all is number". By this motto, however, it was not necessarily meant that everything was some form of a mathematical expression, or that everything was expressible mathematically. Rather, numbers encoded objects, ideas, concepts, and elements. Such mysticism was encapsulated within the tetractys (Figure 2.6). The first dot at the apex of the tetractys is number 1, the Monad, and this represents the geometrical point, unity, and later in the Christian era the one true God. The second row, numbers 2 and 3, corresponds to the Dyad, accounting for the geometrical line, along

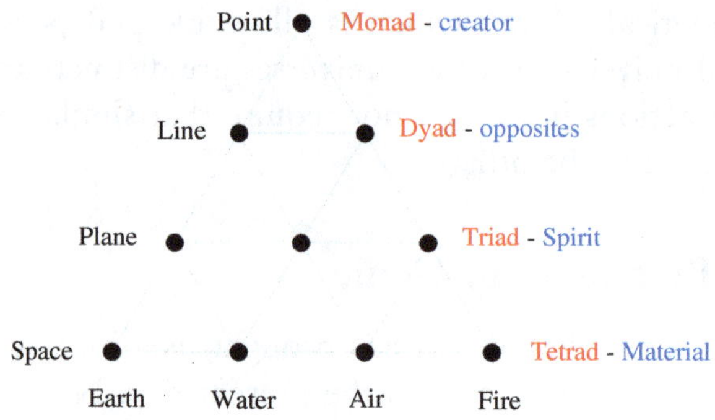

Figure 2.6. The tetractys of the Pythagoreans, being the fourth triangle number T_4. The four rows sum to the Dekad: $1 + 2 + 3 + 4 = 10$.

with the property of twoness and opposites, such as, for example, the Sun and the Moon, or good and evil. The third row is the Triad, representing the 2-dimensional plane, as well as wisdom, and harmony. The last row gives the tetrad, representing the organization of space, as well as justice, the four seasons and the four classic elements: earth, air, fire, and water. The total number of dots in the tetractys is ten, and ten makes the Dekad (corresponding to the 4th triangular number[18]), and this was taken to symbolize new beginnings and limitlessness. Numbers, in the Pythagorean manner, were symbolic, in the sense that they could be subject to mathematical manipulation, but they could also represent concepts (such as wisdom and harmony) that are not amenable to any form of mathematical analysis. Later commentators would greatly add to the mysticism of tetractys, its form being adopted and found in alchemical writings, works on astrology, and medicinal texts.

[18] These are the numbers that can be arranged, in the form of dots, as an equilateral triangle. The sequence of triangular numbers begins as 1, 3, 6, 10, 15, ... The sequence increases as $T_n = n\,(n + 1)/2$, where $n = 1, 2, 3, \ldots$ T_1 is clearly a special case since it corresponds to a single dot, rather than a triangle. The great mathematician Fredrich Gauss discovered in 1796 that every integer number can be expressed as the sum of three triangular numbers — provided, that is, one accepts the notion that $T_0 = 0$ is a triangular number.

The Pythagorean cosmos (recall Figure 2.1) was constructed in an attempt to put some form of numerical order into creation. It accommodated the Earth, where we live, the Moon and Sun, the five wandering stars (that is, the planets Mercury through Saturn), the outermost sphere of the stars, a counter-Earth, and the central fire. It is a combination of the observed and the imagined, the real and the unreal, the moved and the unmoved. The Central-Fire, as described earlier is essentially the motivating *spirit* of the cosmos — it provides cosmic order and it drives cosmic motion. The counter-Earth is a purely aesthetical insert — a brazen mathematical imposition (or blindness) — but for all this, an intended real planet. Not for another 2000 years, and more, would humankind be brave enough to magic a planet out of the machinations of numbers. The reasons for invoking this new planet, planet Neptune, was not through a belief in pure number properties, however, but it was because of an unshakable belief in Newtonian dynamics (but more on this in Chapter 6).

Based upon objects moving at a constant speed in circular orbits (about the central fire), the Pythagorean cosmos is envisioned as a simple geometrical structure. Problematically, however, it is far too simple in design. Indeed, it is not complicated enough. The Pythagorean cosmos might be ordered and complete, from an ancient Greek perspective, but it fails badly from the perspective of reality. It is admittedly a little unfair to criticize the model for this latter failing since it was never suggested that it was capable of explaining observed astronomical phenomena. Rather the model was an ordering principle. The model accounted for the supposed ordering of the planets, but it was never intended that it account for, say, eclipse phenomena, especially lunar and solar eclipses. Likewise, it was known a thousand years before the Pythagoreans that the motion of the Sun, with respect to the background stars, did not take place at a constant rate. Nor, for that matter, was the motion of the superior[19] planets across

[19] I am using modern terminology here — introduced, in fact, by Copernicus. Inferior planets are those that orbit the Sun interior to Earth's orbit, while superior planets orbit the Sun outside the Earth's orbit.

the sky constant; these planets showed direction reversals, or retrograde loops (see Chapter 4), that are obvious to anyone who diligently watches the sky for just a few years. Indeed, accounting for the retrograde motion was one of the key problem points to account for in the early descriptions of planetary motion. The ancient Greek astronomers ultimately solved the problem of retrograde motion in terms of epicycles (as will be seen in Chapter 4), but it was Copernicus, in the 16th century, who eventually re-shaped the problem — and to do this, he had to metaphorically move both Heaven and Earth (as will be seen in Chapter 5). The Copernican revolution, however, was not predicated upon a revolution in mathematical thinking, rather it was predicated on the notion that mathematical models could actually represent reality.

The Pythagorean cosmos was not, so much, concerned with practical astronomy, and it was completely incapable of acting as a predictive tool. It can be argued, however, that while there was a mathematical underpinning to the structure of the Pythagorean cosmos, there was no physical reality attached to it (or, being more generous, not much physical reality) — it was number without substance. This, again, is perhaps too harsh a criticism, since the required instrumentation, technology, and observational data to construct a physically realistic model, capable of explaining an astronomical phenomenon, did not exist for the Pythagoreans. Importantly, however, the Pythagoreans introduced the expectation that the universe, the real universe in which we live, was knowable through mathematics and number. This mathematical outlook, however, had, as argued earlier, unexpected consequences, letting loose monsters: infinites big and small, and inexpressible quantities. The Pythagoreans also released a philosophical demon — a demon that has lived and grumbled its way down the centuries, and this demon begs the question, is the observable universe the same as the mathematical universe, or are the two universes distinct domains that by pure chance have useful cross-over points and commonalities.

While the Pythagorean cosmos tells us very little about physical astronomy and/or astronomical phenomena, it is very definitely finite — large, yes — but finite in extent. We are given no specific details as to what might reside beyond the sphere of the fixed stars. To this extent, the Pythagoreans did not apply a mathematical infinity to the heavens. This is in contrast to the earlier teachings of Anaximander, who lived during the 6th century BC, and who postulated that the Earth (oddly shaped as a cylinder 1/3rd as deep as it was wide) floated in infinite space, neither moving up or down nor back nor forth, since it had no specific compunction[20] to move in any direction within the infinite realm. Indeed, Anaximander made a great play of the concept of *aperion*, meaning boundless or unlimited, in his cosmology, arguing that everything was created from *aperion* and would eventually return to it. Later, Aristotle firmed up this concept of boundlessness and introduced the notion of actual and potential infinities. Indeed, Aristotle argued that physical infinities could not exist, since that would allow for the possibility of something being bigger than the heavens themselves. The mathematical or potential infinite was allowed, however, since this was something that could never be physically realized.

That the universe must be finite in extent is an idea incorporated into the notion of the celestial sphere, where all the stars reside in a spherical shell centered on the Sun. While Christian scholars in the medieval era (a thousand years on from the Greeks) adopted wholesale the ancient cosmological model (as eventually refined and assembled by Ptolemy — see Chapter 4), they usurped the infinite space beyond the celestial sphere to accommodate the kingdom of God. The first modern-sounding suggestion that space might actually extend beyond the domain of a closed shell, with stars being distributed at different distances from the Sun, was introduced (albeit tentatively) by mathematician Thomas Diggs in 1576. Presenting his thoughts in what was the first English translation of the Copernican

[20]This idea of having no compunction to move one way or another was later incorporated into the story of Buridan's donkey. Caught equidistant between two drinking troughs, the hapless donkey, it as argued, will eventually die of thirst since it has no clear choice in which specific direction to move. This story, of course, tells us more about the failing of philosophers to understand the actual universe and basic animal behavior.

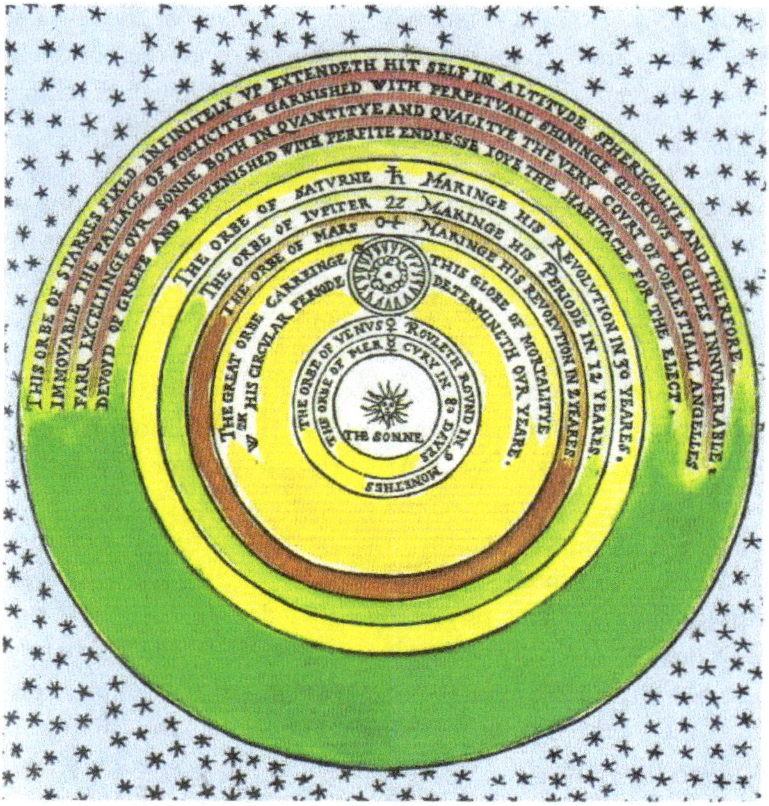

Figure 2.7. The universe as described by Thomas Diggs (1576). Oddly bounded and infinite at the same time, the stars are distributed through space at varying distances from the Sun.

thesis, Diggs set an infinite number of stars adrift in the infinite depths of space — defusing them amid the fixed palace of heaven: "the court of celestial angels, devoid of grief and replenished with perfect endless joy, the habitat of the elect" (Figure 2.7). Having liberated the stars from the shackles of the celestial sphere, Diggs let loose a new and unintended problem: the dark-sky paradox. Diggs, and just a little later in time Kepler, realized that there was a problem in positing the existence of an infinite number of stars within the universe, even an infinite universe. The problem was a blindingly obvious one, visible to all observers (high and low brow) as soon as the Sun set below the horizon — the night sky is dark. More commonly known as Olbers Paradox in the modern era (after astronomer Heinrich Olbers who discussed the problem in 1823), the paradox of the dark sky revolves around the notion that every possible line of sight in the night sky should intercept a star, and accordingly, the sky should simply glow with a uniform brightness — something that it clearly does not do. The full resolution of this paradox (not to be achieved until the 20th century) required the profound realization that the universe cannot be infinitely old, nor infinitely large. There are many stars in the universe, but their number is not infinite.

2.7. Archimedes Poppy Seeds

Counting is easy — it's as simple as $1 + 1 + 1 + 1 = 4$. In most branches of life, both in the past and in modern times, we don't generally have to deal with particularly large numbers, rarely having to physically count beyond a few tens or maybe hundreds. Larger numbers are tabulated on paper and counted by machines or statistical methods. In astronomy, however, numbers can be, well, astronomical in size, and we (as humans) very soon lose contact with any intuitive meaning of their magnitude, especially so when it comes to the sizes of the planets, the distance to the stars, the vastness of galaxies and the scale of the cosmos. They are just too big for our senses to comprehend, and putting a number next to such scales carries no improvement in understanding or emotional attachment. For astronomers and mathematicians, of course, the number, whether big or small, is vitally important, and at some stage, a system of expressing very large numbers needs to be put in place. This, of course, is just a convenience to aid in writing down and then manipulating extreme numbers. We have already made much use of the powers of ten notation in this book, using the textual compaction that it provides to express small and large numbers. Greek mathematician (and ace weapons designer) Archimedes introduced a system for the compaction of very large numbers but used the unit of the myriad, where 1 myriad = 10,000 = 10^4, and a myriad of myriads = 10,000 × 10,000 = 10^8, and so on in increasing order. Indeed, Archimedes, who lived in Syracuse in the 3rd century BC, was prepared to deal with numbers much, much larger than could possibly be encountered in everyday life. His system of large numbers is described in the *Sand Reckoner* — a work of mathematics, astronomy, and pure unadulterated fantasy.

The *Sand Reckoner* is addressed to King Gelo II of Syracuse, and within its text, Archimedes provides the solution to a marvelous fantasy question. "There are some", it begins, "who think that the number of sand [grains] is infinite in multitude", but the reality, Archimedes goes on to argue, is that they can all be accounted for. Indeed, if the entire celestial sphere was filled with sand grains, Archimedes explains, then their number would still be finite and amenable to investigation. To obtain this understanding, Archimedes turns to astronomy in order to determine the diameter of the celestial sphere, and to mathematical notation, in order to develop a scheme of enumeration for very large numbers.

The formula that Archimedes uses to determine the size of the celestial sphere is based upon a Sun-centered cosmological model that he attributes to Aristarchus (this latter work now being long lost), and it is cast in terms of the Earth–Sun separation (what we now call the astronomical unit) and the diameter of the Earth. Accordingly, the expression to evaluate is as follows:

$$\frac{\text{Celestial sphere diameter}}{\text{Earth} - \text{Sun distance}} = \frac{\text{Earth} - \text{Sun distance}}{\text{Earth diameter}} \tag{2.5}$$

No specific justification for this formula is given by Archimedes, and it is really an imposition that enabled a calculation to be made. In modern terms, however, the expression builds upon the notion that the stars (situated within the shell of the celestial sphere) must be sufficiently far away that they show no measurable parallax. Within equation (2.5), the size of the Earth was taken as being 100 myriad stadia (10^6 stadia), and the Earth-Sun distance was deduced as being 10^{10} stadia. This latter result was based upon a set of astronomical measurements requiring knowledge, that is the measurement of, the Sun's angular diameter — this Archimedes found to be about ½ a degree on the sky. The stadion was not a universally fixed unit of measure, and various values were adopted in the ancient world, but roughly speaking, 1 stadia is equal to 1/5th of a kilometer. In this manner, the Earth is deemed to be about 200,000 km across (which is about 15 times too big by modern standards), and the astronomical unit comes out as 2 billion km (which is about 13 times too big). With these estimates, the diameter of the celestial sphere deduced by Archimedes comes out to be 10^{14} stadia — which is equivalent to a distance of about two light years. By modern standards, the universe of Archimedes is very small and entirely claustrophobic and would comfortably sit within the confines of the solar system's Oort cloud of cometary nuclei. For all this, it is still a staggeringly large number.

Having derived an estimate for the size of the cosmos, Archimedes then sets about filling it. To do this, he begins by noting that 40 poppy seeds laid side by side correspond to a distance of 1 dactyl (that is one finger width or about 2 cm). He then states that each poppy seed contains a myriad of sand grains — this introduces some confusion in our times since a sand grain is generally taken to be about the same size as a poppy seed. Something that is 10,000 times smaller than a poppy seed (which is about ½ a millimeter across) is not visible to the human eye. Nonetheless, with these assumptions in place, Archimedes determines that within a sphere of radius 1 stadia, there are some 10^{21} sand grains, and with the volume of the celestial sphere being some 10^{42} cubic stadia, a total of 10^{63} sand grains are required to fill it. A large number indeed but one very much expressible and finite.

The whole idea outlined in the *Sand Reckoner* is fantasy, but it is fantasy with a strong philosophical purpose. It espouses the audacity that the scale of the entire universe can be measured and is knowable. Furthermore, it expresses the idea that once the scale of the universe is known, then we can fill it with all manner of imagined things. To Archimedes, nothing was too big to be comprehended, and even the great expanse of the universe was something amenable to expression in numerical form.

2.8. Counting the Stars

The Yale Observatory *Bright Star Catalog* tells us that there are 8692 stars brighter than magnitude 6 (the limit of human perception) in the sky as seen from the Earth. Given that our binocular vision enables us to observe, from a fixed location, and without moving our head, about 1/3rd of the celestial vault, so at any one instant, under perfect viewing conditions, no more than about 2900 stars might be perceptible to the human eye.[21] This is a sizeable number of stars to count by hand, as it were, without, say, photographic aid, but, in principle, it is the sort of number count that most humans could deal with — if they had to. In spite of the oft-used aphorism that there are innumerable stars in the sky, their number is, in fact, relatively small. Figure 2.8 shows the justly famous Hubble Deep Field image, obtained by the Hubble Space Telescope in 1999. Pretty much every object seen in the image is a galaxy, and analysis of the image data indicates a total of some 1600 galaxies out a distance of about 6 billion parsecs. The area covered by the Deep Field image amounts to just 460 millionths of the entire sky. Taking the Deep Field image to be typical of that for any deep image of the sky in any direction, the number of galaxies in the visible universe amounts to about 35 billion. (If we were counting money here, as opposed to galaxies, at a net worth of 35 billion dollars you would rank number 40 on the *Forbes* 2021 list of wealthiest people.) On average, astronomical survey work indicates that each galaxy contains about 300 billion stars (some more, some less). Accordingly, our estimate for the total number of stars in the observable universe turns out to be about 10^{23} — a number much, much smaller than the number of sand grains required to fill the claustrophobic universe imagined by Archimedes.

One can take the Hubble Deep Field calculation a little further, and estimate the mass of the observable universe (in the form of

[21] This number is probably a factor of two too big since night sky viewing conditions are never uniformly perfect, the field of view is typically obscured by foreground objects, and atmospheric extinction will reduce the observable count. Likewise, the acuity of the human eye varies from one individual to the next.

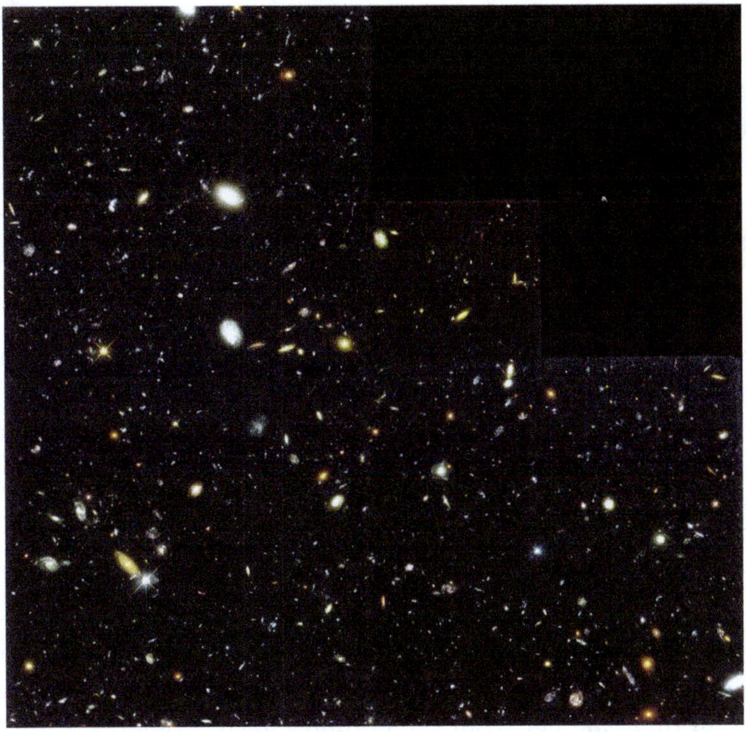

Figure 2.8. The Hubble Deep Field. Virtually everything visible in this image is a galaxy. Image courtesy of NASA.

stars). The typical stellar mass is about ½ that of the Sun (such stars being the predominant spectral type M dwarfs), and since the Sun weighs in at some 2×10^{30} kg, the total mass of material in the form of stars in the observable universe is of order 10^{53} kg. How does this compare with the numbers derived by Archimedes? Taking a dust grain (as envisioned by Archimedes[22]) to have a mass of some 3×10^{-11} kg, so, Archimedes sand-packed universe would have a mass of some 3×10^{52} kg — a number remarkably similar to (within a factor

[22]Here, I use a snippet of Wikipedia wisdom that 3300 poppy seeds constitute a mass of 1 gram. Taking each poppy seed to contain a myriad of sand grains, as per Archimedes description, then each sand grain has a mass of 3×10^{-11} kg.

of 3) the modern-day number for the mass of stars within the observable universe. Coincidence? Yes, very much so, but it gives one momentary pause for thought.

Having estimated the number and mass of stars in the observable universe, we can now estimate the number of atoms in the universe. Assuming, fairly safely as it turns out, that stars are mostly composed of hydrogen, and taking the mass of the proton to be 1.7×10^{-27} kg, the number of protons in the observable universe is of order 6×10^{79}, there being an equal number of electrons to go with this number of protons to preserve charge neutrality. Now this is a number that Archimedes would have surely been proud of. Indeed, the number just derived is often called the Eddington Number, after astronomer Arthur Eddington, and it will be further discussed in Chapter 13.

The premise of the *Sand Reckoner* is fantasy — it is not possible to physically count the number of sand grains required to fill the universe (even if we wanted to). Likewise, the Eddington number is fantasy (although Eddington believed in it) and we cannot hope to physically count the number of protons in the universe. The important point that emerges from such fantasy ideas, however, is that the human mind is able to *see*, *enumerate*, *express*, and *determine* the quantities involved. Our minds can take us outside of physical reality, and what we call mathematics is able to follow the human mind into fantasy domains.

2.9. Computing the Universe

Electronic computers are ubiquitous in the modern era, and today we can hardly imagine life without them. They run the world for us, and they perform a stupendous number of operations per second. Indeed, much of mathematics is the straightforward evaluation of numerical algorithms of one form or another, and this is a process that electronic computers can perform very efficiently, and at great speed.

A measure of the number of operations N_{ops} that all desk-top-style computers might have made over the last 10 years T (with $T \approx 3 \times 10^8$ seconds) can be gauged by adopting a typical clock rate R

and characterizing the number of the logical operations performed per clock cycle as N_L. Given that at least 2 billion computers have been manufactured, so the number of operations performed by artificial (electronic) computers over the last 10 years is of order: $N_{ops} = 2 \times 10^9 R N_L T \approx 6 \times 10^{31}$, where we have taken a clock rate of one billion Hertz and assumed that 100,000 logical operations are being performed per clock cycle. This same number of operations provides a limit to the number of mathematical calculations that artificial computers have made in the same 10-year timeframe. Many calculations have been made by computers, but it is still a finite amount of mathematical processing that has taken place.

Given that (human-made) electronic computers have performed something like 10^{32} logical operations over the past 10 years, what, in comparison, are the theoretical processing capabilities of the entire universe? This latter number corresponds to the computational capacity of the universe and is something that physicist Seth Lloyd (MIT) has investigated. Writing in the early years of this century, Lloyd argued, "merely by existing, all physical systems register information. And, by evolving dynamically in time, they transfer and process information" — and they do all this transfer and processing according to the laws of physics. The basic limit on the amount of information processing that might be performed by a physical system is set by the energy E available. This simply addresses the point that energy is required to change the state of a system — such as the spin state of an electron. According to a theorem developed by Norman Margolus and Lev Levitin, the time required for a quantum system to undergo a state change is $\Delta t = h/4E$, where h is Planck's constant. This constraint dictates that the maximum number of elementary operations that a given system can perform per second N with the energy E available is $N \approx 4E/h$. Additionally, the maximum speed at which information can be moved around within a system is set by the speed of light c. In terms of estimating the calculating capacity of the universe, we need to consider the total number of operations that might be performed in the spherical volume $V = 4\pi (cT_u)^3/3$, where $T_u \approx 13.8$ billion years is the age of the universe. Taking the matter density of the universe to be $\rho \sim 10^{-27}$ kg/m^3 (equivalent to an average of 1 proton per cubic meter of space), the number of operations that the universe could in principle have made since its formation is $N \approx \rho c^2 V(4/h)T_u \sim 10^{120}$. Of course, the universe is not a computer, and it hasn't been computing anything since it formed, but if it had been running some vast, pre-ordained, and possibly endless computer program, then it could, in principle, have made many calculations, but only made a finite amount of headway with respect to its potential outcome. Indeed, if the purpose of the entire universe was to compute the full expression of $\sqrt{2}$ then it would not yet have achieved this result, and indeed, it could never actually achieve such a goal. There is, apparently, only a finite amount of mathematics that even the entire universe can actually perform.

Chapter 3

Plato's *Timaeus*

Plato, in the mid-4th century BC, moved away from the numerical mysticism that enshrouded the Pythagorean school, but he still enforced numerical order upon the formation and structure of the heavens. The universe had a creator and a demiurge to perform the hard graft of construction, but even the creative demiurge was constrained by number relationships when it came to organizing the structure of the cosmos. The details of Plato's cosmos are described in his *Timaeus*.

The *Timaeus* is generally considered to be one of Plato's most important works, and it relies to a large extent upon mathematics, rather than pure dialog, to make many of its points about the origin and structure of the cosmos. Within the *Timaeus*, Plato provides an account of the formation of the universe and he explores the reasons for its order and beauty. Indeed, the cosmology in *Timaeus* is a teleological account in that it seeks to explain the cosmos as being structured toward an end — an end, that is, that makes it as good as it can possibly be.

Unlike the Pythagoreans, Plato placed a stationary Earth (both in the sense of orbital motion and spin) at the center of the cosmos, and he did away with the central fire and the counter-Earth. Plato additionally animated the celestial sphere, requiring that the vast outermost shell of stars lumbered around the Earth once every 24 hours. Indeed, Plato undertook a radical re-thinking of the structure of the cosmos, but for all its newness and innovation, it was still no better at accounting for actual astronomical observations than that presented by the Pythagoreans (recall Figure 2.1). This being said, Plato's cosmos did distinguish between the celestial equator and the ecliptic (Figure 3.1). These two circles, that trace across and around the sky, account for the division of the celestial sphere into two equal halves (this is the celestial equator) and the plane of rotation of the Moon, Sun, and planets (this is the ecliptic). The two circles are inclined to each other by some 23.5 degrees, and in modern terminology, this angle is called the obliquity of the ecliptic.

There was by necessity a need for an activating *spirt* in Plato's cosmos, and this order was attributed to the world soul. Indeed, the cosmos was considered to be alive and intelligent, and since it was intelligent, it must have a soul. Like the central fire of the Pythagoreans, the world soul permeated and enlivened the cosmos, and it provided

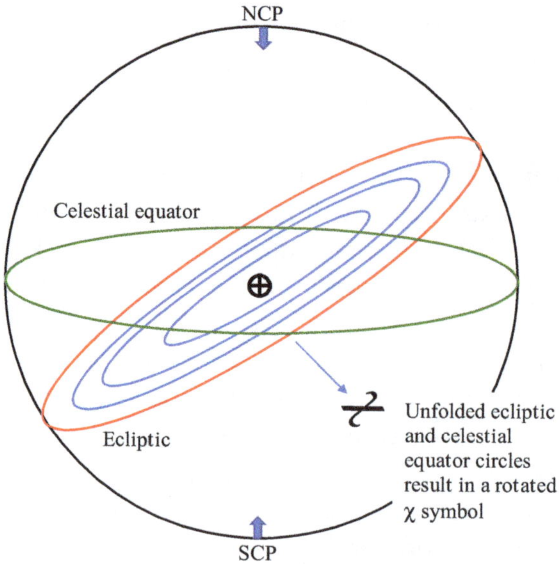

Figure 3.1. The χ segments of the world soul are composed of the unwound circles of the ecliptic and the celestial equator. The planetary orbits are illustrated in blue and all lie within the ecliptic. The Earth is located at the center of the cosmos, and the entire outer shell of stars rotates about the spin axis joining the North (NCP) and South Celestial Poles (SCP).

for an order-controlling essence. Indeed, Plato explains in his *Timaeus* that it was by manipulating *soul stuff* that the creating demiurge brought the cosmos into existence. The demiurge, in fact, started by taking two strips of *soul stuff* and crossing them over to form the shape of the letter chi (χ). The ends of each strip were then curved around so as to make for two intercepting circles. These two circles were the celestial equator and the ecliptic, and their interception points (on the sky) define the direction of the vernal and autumnal equinoxes. Once formed, the celestial equator was then given what Plato called the motion of the same, that is, circular motion at a constant speed. It was this perfect "same" motion that drove the celestial shell around the Earth once every 24 hours. The strip of *soul stuff* accounting for the ecliptic, however, was split into numerous bands and given motion of difference. It was this motion of difference that allowed the planets to move around the ecliptic at different speeds, so as to account for their observed motions: the Moon moving most rapidly around the sky (once every 27.5 days) and Saturn most slowly (once every 29 years).

No physical scale is given for the χ-cosmos, but Plato does provide a detailed (but complex) account for the distance ratios between the planets. The ratios are based on the numbers derived from Plato's lambda (Figure 3.2), which shows two numerical sequences. The first sequence runs 1, 2, 4, 8, while the second sequence runs 1, 3, 9, 27, and they are the first four terms in the (potentially infinite) square and cube number series 2^n and 3^n, where $n = 0, 1, 2, 3$. These

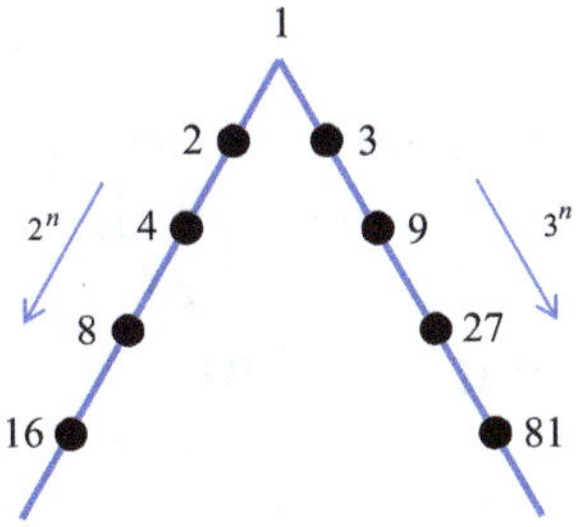

Figure 3.2. Plato's lambda with one branch corresponding to the sequence 2^n and the other to 3^n, where $n = 0, 1, 2, 3, \ldots$

Figure 3.3. The Christianized Platonic cosmos. From Peter Apian's *Cosmographia* (1539). The Earth is at the very center of the cosmos, and it is surrounded by the concentric domains occupied by the Moon and the planets (including the Sun). The outermost astronomical shell contains the stars (*Firmamentum*). This shell is further surrounded by the first-moved (*Primum Mobile*) region, from which all lower celestial motion is derived. Below the Moon's sphere are the four domains of matter: water, air, earth, and fire. Beyond the outermost shell is the heavenly realm: the imperium of the *Electorum*.

two number sequences were taken to be representative of the plane (2-dimensional) and solid (3-dimensional) figures, and these in turn were seen as the basic building blocks of nature. It was by constructing various number ratios (plucked from the lambda sequences) that the demiurge was able to create and order the cosmos.

Plato explains that the Earth is located at the center of the cosmos and that the order of the celestial bodies runs from the Moon, Mercury, Venus, Sun, Mars, Jupiter, Saturn, and the celestial sphere (Figure 3.3).

Plato further writes that the *soul stuff* that makes up the ecliptic is split into seven unequal circles according to the numbers 1, 2, 3, 4, 8, 9, and 27. The picture that Plato develops for the cosmos is an entirely symbolic one, and with respect to planetary spacings and motions, no physical and/or practical measure can be read into what is described. Indeed, to understand the motion of the planets, Sun, and Moon, entirely different geometrical constructions were required.

The cosmos that Plato describes in his *Timaeus* is not an obviously better description of the heavens than that given by the Pythagoreans. Some aspects are arguably simpler in design, and he does distinguish between the ecliptic and the celestial equator, but it effectively does little more than replace one set of assumptions with another set of assumptions and mathematical impositions (the lambda sequence). For all this, the Pythagorean and Platonic models set in place a series of ideas, one can even say dogmas, that influenced, as well as constrained, the outlook of those philosophers that contemplated heavenly structure. Three dictates in particular were set in stone, and for a time span amounting to something like 50 human generations,[1] they were accepted and applied without question. Indeed, their emplacement as absolute requirements rippled down the centuries until the late 16th century and the emergence of the new astronomy developed by Nicolaus Copernicus and Johannes Kepler. Based upon the aesthetics of ideals, and the perfection of geometry, the dictates (here labeled with a P for Plato) were as follows:

Plato's Cosmological Dictates:

P0: The Earth is a sphere located at the exact center of a spherical universe.

P1: All heavenly motion is circular.

P2: All heavenly motion takes place at a constant speed.

Given these absolute edicts (one could even say, formal axioms) concerning the structure of the cosmos, the role of the mathematicians

[1] Here, I take a human generation to be equivalent to 35 years.

was thereafter to save the appearance of the heavens as described by the astronomers.

Neither the Pythagoreans nor Plato offered any clear statements about actual planetary spacings and motions; their knowledge only extended as far as an ordering sequence based on their observed sidereal periods and specific harmonic ratios. What the ancient philosophers had achieved for posterity, however, was a mathematical expectation; the universe and the planetary realm, irrespective of appearances, must surely be understandable through geometry and number. And yet, for all this, the mathematics of Plato's cosmos was essentially impotent: it made no predictions, even of simple astronomical phenomena, and it openly contradicted what was observed with respect to planetary motion. While the impotency of the cosmological models presented by the Pythagoreans and Plato remained in place for the next 1500 years, mathematicians were able to successfully save the appearance for isolated components. Geometry was key to making sense of the observed motions — that and a lawyer's eye for identifying loopholes within Plato's three edicts P0, P1, and P2. In saving the appearance, however, there was no specific expectation that the mathematical model must represent physical reality. This latter attribute, the representation of reality, was only something imposed upon the mathematical models at a much later time. All the mathematical model had to do for the ancients was to be consistent with the actual observations. To a certain extent, giving the right result, with respect to the observations, is all that is required of modern mathematics as well, but in the modern era, there are more stringent conditions placed upon the structure of the underlying physical principles.

3.1. Tracking the Sun

Central to the development of observational astronomy, and indeed, central to the idea of developing a practical astronomy, is knowledge of the Sun's motion across the backdrop of the celestial sphere. The motion of the Sun, with respect to an observer's horizon, has, since the most ancient of times, been used to measure the progress of the seasons and establish its natural one-year cycle. The southern and northern-most limits of the Sun's rising (and setting) locations are used to determine the times of the winter and summer solstices — corresponding to the shortest and longest hours of daylight, respectively. Between these are the equinoxes (indicating the onset of the spring and the fall seasons), when there are equal hours of daylight and nighttime. The Sun is indeed the celestial clock of the seasons, but its angular rate of motion along the zodiac (along the ecliptic) is not constant, alternately speeding up and slowing down in its progression on a periodic six-monthly cycle. Such variable motion (the solar anomaly) is inconsistent with Plato's cosmic dictates, and the question arises as to how might the observations be described. Here, the idea is to account for the observations rather than adjust the fundamental underlying cosmic model. It is the mathematical description of the phenomenon that is important, with no specific emphasis being placed upon the representation of reality.

Hipparchus (c. 190–c. 120 BC) found a highly practical way of reconciling the observed non-constant motion, as measured with respect to the background stars, of the Sun and Plato's dictates: he saved the appearance of actuality against Plato's imposed ideals [1]. Specifically, he let the Sun move along a perfectly circular orbit at a perfectly constant speed, but with a lawyer's smirk, he then offset the position of the Earth from the center of the Sun's orbit (or deferent). The constant motion of the Sun is now, from the offset Earth, perceived as being non-constant: the angle swept out in a given time interval being larger when the Sun is closer to the Earth and slower when further away (Figure 3.4). Effectively, Hipparchus developed a two-parameter theory to describe solar motion, with the two parameters being the displacement of the Earth from the center of the Sun's deferent (OE) and the longitude of apogee (*A*). In this manner, P0 is preserved, since Earth is still taken to be at the center of the celestial sphere, and no one said that all the centers of the circular orbits had to be in the same location. Furthermore, P1 and P2 are preserved for the Sun since it is taken to move around its circular orbit at constant

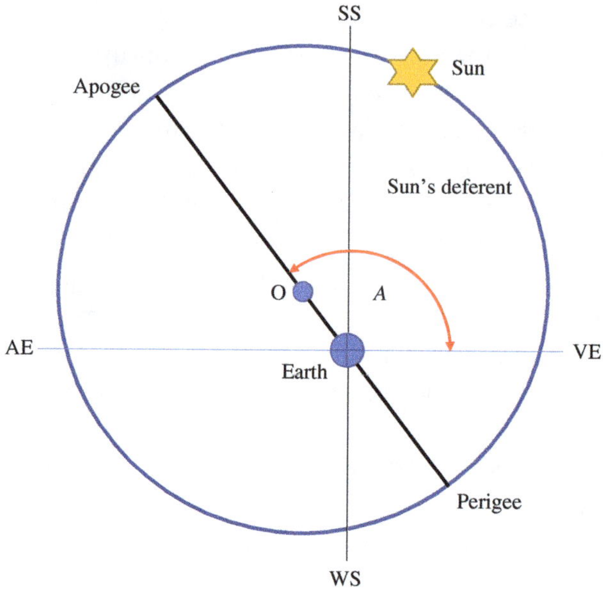

Figure 3.4. The solar anomaly theory of Hipparchus. The outer circle (deferent) indicates the circular track of the Sun, with center O, along which it moves with a constant velocity. The symbols WS, SS, AE, and VE correspond to the location of the Sun at the times of the winter solstice, summer solstice, autumnal equinox, and vernal (spring) equinox. Apogee and Perigee correspond to the greatest and least distance between the Earth and Sun.

speed. The model is adjusted and orientated in accordance with the direction of the vernal equinox[2] (VE) on the celestial sphere.

The solar anomaly model of Hipparchus is based on the tropical year and the counting of days between each successive equinox and solstice. If the time interval between each equinox and solstice was

[2]The direction to the vernal equinox is also identified as the first point of Aries and is the point on the sky where the great circles of the ecliptic and celestial equator intersect each other. At this location, the Sun is moving from below the celestial equator to above; at the other intersection point (the autumnal equinox), the Sun is passing from above the celestial equator to below. Because of precession, the first point of Aries is not fixed on the celestial sphere but moves westward with respect to the ecliptic at a rate of about 1 degree per century. Currently, the first point of Aries is in the constellation of Pieces.

exactly ¼ of a year, then they should be separated by a time interval of just over 91.3 days. Observations indicate, however, that the time interval between the time of the winter solstice and that of the vernal equinox is shorter than the time interval between the vernal equinox and the summer solstice. A quick check with a modern-day calendar, for example, will reveal that the time interval between the vernal equinox and the summer solstice is 93 days, that between the summer solstice and the autumnal equinox is 93 days, that between the autumnal equinox and the winter solstice is 90 days, and that between the winter solstice and the vernal equinox is 89 days. It is these time intervals that are mapped onto the deferent (with center O, see Figure 3.4) and the intersection point of the lines joining the two equinoxes and the two solstices indicate the position of the Earth. The line drawn through the center of the deferent O and Earth (see Figure 3.4) corresponds to the line of apsides, which intercepts the deferent at those locations where the Sun is moving with its greatest and least angular speeds, as seen from the Earth. At perigee (the point of closest approach), the Sun has its greatest angular speed; at apogee (the point of greatest separation), it is moving at its slowest angular rate.

The angle of longitude (A) in the solar anomaly model is not actually constant but gradually increases at a rate of some 1 degree every 72 years: an effect due to the precession of the equinoxes. This progression is a slow enough change that it need not be worried about on timescales shorter than a century or so, but it does steadily increase and ultimately adjustments to any tabulated results will need to be made. It is generally taken that Hipparchus discovered the gradual shift (on the celestial sphere) in the equinox locations, but none of his actual writings on the topic have survived.

The solar anomaly model presented by Hipparchus is a purely geometrical construction, but once constructed (see Chapter 4), it works extremely well, enabling a highly accurate description of the Sun's motion along the ecliptic to be made. This same basic geometrical approach, using circles with centers off-set from the Earth, was later expanded and applied to the description of planetary positions, culminating in the great encyclopedic work of Claudius Ptolemy, in the 1st century AD, and these will be discussed in Chapter 4.

3.2. The Platonic Solids

Central to the theme of the *Timaeus* is a description of the four elements out of which all substances are made. These elements corresponded to earth, wind, fire, and water, with the idea that such elements could be combined to explain such structures as mud, being a combination of earth and water, or steam, being a combination of air and fire. In the *Timaeus*, these elements are described in terms of the simplest, most basic, planar figure: the triangle. With this idea in place, Plato then explains how various solids can be constructed from bounding faces made up of triangles or composite triangular faces. The cube, for example, is composed of six square faces, with each square face being composed of two equilateral, right-angled triangles joined along their hypotenuse. Of the many possible solids that can be made, Plato paid particular attention to those solids that have identical faces. These are known as the Platonic solids [2].

The Platonic (or regular) solids (Figure 3.5) are special, and they compose a unique set of exactly five polyhedra. These particular

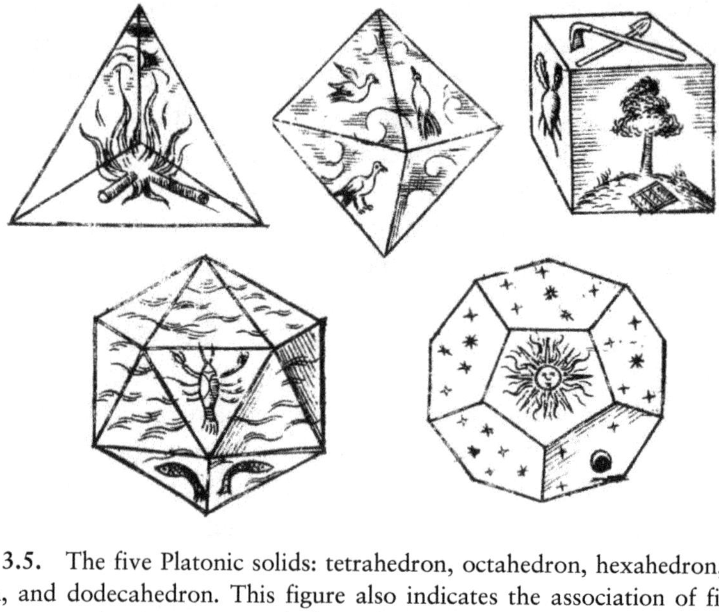

Figure 3.5. The five Platonic solids: tetrahedron, octahedron, hexahedron, icosahedron, and dodecahedron. This figure also indicates the association of fire, air, earth, water, and quintessence with the various polyhedrons.

solids are familiar to us even today (and especially so to anyone who plays the *Dungeons and Dragon* role-playing game). The Platonic solids correspond to the triangular pyramid (the tetrahedron), made of 4 equilateral triangles, the octahedron, made of 8 equilateral triangles, the cube (or hexahedron), made of 6 squares, the icosahedron, made of 20 equilateral triangles, and the dodecahedron, made of 10 pentagons.

The special characteristic of the Platonic solids is that every face of the solid is the same 2-dimensional shape (a triangle, a square, or a pentagon). Other shapes (polyhedra) can be constructed from combinations of triangles, squares, pentagons, and so on, but these are not Platonic polyhedra. A more formal definition of a Platonic solid might run as follows:

Definition PS: A Platonic solid satisfies the following conditions:

1. All its faces are identical planar polygons.
2. Each vertex (corner) has the same number of incident edges.
3. It is convex.

In this definition, we have snuck in an additional term, convex, that requires a little explanation. A shape (polygon or polyhedra) is said to be convex if when drawing a line from any point on the shape's boundary to any other point on the shape's boundary, the line is always interior to the boundary. So, for example, a circle is convex, a square is convex, but a star shape (the pentagram, say) is not convex (see Figure 3.6). The Platonic polyhedra also satisfy other important conditions, such as having a perfect geometrical symmetry, meaning that all the vertices are the same distance away from a common center and that the vertices are evenly distributed around the center. In this sense, one could imagine enveloping a Platonic solid within a sphere, with all the edges just touching the sphere's surface — this is the so-called circumscribed sphere. Likewise, one could also imagine placing a sphere within the interior of a Platonic solid so that the sphere's surface just touches the central point of each face: this is the circumscribed sphere. Johannes Kepler, as we

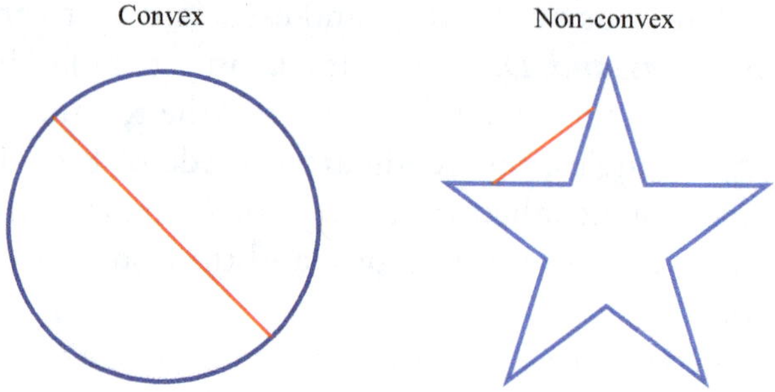

Figure 3.6. Illustrating the idea of convexity. A line drawn between any two points on the boundary of a circle is always interior to the boundary, and the circle is accordingly said to be convex. For a pentagram, two points on its boundary can be connected by a line passing outside of the boundary — the pentagram is therefore non-convex.

shall see in Chapter 6, made great use of this sphere-defining characteristic of the Platonic solids.

While studied by the mathematically inclined for millennia, perhaps one of the most intriguing characteristics of the Platonic solids is that there are only five of them. This is a rather remarkable outcome when we think of the fact that there are in principle an infinite number of 2-dimensional shapes that we could draw, and yet, only three of them (the triangle, the cube, and the pentagon) can produce closed polyhedra satisfying Definition 3.1, and then only five such polyhedra can be constructed from these generating surfaces. Such a result demands proof. For indeed, how do we know that there are five and only five such solids? Could it be, one might ask, that some additional Platonic solid, with faces made up of a polygon having 48 sides (say), might exist and that we simply haven't constructed or experienced such a solid in real life? This is a fundamental question, since it soon becomes reasonably clear through a little experimentation (say with card and tape in the modern era) that five solids can be constructed from triangles (the tetrahedron, octahedron, and icosahedron) and squares (the hexahedron or cube) and the pentagon (the dodecahedron), as soon as we introduce shapes with more than five sides, no solid shape, without gaps, can be constructed. Such questions, of course, inspire the mathematician, and mathematicians are good at proving specific statements or theorems. Let us state the problem in mathematical terms:

Theorem 5PS: There are five, and only five, Platonic solids.

What is needed now is a proof of theorem 5PS. Let us begin this proof by making a few straightforward statements and then (because we can) grab one mathematical result due to the great Swiss mathematician Leonhard Euler:

(1) The minimum generating face for any solid is the triangle — anything with fewer sides cannot be joined together to make a 2-dimensional shape.
(2) Leonhard Euler (1707–1783) has shown by a straightforward counting scheme that for any solid having V vertices (or corners), E edges, and F faces that

$$V - E + F = 2 \tag{3.1}$$

(3) By straightforward counting, we also find the following relationships for any solid made with faces that have p edges and in which q faces meet at each vertex: $pF = qV = 2E$. To see a demonstration of this rule, take a look at the hexahedron. A cube is made up of $F = 6$ faces, each of which is composed of a square ($p = 4$) with three faces meeting at each vertex ($q = 3$). By simple counting, we have $V = 8$ vertices and $E = 12$ edges. Now, straightforward multiplication yields $pF = 4 \times 6 = qV = 3 \times 8 = 2E = 2 \times 12 = 24$, and this demonstrates our result.
(4) Substituting the results from (3) into Euler's formula, equation (3.1) then gives us, after a bit of algebraic rearranging, the result that for any Platonic solid,

$$\frac{1}{q} + \frac{1}{p} = \frac{1}{2} + \frac{1}{E} > \frac{1}{2} \tag{3.2}$$

with the last, >½ step, simply accounting for the fact that E (which might in principle be very large) cannot be less than 3 (corresponding to the triangle and the generation of the tetrahedron: the minimal Platonic solid).

The final step builds upon the fact that the smallest building unit is a triangle and that p and q must therefore always be greater than or equal to 3. The rest of our proof can be done graphically. The graph of p versus q, which must always be integers, is shown in Figure 3.7. The region where $1/p + 1/q < ½$ is to the upper left in the diagram (above the dotted line), and this indicates the region in which Platonic solids are excluded — remember, Platonic solids can only exist for (p, q) combinations (points) that fall below the dotted line. So, by inspection, the only (p, q) combinations that satisfy Euler's equation are (3, 3), (4, 3), (3, 4), (5, 3), and (3, 5) and these configurations correspond to the tetrahedron, the cube, the

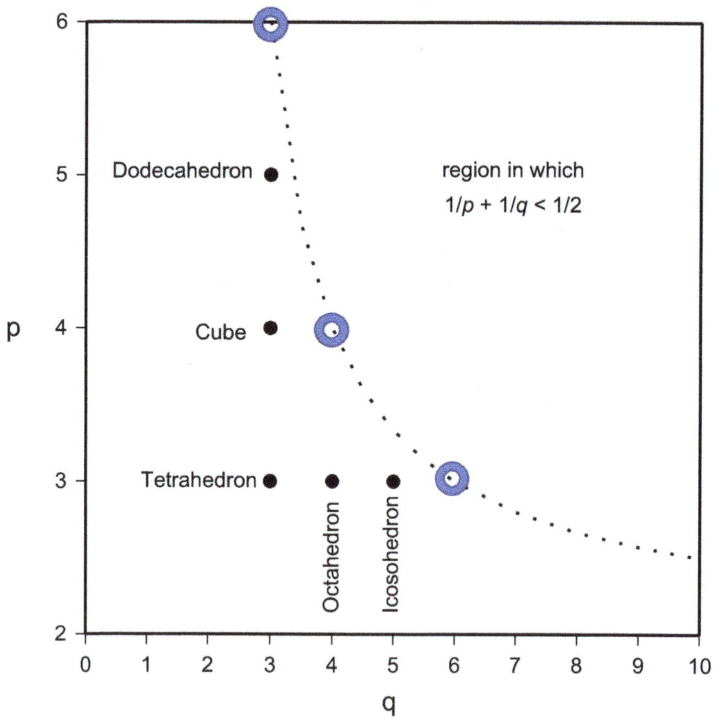

Figure 3.7. The (p, q) combinations that pick out the Platonic solids. The circled points pick out those polygons that tile the plain.

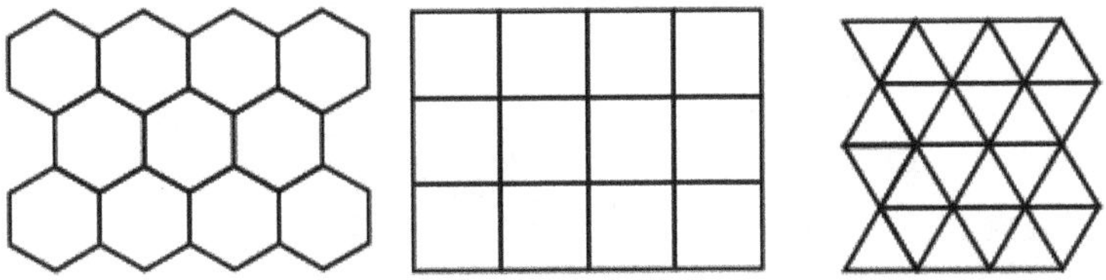

Figure 3.8. Regular plain tiling as generated by the hexagon, the square, and the equilateral triangle.

octahedron, the dodecahedron, and the icosahedron, respectively. There are no other combinations of (p, q) — in spite of their being an infinite number of integers — that will satisfy the condition that $\frac{1}{q} + \frac{1}{p} > \frac{1}{2}$ once p and q are greater than 5.

The dotted curve in Figure 3.7 corresponds to the special case where $1/p + 1/q = ½$, and this is satisfied for the (p, q) combinations (3, 6), (4, 4), and (6, 3); these points are circled on the graph. These combinations correspond to what is known as regular plain tiling — literally shape combinations that can cover the entire 2-dimensional plane without any gaps (Figure 3.8).

As an alternative "observational" proof that there are only five Platonic solids, we need only make recourse to the natural world. If there were more than five such solids, one strongly suspects that evolution would have found them by now. That is, through the amazing structures exhibited by radiolarians (Figure 3.9): protozoa that are 0.1–0.2 mm in size. These incredible lifeforms have mineral skeletons that take up the shape of the five Platonic solids. This, of course, is not a mathematical proof, but evolution is very thorough when it comes to teasing out available niches and geometries, and it has additionally had many millions of years in which to experiment and try out all possible combinations.

The finite completeness of the Platonic solids, that there are five and only five of them, captured the imagination of the Greek philosophes. There was a provably finite group of objects, within the infinitude of possible objects, that had a fixed set of attributes (that are satisfied by definition PS) and clearly, by this property of

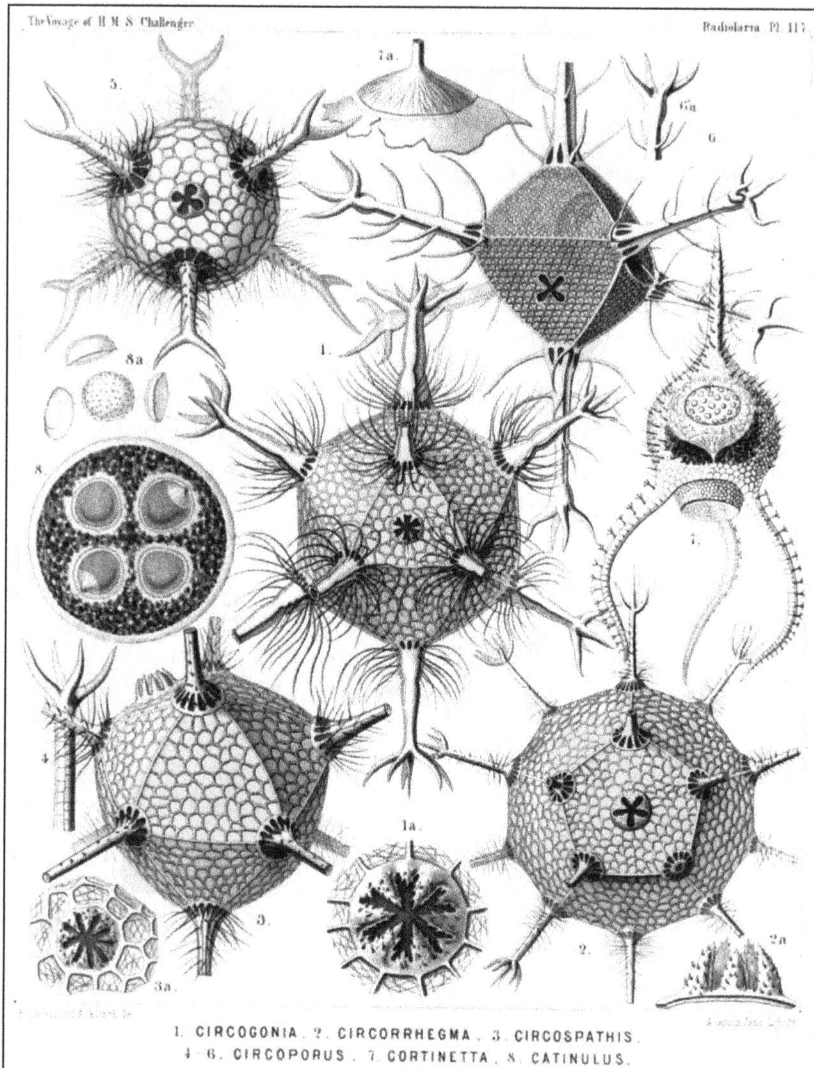

Figure 3.9. Radiolarians that have adopted shapes corresponding to the Platonic solids.

completeness, must be special. Indeed, we have a clear example of the human mind imposing importance on a specific group of objects purely because of their mathematical attributes. It is not the mathematics that is important (although it is intellectually stimulating) but the human association and imposition of importance to a mathematical result; this is something that Johannes Kepler championed in the late 16th century (see Chapter 6). In the modern era, this imposition of mathematics upon reality still dominates much thinking, and while physicists have moved on from the Platonic solids, they still struggle to define many aspects of subatomic structure in anything other than complex mathematical terms. This does not mean, of course, in the sense of putting the cart before the horse, that the structures so developed are mathematical realities. The ever-growing complexities of physical models (e.g., string theory) may not be a result of the complexities of reality but a result of the fact that physicists have started their search for reality in the wrong place [3, 4]. As the old saying goes, "you can't get there from here".

Plato, in the *Timaeus*, attaches specific attributes to the five-regular polyhedra; indeed, they are taken to be the very building blocks of the cosmos (recall Figure 3.5). The solids are the embodiment of the five elements, air, earth, fire, water, and quintessence, and the very fact that there are five and only five Platonic solids means that everything in the universe must be explainable in terms of a finite set of building blocks. This is clearly a powerful idea, although, by modern standards, entirely wrong. And yet, for the ancient philosophers, the idea was rich with possibility, and it satisfactorily explained the world as they knew it. In modern times, the methodology of association is no different, we just have an expanded, but still finite, set of basic building blocks (atoms), and while it is no longer *shape* that is important, it is the enumeration of atomic mass and the number of associated electrons. From the finite list of atoms in the periodic table of elements, all chemistry flows, and all structures are made.

3.3. The Platonic Universe

The importance, with respect to ancient philosophy, that was attached to the Platonic solids was something imposed by the human mind: it was not a mathematical result dictating human understanding of the universe. To echo and paraphrase a comment once made by pioneering astrophysicist Arthur Eddington, the first key idea was philosophical (that is, it was a human construct) and the mathematics wasn't there and made relevant until someone put it there. There is, in some quarters, a trend in the modern era to project mathematics as being somehow fundamental to the structure of the observable universe. This, of course, is a delusion, and what we call mathematics is really a flow of inventions constructed from an imaginary *realm* that are artificially (and artfully) mapped onto observable phenomena. In effect, the proponents of the universe being composed of a series of mathematical structures perversely confuse the mathematical model with actuality. For all this philosophical bickering, and it is an argument that has run for thousands of years, mathematics is an extremely useful tool for describing the way in which the universe appears to work, but it is nothing more than a useful tool. We attribute importance to mathematical results, not because the universe is some mathematical construct, or because it was designed by or run by some mathematician, but because problems and results that can be cast in idealized mathematical terms are useful *entities* for the human mind to work with. In other words, mathematics is imposed upon the universe by human beings.

The idea that there is some fantastic place, distinct from the observable universe, in which all mathematical theorems, proofs, and infinite forms exist in some ready-made, timeless stasis is due to Plato. The Platonic realm contains all of the ideals for the shapes, geometry, and numbers that the mathematician manipulates. In this realm, we are told, there are such entities as perfect infinitely small points, perfect triangles in which the angles really do add up to exactly 180 degrees, and where the full infinite expansions of π, e, and $\sqrt{2}$ really do exist. It is a realm in which time has been expunged, everything is and nothing changes; indeed, there is nothing new in the Platonic realm since it contains the infinitude of all possible things. There are essentially two ways of looking at the Platonic realm (Figure 3.10): one way is to bestow upon it an independent, distinct, but real existence and then consider mathematics to be the method by which it can be accessed and explored by the human mind; in this way, mathematical results are discovered because they already existed in the Platonic realm. A second way is to think of the

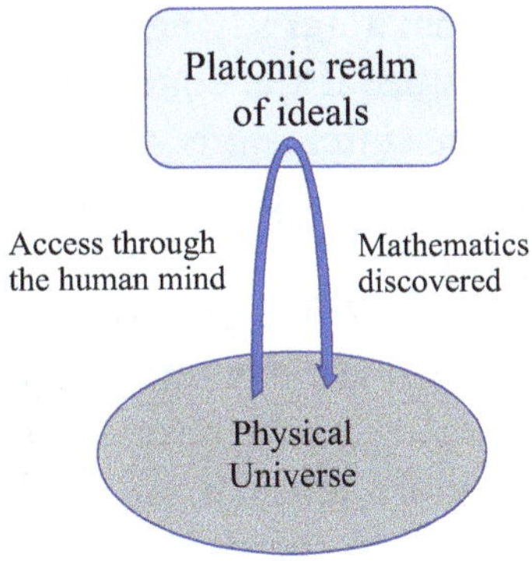

Figure 3.10. The two channels by which the physical universe and the Platonic realm could be connected. In this interpretation, the Platonic universe is a real, timeless, but physically distinct realm, only accessible by the human mind. In this manner, all mathematical results are interpreted as being discoveries rather than inventions.

Platonic realm as an idealized domain, with no actual existence, constructed through human abstraction and need. In this manner, the Platonic realm emerges from the physical universe (through human abstractions) and any mathematical results found there are invented with no prior existence being implied.

For all of its unreality, the Platonic universe provides a comfort zone: a place where infinities, monsters, and mathematical strangeness can exist, somehow, and somewhere other than in the observable universe. Many mathematicians buy into the idea that their results are discovered and that the Platonic realm is a real place that exists independently from and outside the observable universe (see Chapter 12). There is, of course, absolutely no proof that this is the case, and the question which asks if mathematics is discovered or invented remains profoundly vague and probably has no specific answer (see, however, Chapter 11).

3.4. The Sorites Paradox

Eubulides of Miletus, who lived in the 4th century BC, it is said, was worried about paradoxes and vagueness. He is perhaps best remembered for his exposition of the liar's paradox, in which a philosopher makes the statement, "what I am saying is a lie". This presents the listener with a paradox, since if the statement is true, then the philosopher is lying, but if the statement is a lie, then the philosopher is telling the truth. Among his other paradoxes, the one of interest here is the sorites (that is heap) paradox. This paradox cuts to the idea of vagueness and runs as follows: imagine that you have a heap of a million sand grains, and you remove one grain from the heap. This surely makes little difference and the heap of sand remains a heap of sand. You remove one more grain from the heap, and again this makes next to no difference to the collection, and the heap of sand remains a heap. Eventually, you have removed so many grains from the heap that only one grain is left. Is then this single grain still a heap or not? And, if the single grain is not a heap, with the removal of which specific grain in the sequence did the heap of sand change into something else? The paradox essentially arises according to the vagueness of the word "heap."[3] Indeed, vagueness is found everywhere within spoken language, and the basic philosophical problem is that there is no way to gauge, beyond some arbitrary set of definitions, where and/or when the vagueness ends. Logic alone, for all of its cut-and-dry powers, cannot resolve the vagueness problem.

The boundary, or borderline, beyond which a heap of sand is no longer a heap of sand, is both unclear and arbitrary, and this runs counter to our expectations of the world, and the phenomena within it, being precisely defined. Indeed, we expect that ultimately the behavior and description of the physical world should be expressible in precisely defined physical laws and mathematical terminology. Ontic vagueness, therefore, is something beyond the reach of mathematical description, and it is presumably something that will not apply to laws of physical reality, or is it? Philosopher Eddy Keming Chen believes that ontic vagueness could, in fact, be something that

[3] A similar paradox is that known as the ship of Theseus. In this case, it is envisioned that over time the various parts of a ship that have rotted or no longer function properly are faithfully replaced with a new component. After a while, the entire ship has been replaced although its appearance has not changed, and the question becomes as follows: Is it still the original ship? Further, if it is not the original ship, then when did it become a new or different one?

permeates even physical laws (and not just language) [5]. What if, Chen questions, some fundamental laws are inherently vague, not as a result of some specific lack of knowledge or data but because they are intrinsically so? Such laws would act in a fundamentally different way to nomic exact laws (such as those found in classical mechanics, quantum mechanics, and general relativity) since they imply a fundamental limit to exact knowledge and they indicate the existence of a realm entirely unreachable through mathematical analysis. Ontic vagueness, if it genuinely exists, implies that some indeterminacy must exist within the observable universe and that such indeterminacy is beyond the reach of mathematical ideals; indeed, vagueness is not something that is deemed to exist in Plato's universe. Any law that is susceptible to a sorites-like paradox, Chen argues, could be an indication of nomic vagueness at play, and by way of example, he has investigated the implications of such ideas on the cosmological Past Hypothesis. This hypothesis is a fundamental law that reflects upon the origin of the universe and the arrow of time. The Past Hypothesis postulates that the observable universe must have begun in a low entropy state with a very high degree of order. This postulate is an attempt to account for the fact that we do sense a directionality to time, moving from the very beginning (the Big Bang), through the distant past, to the present day, and onwards to the future. This directionality of time is apparent even though the world is governed by time-symmetric physical laws. The Past Hypothesis accounts for the flow of time through the 2nd law of thermodynamics, which states that in a closed system, the overall disorder, quantified statistically by a term called the entropy of the system, must always increase. This result essentially allows for the fact that there are more disordered states that the system could be in than ordered ones. In this picture, the arrow of time is created by the continued increase in disorder. The full details of Chen's arguments concerning the ontic vagueness of the Past Hypothesis need not concern us here; the point, however, is that if his arguments hold true, then it means that mathematics will never be able to fully describe the objective order of the observable universe. The final theory of physics, Chen therefore argues, might not be entirely expressible in

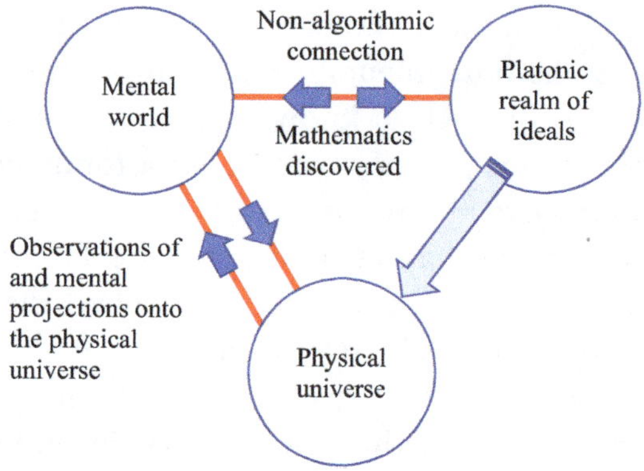

Figure 3.11. Penrose has championed the idea of three interconnected realms. The mental world, which operates within human consciousness, is able to access in a non-computational manner, the perfect and timeless realm of Platonic ideals. In turn, the physical universe is taken to be a corrupted simulacrum of the Platonic realm. Through a tangible web of observations and mathematical discoveries, the human consciousness begins to make sense of the physical world.

mathematical terms. Nomic vagueness does not deny the great power of mathematics to explain the physical universe; it simply argues that mathematics will never fully capture all of its machinations.

The admittance of nomic vagueness provides for a situation in which the workings of reality might escape the grasp of full mathematical description. Furthermore, Nobel Laureate Roger Penrose has also suggested that the cognitive processes in the mathematician's (indeed, all human) brain might be something beyond the grasp of mathematical description (Figure 3.11) [6]. Indeed, Penrose's argument cuts to the very core of artificial intelligence, and the claim is made that the mind (consciousness) acts in a non-algorithmic fashion. In other words, thinking and inventiveness (the very stuff of mathematics) are not equivalent to running some very complicated, step-by-step, computer program. Penrose writes, "mathematical truth is not something that we ascertain merely by the use of an algorithm."[4] Does this non-algorithmic property of consciousness

[4] Algorithm is taken here to mean some calculational procedure such as can be carried out on a computer.

imply that mathematics must be invented? In many ways, yes, it does, but Penrose maintains that somehow or other, human consciousness is still able to make contact with the Platonic realm of ideals (viewed as a very real but physically inaccessible place). In this manner, mathematics, according to Penrose, is discovered and the physical universe is seen as some form of shadowy projection of an ideal and timeless Platonic world. For all this, the idea of a real Platonic universe continues to remain vague and, for many, unconvincing; it is a convenient philosophical fallback that provides no physical answers about the reality in which we live. Just like the many-worlds interpretation of quantum mechanics, the cosmological multiverse, or the multiple extra-dimensions invoked in string theory, the idea that the Platonic realm is somehow real, physical, and influential is formulated on little more than hope and theoretical moonshine.

3.5. Aristotle's Alternative

There is something both liberating and depressing about the idea of a timeless and perfect Platonic realm. The liberation allows for the mathematical mapping of ideal forms into our shadowy and time-ravaged (observable) universe. The depression resides in the fact that we will never see or achieve perfection on Earth or, indeed, find it anywhere in the observable heavens. Well, so be it. There is no law that says humanity (and our distant descendants) are deserving of finding perfection. Of course, the Platonic realm is just one of many philosophical constructs that have been developed (by humans over many thousands of years) to explain the appearance and actions of the physical world. Perhaps its nearest philosophical rival is the approach promoted by Aristotle, and in some sense, it simply turns the concept of the Platonic realm upon its head. The Platonic tradition essentially sees the universe as being simple, and ordered in its form, with a timeless mathematical underpinning. In contrast, the Aristotelian viewpoint is that the universe, at heart, is complicated, random, and fundamental temporal in nature. In the Platonic approach, the idea is to identify the timeless and unchanging laws of nature (and these are expressed through timeless mathematics in the form of symmetries). In contrast, the Aristotelian approach is to place emphasis on the actual observations rather than the unobservable rules that underlie them. The world, according to this approach, is messy, dirty, fully alive, and kicking, and there is no underlying assumption that the complexities are explainable in simple mathematical terms. In essence, in the Aristotelian perspective, we do not observe the laws of nature but only the outcome of those laws operating on actual objects. It is not so much the origins, therefore, but the outcomes that inspire the Aristotelian approach. The deduced and repeated outcomes from the observations can still be expressed mathematically, but echoing the sentiments of Arthur Eddington (recall Chapter 1), the mathematics is not there until we (that is humans) put it there. The laws of nature, as such, are the compression, distillation, and fermentation of the observations that are actually made. The proof, according to Aristotle, is in the tasting of the pudding rather than in its recipe.

One field of modern research that sports aspects of both Platonic and Aristotelian interpretation is that of quantum mechanics. It is Platonic in the sense that it is formulated according to a mathematical wavefunction, which is a complex-valued probability amplitude. One cannot see a wavefunction, however, it is in essence a wave of information, but its mathematical properties enable calculations to be made. The Aristotelian aspect resides in the fact that the outcome of quantum mechanical calculations (derived according to the fully deterministic — that is classical world — Schrodinger equation) can be understood even though we have absolutely no idea what it is that nature is actually doing. The probabilistic and uncertain nature that is inherent to any quantum mechanical calculation only appears when an observation is made, and it is the observation that attempts to extract specific classical information (e.g., the speed, location, and direction of travel of an electron).

Interpreting quantum mechanical outcomes in terms of a many-worlds outlook, which argues that all possible outcomes are realized in an infinity of other universes that are causally disjoint from our own, is highly Platonic in its outlook, but it basically seeks to sweep unwanted problems of interpretation into inaccessible realms. Physicist Richard Feynman famously commented that one should not try to understand quantum mechanics (madness lies down that path), and a utilitarian mantra of "shut up and calculate" has effectively evolved within the user community. We might not know what is actually going on deep down, but we do have sets of equations and mathematical functions that can be solved in order to make predictions, in some cases to exquisite accuracy, about observable outcomes. And, in an Aristotelian, practical sense, the point to this is that it works, and what more, therefore, does one want? Poet Samuel Coleridge once remarked that every man is born either a Platonist or an Aristotelian, and around this division has grown the idea that the Platonists are reasoners, artists, poets, mathematicians, and symbolists, while the Aristotelians are logicians, classifiers, and encyclopedists. Coleridge is only partly right, and there are many different philosophies that one can espouse, but this is another story altogether.

The pragmatic Aristotelian outlook of "shut up and calculate" provides a new way to study and interpret the heavens. It is the observations that need to be accounted for, and this can be achieved by any means that are legally allowed mathematically. Indeed, Claudius Ptolemy, in the 1st century BC, under the Platonic guise of "saving the appearance", used just such a philosophy to assemble a set of geometrical models capable of describing planetary motion to a high degree of accuracy, and it is this that we look at in Chapter 4.

References

[1] Evans, J. *The History and Practice of Ancient Astronomy*. Oxford University Press, Oxford, 1998.

[2] Cromwell, P. *Polyhedra*. Cambridge University Press, Cambridge, 1997.

[3] Hossenfelder, S. *Lost in Math — How Beauty Leads Physics Astray*. Basic Books, New York, 2018.

[4] Penrose, R. *Fashion, Faith, and Fantasy in the New Physics of the Universe*. Princeton University Press, Princeton, 2016.

[5] Chen, E. Fundamental nomic vagueness. *The Philosophical Review*, **131**, 1–49, 2022.

[6] Penrose, R. *The Emperors New Mind — Concerning Computers, Minds, and the Laws of Physics*. Vintage, London, 1989.

Chapter 4

Ptolemy's Solutions

One of the most important realizations to emerge from the musings of the ancient Greek Philosophers was that the universe is amenable to rational understanding. The universe, the Earth, and the everyday world, all can be explained by *natural* rules based upon physical cause and physical effect. The universe is not powered by magic or mystery, and human beings can comprehend the complex machinations behind its workings. What a wonderful, liberating, realization this was — which is not to say, of course, that the universe isn't full of seemingly magical and mysterious phenomena. Incredibly, however, we can be sure, on the basis of this ancient rationality, that any mysteries that appear to exist within our present understanding of the universe are only provisional. There is absolutely no need within any physical (that is rational and scientific) theory that we, as human beings, might develop to ever use the phrase: "insert miracle here". Capricious gods, irrationality, and wondering spirits do not animate the world nor do they dictate or control the circumstances of our lives. That the world is understandable according to rational thought, and that it can be investigated and interpreted according to rational explanations, is a truly remarkable discovery, and there is, of course, no reason that it had to be so. The universe, the Milky Way galaxy, the solar system, and planet Earth, all are entirely indifferent to our (that is humanities) existence, and it is not at all obvious that we, as *homo sapiens*, should have any ability to understand how it is that the whole *thing* works. And yet — and yet — in spite of this innate indifference, there is a very real sense in which it can be argued that we do understand some of the physical workings of the universe. This understanding is one of the great triumphs of adopting a scientific outlook — an outlook predicated, nurtured, and honed on the realizations of the ancient Greek Philosophers, and the denial of miracles and spiritual influence. Writing in the early 17th century Galileo insisted that the universe (and nature) could only be interpreted through the language of mathematics, and by mathematics Galileo essentially meant geometry. In this sense, Galileo was buying wholesale into the astronomy of the ancient Greeks, and, by default, this was effectively the works of Ptolemy, although his allegiance was purely *in spirit* rather than in *nature*.

4.1. The *Syntaxis*

Claudius Ptolemy lived in Roman-controlled Alexandria during the 2nd century AD, and he was a great (and talented) compiler of information and knowledge. His name has rippled down through history largely because of his production of three great compendiums: the *Tetrabiblios*, *Syntaxis Mathematica* and *Geographia*. The first of these works is concerned with astrology, the second astronomy, and the third geography. Each of these compendia had its specific innervations, and each has been historically controversial in one way or another. Of interest to us here is the *Syntaxis Mathematica*, which was a 13-volume compendium on all astronomy. Book I is concerned with Aristotle's cosmological model, and sets out the idea of a spherical Earth located at the center of a spherical celestial sphere. The remaining books considered the motion of the stars through the sky, along with the motion of the Moon, Sun, and Planets through the celestial backdrop. The *Syntaxis* was intended to be a practical astronomy textbook, and through it, and a later text, the *Planetary Hypothesis*, all the theories and tools necessary for predicting planetary positions are presented. In terms of its aims, as a teaching reference, the *Syntaxis* was supremely successful, and it became the go-to book on astronomy for the ensuing 1200 years. For all this, it had many detractors, who produced commentaries criticizing one part or another, but none could find an alternative to the Ptolemaic model — that is, until the epoch-changing work of Copernicus in the mid-16th century (see Chapter 5). So successful was the *Syntaxis* that after its appearance, older astronomy texts were essentially abandoned. Indeed, while the *Syntaxis* was widely copied, it was largely preserved in Arabic translations, first made in the 9th century. Under the Arabic title of the *Almagest* (meaning, the greatest) Ptolemy's text eventually found its way into European universities during the 12th century.

4.2. The Marvelous Equant

Geometry was at the heart of the *Syntaxis*, although it also included a star catalog (listing some 1022 stars), original observations, and mathematical tables. For all this, the *Syntaxis* carefully outlines how observations of the heavens can be made and interpreted, and how the planets move through the constellations. With respect to the latter issue, Ptolemy extended the search, begun by Hipparchus (recall Figure 3.4), for legal loopholes to Plato's dictates of planetary motion (recall Chapter 2). To this end, not only was the Earth displaced away from the center of the deferent (*carrying* circle), the planet was envisioned to move upon a secondary epicycle, the center of which moved at a constant angular rate about the equant — the equant point itself being displaced from the center of the deferent — the essential geometry is shown in Figure 4.1. The Ptolemaic model is effectively a 6-parameter model, and the key variables are the displacement of the Earth and the equant point from the center of the deferent,[1] the angular rate at which the center of the epicycle moves along the deferent (as seen from the equant point), the radius of the deferent, and the radius of the epicycle. There is also a constraint related to the position of the (so-called) mean Sun. Each of the main parameters can be changed and adjusted so as to bring the model predictions into agreement with respect to the observations of each of the planets.

[1] This being said, they were usually taken as being equidistant from the center of the deferent.

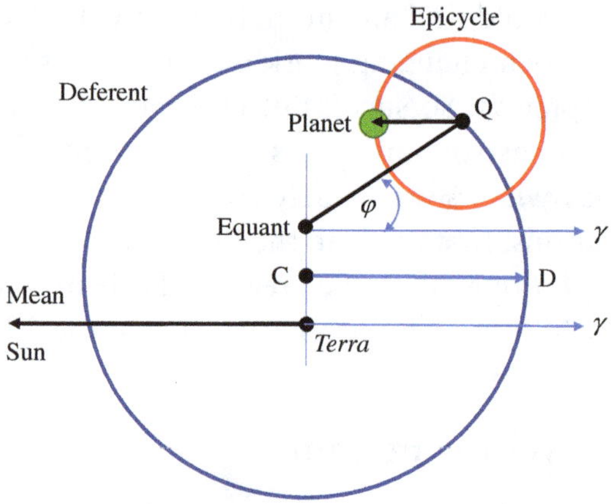

Figure 4.1. The essential planetary model developed by Ptolemy. C is the center of the deferent, Q is the center of the epicycle, and γ is the direction to the first point of Aeries. See text for additional details.

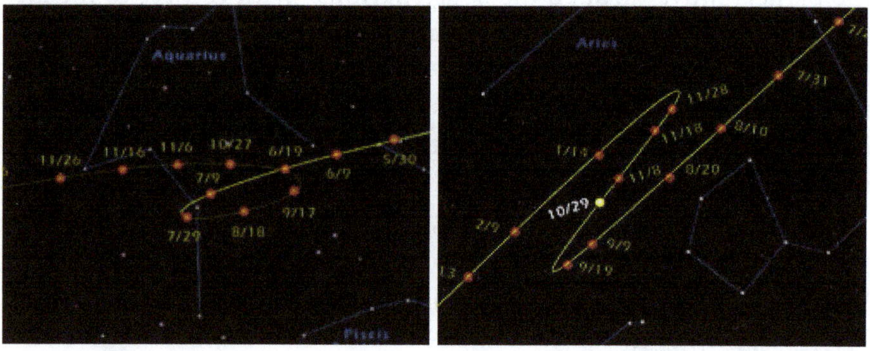

Figure 4.2. Retrograde motion of Mars (left) as seen between May and November 2003, and (right) from July 2005 to March 2006. Image courtesy of NASA.

The epicycle was introduced by Ptolemy for a very specific reason. Indeed, the epicycle is a highly elegant way of keeping true to Plato's dictates and explaining retrograde planetary motion (Figure 4.2) — a motion that is decidedly noncircular in behavior (at least in the sky). The essence of retrograde motion is a reversal in the direction of advancement that a planet makes with respect to the background stars. Generally, planetary motion is westward with respect to the fixed stars, but at regular intervals, the progress slows and reverses (moving eastward), before once again slowing and reversing again to continue a steady westward drift with respect to the constellations. This is not the kind of motion that one would expect of a planet moving in a circle at a constant speed about the fixed Earth. To explain this motion, a fix to the pure circular motion idea was required, and the way to do this was first described by Apollonius of Perga in the 3rd century BC. Specifically, Apollonius realized that the retrograde loops could be described through a combined set of circular motions. In this manner, he constrained the planet to move along a secondary, or epicyclic circle, the center of which moved around a primary, or deferent circle. In this manner, the planet is still constrained to move along a circular path. As Ptolemy found, with appropriate adjustments, the epicyclic architecture could be adjusted so as to account for the times at which retrograde motion, for each specific planet, was observed to occur.

The basic parameters that make the Ptolemaic model work are the radius of the deferent (CD in Figure 4.1), the offset of the equant point, and the Earth, and the radius of the epicycle. After this, the model is set in motion with respect to angular variations tapped out in accordance with the motion of the Sun. The key angles to fix are those relating to the rotation of the epicycle (angle φ in Figure 4.1), and the position of the mean Sun — both of which are traced out with respect to a zero-point on the sky located at the first point of Aries (γ). This zero-point is the location on the celestial sphere where the ecliptic and celestial equator intercept each other when the Sun is moving from below the celestial equator to above it (this occurs around March 20th each year). The mean Sun is a fictitious Sun that is imagined to move around the ecliptic at a constant speed. Accordingly, the mean Sun location and the actual Sun locations only coincide exactly twice per year — at the time of the equinoxes when the Sun cuts through the celestial equator. Where the constraint of the mean Sun comes into play is that the radial arm Q-Planet (Figure 4.1) must always lie parallel to the Earth-mean Sun line.[2] Since the model is governed according to successive passages through the equinox (first point of Aries) location, the appropriate year time interval is that of the tropical (or solar) year Y_T corresponding (in modern terms) to 365 days, 5 hours, 48 minutes, and 45 seconds ($Y_T = 365.24219$ days). The tropical year differs from the sidereal year — the time for the Sun to complete a full orbit with respect to the stars, by about 20 minutes. The Gregorian calendar, introduced in October 1582, was specifically developed (with its leap year rules, and its initial 11-day jump forward with the 4th of October being followed by the 15th of October) so that the calendar year and the solar year remain approximately synchronized (at least over time intervals of thousands of years). It is in this way that the important dates of the equinoxes and solstices fall on approximately the same calendar days each year.

With respect to the first point of the Aries reference line (Figure 4.1), the angles swept out by the mean Sun φ_{MS} and the angle φ of the epicycle center Q as seen from the equant are as follows:

$$\varphi_{MS} = (360°/Y_T) \times T \text{ (days)}$$

where T (days) is the time interval since the day of the vernal equinox and

$$\varphi = (360°/\text{planet period}) \times T \text{ (days)}$$

where the planet period is appropriately adjusted to that observed for each planet. Table 4.1 shows characteristic values for the complete Ptolemaic models for Venus, Mars, Jupiter, and Saturn. Details for Mercury and the Moon are not given in Table 4.1 since additional constraints were placed on their motion and these need not directly concern us here.

By way of example, Figure 4.3 shows the position of Mars calculated over a 3000-day interval. In this diagram the radius of the

[2] This rule does not apply to the case of Mercury and Venus, although the center of their epicycle must always fall on the Earth–Sun line.

Table 4.1. Planetary calculator parameters (with reference to Figure 4.1). The distance O-Earth is taken to be equal to O-Equant, and the radius of the deferent has been fixed as 10 units.

Planet	O-Equant	CD	Q-Planet	φ (deg./day)
Venus	0.15	10.0	7.23	0.986
Mars	1.03	10.0	6.56	0.524
Jupiter	0.48	10.0	1.92	0.083
Saturn	0.53	10.0	1.05	0.034

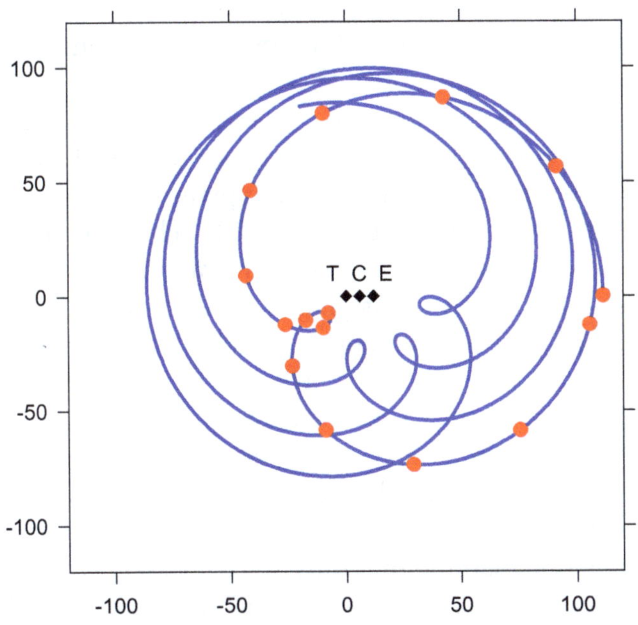

Figure 4.3. The motion of Mars over the course of 3000 days. The scale is arbitrary, with the deferent radius CD being set to 60 units. T corresponds to the position of the Earth (*terra*); C to the center of the deferent, and E to the location of the equant. The red dots pick out the position of Mars at 50-day time intervals for the first 700 days, with the arbitrary starting point being set at the far right in the diagram [at position (111.4,0)] corresponding to the maximum Earth–Mars separation.

deferent is set to CD = 60 units, giving an epicycle radius QP = 39.4 units. The origin (0, 0) in the diagram corresponds to the location of the Earth (T = *Terra*), with the center of the deferent C being at (6.2, 0), and the equant point being at (12.4, 0). The calculation starts at the arbitrary location when Mars is at its greatest distance from the Earth (111.4, 0). Retrograde loops occur at intervals corresponding to 780 days (the synodic period of Mars), with each retrograde loop occurring at a different location (as measured from T) on the ecliptic, with the varying Earth distances experienced during each period of retrogradation indicating a different maximum brightness of Mars. Because the rate at which the angle determining the location of the epicycle center Q on the deferent is non-rational, the specific path followed by Mars and the locations of the retrograde loops never exactly coincide.

Once the geometrical parameters and the angular rate of progress, for each planet, have been tabulated (as per Table 4.1), one can set about the process of determining planetary positions into the past and into the future (as per Figure 4.3). The primary reason, of course, for computing such planetary information was the casting of horoscopes — rather than satisfying any specific astronomical interests. The art of casting horoscopes, was, in its own right, a highly organized body of science, with intricate rules, methodologies, and interpretations being applied to the zodiacal placings of the planets. The idea that planets might be something other than prognostication signposts is very much a modern concept. We now think of planets as distinct objects, with a topology, mass, size, rotation, and specific orbit, but this is something that only started to come about with the work of Kepler and Galileo in the early 17th century (see Chapter 6).

The great structural feature of Ptolemy's model is that it can be fine-tuned to each planet, and that, once tuned, it provides highly accurate positional results. Simply put, Ptolemy's model works, and it works very well. It allows calculations of planetary positions to be made, and, when tested, the predictions agree with the observations. Indeed, it does all that is required of any scientific model — it provides a correct description of the actual observations. Such success, however, was not achieved without its critics. Front and center of the criticisms leveled against Ptolemy's model was his introduction of the equant. The problem, of course, is that it insisted upon a constant angular rate of advancement in the epicycle center (Q in Figure 4.1), a dictate that resulted in the epicycle center moving with a non-constant speed along the deferent — and this, of course, contradicted Plato's constant motion dictate. In a practical sense, the equant is a brilliant mathematical invention that enabled the planetary model to work (with respect to making accurate predictions); philosophically, however, it rankled subsequent commentators. The scorn directed toward the equant[3] was entirely predicated on the *idee fixe* that the heavens must be perfect, and that its perfection was enshrined within the simplest repetitive motion (constant speed around a circular path). At issue was the tension between an assumed human ideal and the workings of a mathematical model. The ancient commentators, right through to Copernicus, were not objecting to the Ptolemaic model, and its use of the equant, because it didn't represent reality, but because it didn't represent their perceived (indeed, philosophically imposed) idea of reality.

4.3. The Analog Heavens

While the deferent and epicycle combination is an inherently geometrical construction, it is also a combination that can be realized mechanically through a linked pair of arms (Figure 4.4). Indeed, at the heart of the Ptolemaic model is the idea that there are two radial arms (of length EQ and QP) linked at Q, that sweep-out angles at a constant rate (ω_E and ω_Q) about the pivot points E and Q. To make the model or linkage work one simply needs to specify the appropriate lengths and rotational angles. Of course, there is nothing actually simple about the choice of parameters, since they have to be fine-tuned to describe the actual observations.

[3] Georg Rheticus (acting on behalf of Copernicus) in his *Narratio Prima* (written in 1540) called the equant a "relation which nature abhors". Copernicus was a little more tactful in his *De Revolutionibus* (published in 1543) and simply referred to the equant as a "defect" in the Ptolemaic model.

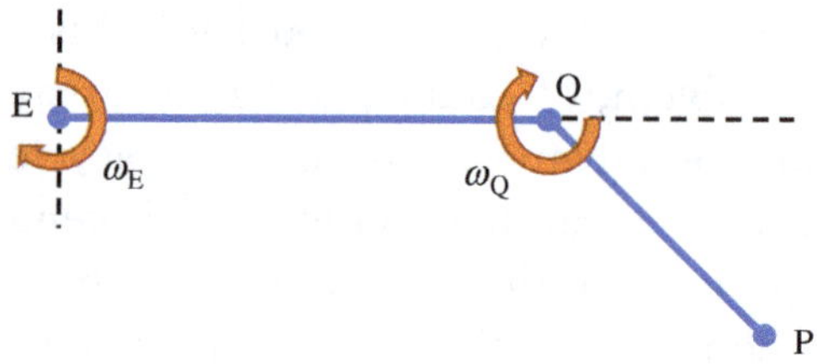

Figure 4.4. Mechanical linkage model for epicyclic motion. Radial arm EQ rotates about E with an angular speed of ω_E, while radial arm QP rotates about Q with an angular speed of ω_Q.

The double-linkage interpretation of the Ptolemaic model readily allows for the construction of mechanical models to determine planetary positions. In this manner the calculation proceeds according to an analog computation. Such models allowed for the development of a tactile sense of heavenly motion. Indeed, such mechanical models provide a satisfying feeling of rationality and stability — the very things that the ancients wanted, and expected, of the heavens. Figure 4.5 shows a medieval volvelle (from the Latin *volvere*: to turn) developed by Peter Apian, in the 16th century. This particular volvelle allows for the determination of the sky position of planet Saturn. The two yellow-colored disks correspond to the deferent and epicycle. The position of the center of the epicycle is set by rotating the orange disk (of which the yellow deferent disk is part). The position of the planet is then determined by rotating the epicycle disk about the center of the small red circle. With an appropriate set of tables to provide starting positions (that is, the known, or calculated, locations of Saturn on certain set dates), the sky position of Saturn can be determined on any specified day into the past or into the future.

The first mechanical representation of the Ptolemaic system, using geared wheelwork, was completed, after many years of construction, by Giovanni de Dondi in 1364 (but see Section 4.4). Called the *Astrarium*, this device sought to show the motion of each planet, via geared wheelwork, mounted within the frame of a seven-sided cage. Consisting of 107 wheels and pinions this device showed the positions of the Sun, Moon, and planets, and it calculated the

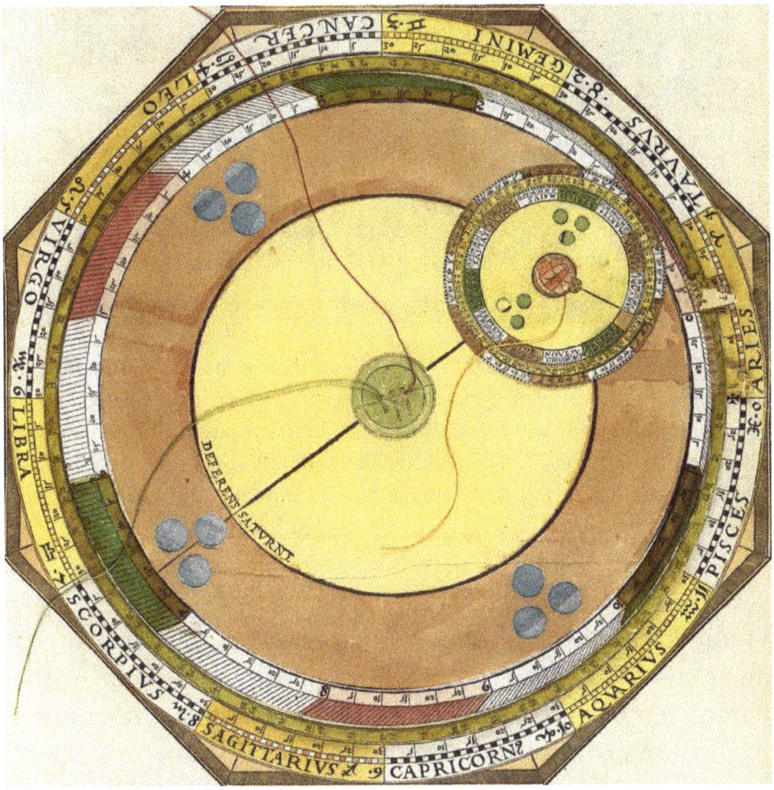

Figure 4.5. The Saturn volvelle from Peter Apian's *Astronomicum Caesareum* (first published in 1540). The various colored strings are used to mark out the direction of the mean Sun and the zodiacal position (depicted in the outermost circle) of the planet as seen from Earth.

dates of the moveable religious feast days. More recently, Stuart Malin designed and constructed (Figure 4.6) a Ptolemaic orrery using 54 gear-wheels [1]. Malin comments that the requirement that the epicycles provide parallel motion to the Earth-Sun line (the Q-Planet line in Figure 4.1), "presented serious mechanical headaches". These "headaches", of course, only come about if one attempts to make an integrated planetary model — something that Ptolemy never intended. The volvelle and de Dondi's *Astrarium* avoid this latter mechanical complication by working with one planet (per disk or clock face) at a time.

Figure 4.6. Mechanical Ptolemaic Orrery designed by Stuart Malin. In this construction, the planetary representations have been split into two groups. To the left is shown the crank and gearwork for the Sun, Mars, Jupiter, and Saturn, while to the right is the wheelwork for the Moon, Mercury, Venus, and the Sun. In each representation, the Earth is at the center of the wheelwork. The orrery base plate is 60 centimeters across. Image courtesy of Stuart Malin.

As noted by Malin, one of the practical drawbacks of the Ptolemaic model is that each planet requires its own set of descriptive parameters (as seen in Table 4.1). Indeed, this non-commensurability feature proved to be one of the critical spurs for Copernicus (in the 16th century) to develop a new architectural description of the planetary system (see Chapter 5). Interestingly, however, within the Ptolemaic system are the seeds that blossomed into its eventual replacement — specifically, the alternative Copernican arrangement championed by Tycho Brahe in the late 16th century. To see this latter connection, we need to first re-iterate the key point that in the Ptolemaic model, the motion of each planet is described according to the ratio of the radii of the deferent and epicycle: $R_{\text{deferent}}/R_{\text{epicyle}}$ = constant. And, while traditionally it was the radius of the deferent that was set to some specific value (i.e., 10 units in Table 4.1), there is no mathematical reason why one cannot set some specific size to the epicycle radius and adjust the size of the deferent appropriately. Under this latter constraint, a translation relationship can always be applied to any deferent — epicycle combination (Figure 4.7). That is, the position of the planet, with respect to the Earth, is the same whether one follows the deferent radial arm and then the epicycle

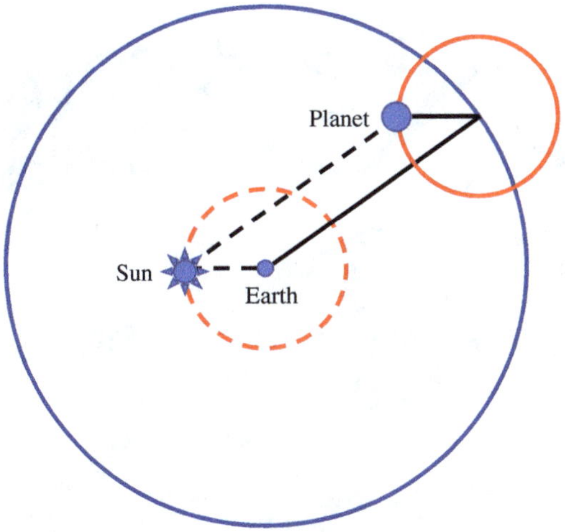

Figure 4.7. Parallelogram theorem relating to the equivalency interchange of the circles for the deferent and epicycle. The proof of the equivalency theorem is by inspection and by simply translating the solid and dashed lines in the parallelogram.

radial arm (solid black lines in Figure 4.7), or *vice versa* in an interchanged system (dashed lines and circle in Figure 4.7). By reversing the order of the deferent and epicycle one can then scale the epicycle to have a radius of 1 unit, which would be equivalent to 1 astronomical unit, and this would represent the motion of the Sun along the ecliptic as seen from the Earth. The deferent radial arms would then correspond to the distance of each planet from the Sun. The minds-eye picture that this reorganization provides is that of a system in which the Earth is fixed, the planets all move around the Sun, and the Sun with its attendant planets moves around the Earth. Indeed, this is the Tychonic planetary system. This equivalency-translation of the epicycle and radiant appears to be a mathematical trick that the ancient astronomers largely missed, and while it cannot be directly transformed into the Copernican planetary model, it contains the suggestion of a possible new ordering: if all the planets orbit the Sun, then why not the Earth as well? This story will be continued in Chapter 5.

Having developed a mental, or geometrical, picture of something, it seems to be a characteristic of human nature to thereafter

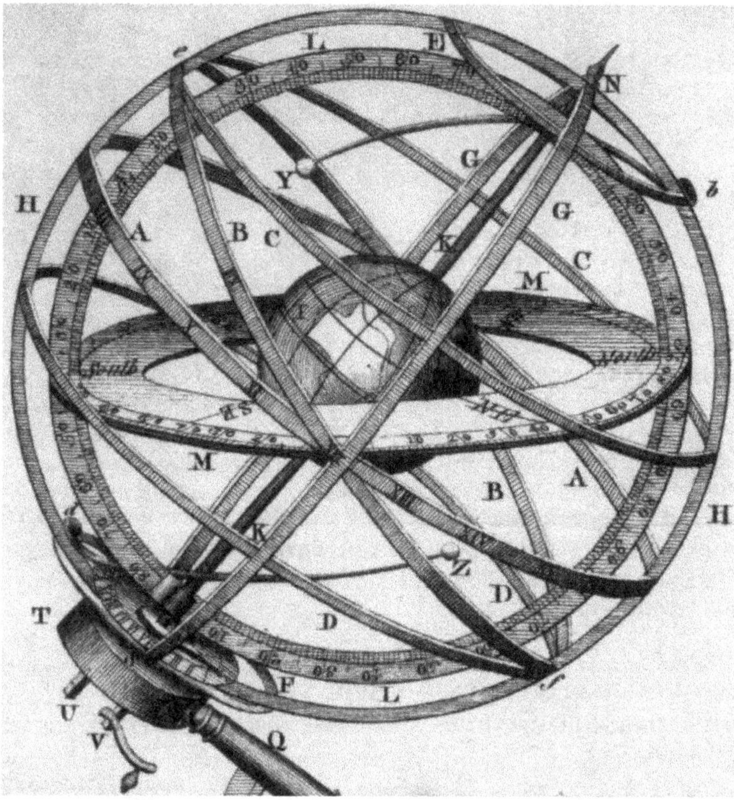

Figure 4.8. The armillary sphere. In this diagram, the Earth is at the center (labeled T), and the Earth's spin axis is labeled K, with N being the north celestial pole. Circle A corresponds to the celestial equator, while circles C and D correspond to the Tropic of Cancer and the Tropic of Capricorn. The inclined circle B corresponds to the ecliptic. The (horizontal) disk M delineates the observer's horizon. Circles E and F correspond to the southernmost limits of the Arctic Circle and the Antarctic Circle, respectively.

make a physical model of it — transforming, thereby, intangible thoughts into the tangible and tactile ones. In the case of the celestial heavens, this transformation of mind stuff into solid matter resulted in the construction of the armillary[4] sphere (Figure 4.8). Such spheres

---

[4]The term armillary relates to the banded or hooped (amulet) construction of the sphere.

allow for a view of the cosmos from outside of the cosmos — indeed, the armillary sphere provides an impossible view of the cosmos. Furthermore, the armillary sphere is a physical model of a human-created concept, not a model of reality. For all this, the armillary sphere is a brilliant teaching tool, and it provides a perfect 3-dimensional representation with which to learn about the practical side of the heavens. Indeed, while mention of the first armillary spheres dates back to the time of the ancient Greek philosophers, they are still demonstration models found within astronomy labs at universities to this very day. The point is, they simply work, and that they are simple to work. Not only are armillary spheres still a standard part of the astronomy curriculum, but they provide a description of the coordinate system used by astronomers to map out the heavens.

Many different forms of the armillary sphere have been made, but the basic form is that in which the planets have been removed from the picture, with the key components being the circles of the ecliptic and celestial sphere (recall Figure 3.1). At the center of the armillary sphere is a stationary Earth, and related to this is a third (movable) circle that represents the observer's horizon. It is the horizon, of course, that divides the celestial sphere into the visible heavens (above the horizon) and the invisible heavens (below the horizon). By changing the angle of the horizon circle, which is equivalent to changing the observer's latitude on Earth, the motion of the heavens can be mimicked as the celestial sphere itself rotates about the spin axis that joins the north and south celestial poles.[5] Such adjustments will reveal, for example, that for an observer located at either of the Earth's poles, all stars rotate around the sky parallel to the horizon, none rising and none setting, with the celestial pole being located in the observer's zenith. For an observer located on the Earth's equator, the celestial poles will be right on the northern and southern points on the horizon, and all stars will rise and set perpendicular to the horizon. This practical demonstration,

[5] From the modern perspective, the rotation axis of the celestial sphere corresponds to the Earth's spin axis.

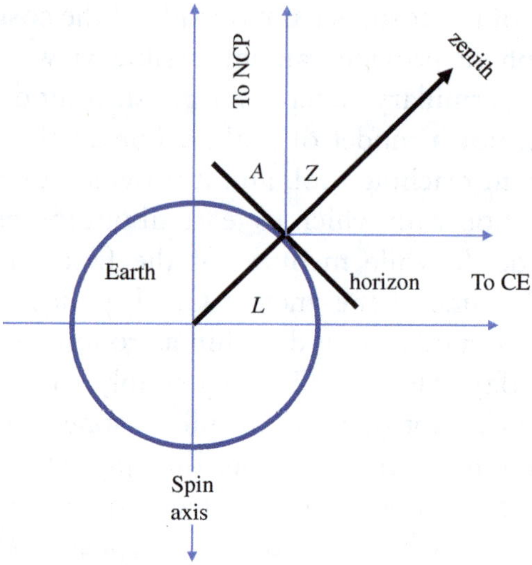

Figure 4.9. Geometry underlying the proof of Theorem ACP.

can in turn be incorporated into a mathematical theorem that sits at the very heart of terrestrial navigation.

Theorem ACP: The altitude of the visible celestial pole corresponds to the observer's latitude on Earth.

Proof of ACP: The proof is straightforward and is illustrated in Figure 4.9. An observer at latitude L on the Earth observes the (north) celestial pole at an altitude of A degrees above the horizon. The line from the center of the Earth, through the observer's feet, and on into the sky defines the local zenith. By definition, at right angles to the zenith line, through the observer's feet, is the plain of the horizon. Now, using the right-angle conditions, we have angles $A + Z = 90°$. Also, $L + Z = 90°$. Substituting for Z in the first relation provides the result (and proof) that $A = L$.

The theorem and proof of ACP just presented makes the assumption that the celestial sphere is very much larger than the Earth. In this way, the sightline along which the altitude A is measured is actually parallel to the Earth's spin axis, and the sightline at 90° to the spin axis runs parallel to the celestial equator. That the Earth was spherical was well established by Ptolemy's time, and indeed in the 2nd century BC. Eratosthenes had estimated the Earth's radius to be within a few percent of the modern-day value. That the Earth must be spherical was further established by theoretical and observational arguments offered up by Aristotle.[6]

The armillary sphere is not a particularly useful observational instrument, although it can, and has been used in such a fashion — Ptolemy used it to convert between coordinate systems on the ecliptic and the celestial equator, and so to did Tycho Brahe in the late 1500s. Rather, the armillary sphere is a model of a concept of the heavens, and it illustrates how the key components of the celestial sphere appear to Earth-based observers. The armillary sphere, however, can be transformed into a portable, highly useful, multifaceted instrument. The secret is to systematically project the celestial sphere onto the 2-dimensional plane that passes through the celestial equator. Such a transformation turns the armillary sphere into an astrolabe (from the ancient Greek for *star-taker*).

The origins of the astrolabe trace back to the ancient Hellenistic philosophers, but they were brought to engineering (and artistic) perfection by Arabic instrument makers in the 9th and 10th centuries. Traditionally made of brass, the astrolabe (Figure 4.10) is a thoroughly beautiful instrument, and with it, one can determine the rise and set times of specific stars on any given day during the course of the year. It could also be used to gauge angles in the sky, or determine heights for surveying purposes, and, according to a specific design, it provided scales to convert unequal hours, of the day and night,[7] into equal hours. The main body structure is the circular *mater* (or mother) plate, the center of which corresponds to the

[6] The most practical observation that demonstrates the curvature of the Earth is that of ships disappearing hull down over the horizon.

[7] While the division of the day into 24 hours has a long and ancient history, the initial convention was to divide the daytime and nighttime into 12 hours each. In this manner the actual minute length of hours would change according to the time of year — there only being 24 equal hours (each 60 minutes long) on just two days each year — corresponding to the days of the equinoxes. The adoption of a pure, every hour is 60 minutes long, system only took hold once mechanical clocks were invented in the early 14th century.

Figure 4.10. A brass astrolabe from the early 14th century. Height 13.2 centimeters, thickness 1 centimeter. This image shows the filamentary *rete*, which indicates specific star positions, against a background *climate* plate for a latitude of about 50° north of the equator. The center of the *rete* corresponds to the north celestial pole, the small inner circle within the *rete* corresponds to the ecliptic, the outer diameter of the *mate* corresponds to the Tropic of Capricorn, south is to the top (indicated by the suspension ring attached to the *throne*) and East is to the left. Arcs and circles (centered on the observer's zenith) of constant altitude (almucantars) can be seen on the *climate* plate. Image courtesy of the British Museum.

north celestial pole (NCP). Seated within the *mater* is a *climate* plate that shows circles and arcs representing equal altitude positions for a specific latitude on Earth — with these circles and arcs being centered on a point corresponding to the observer's zenith (Z). On top of the *climate* plate sits, the *rete* (literally, *net*), which, through a series of filamentary arcs and pointers, indicates the relative positions (typically 20 or so) of prominent fixed stars. The *rete* can rotate about a central peg. According to whether a specific star-pointer is above or below the horizon (as indicated on the *climate* plate) so the star itself can be either seen, or not, from the specific latitude and time of observation. On the back of the astrolabe is a calendar for establishing the zodiacal location of the Sun on any given day of the year. This was established by engraving two centrally offset circles in the manner described in the solar model of Hipparchus (recall Figure 3.4). A rotating arm enabled the reading of the calendar scale. Many astrolabes also had an *alidade* for gauging angles when the device was hung vertically from its support ring.

Key to the design of the astrolabe is a stereographic projection from the location of the south celestial pole (Figure 4.11). Such a projection preserves angles and circles around the celestial sphere when they are reproduced on the surface of the astrolabe. The projection plane is that corresponding to the celestial equator. By running lines between features on the celestial sphere and the south celestial pole (SCP), so their interception locations on the celestial equator can be determined. Figure 4.11 shows how the circles of the ecliptic, the Tropic of Cancer, and the Tropic of Capricorn are located. While numerous examples of medieval astrolabes still survive, very little is known about the craftspeople[8] who designed and made them. Clearly, however, a skilled workforce capable of forging and producing flat brass plates, along with artisans able to cut and stamp remarkably accurate scale divisions, and scribe marks, must have existed, but their story is mostly lost to the annals of history. For all this, among the very first books on astronomy written in the English language was Geoffrey Chaucer's *Treatise on the Astrolabe* (first published in 1391). Better known, perhaps, for his prose work, Chaucer's *Treatise* is a highly detailed account,[9] and not only does

[8] Craftspeople is the appropriate word since it is known that women (nuns in particular) were involved in the design of astrolabes and in the computation of astronomical tables.

[9] Chaucer's *Treatise* is not an original work, as such, and it draws heavily upon earlier manuscripts that had been published in Latin. Radically in the 14th century, Chaucer felt that it is easier to lean and understand matters when they are written in one's familiar language — this notion begins to explain why the *Treatise* opens with an address to a young boy, "Lyte Lowys my sone" — not that Chaucer actually had a son called Lewis.

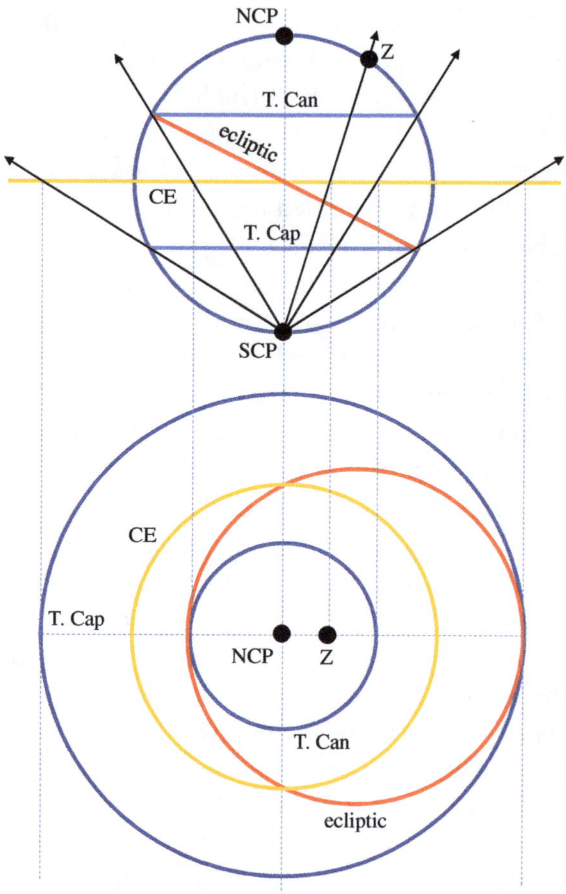

Figure 4.11. Stereographic projection of the celestial sphere from the south celestial pole (SCP). The top image is a side-on view of the celestial sphere. The lower image is a plan view of the celestial sphere projected onto the celestial equator (astrolabe). The outer perimeter of the astrolabe corresponds to the circle of the Tropic of Capricorn, with the smaller concentric circles corresponding to the celestial equator and the Tropic of Cancer. The ecliptic is again a circle, but its center is displaced away from the central point corresponding to the north celestial pole. (NCP). The zenith point Z (for an observer located at a latitude of about 50° north of the equator) is indicated and appropriately displaced away from the NCP in accord with Theorem ACP.

he explain how to make one's own astrolabe (out of card), but he also explains how to use it.

While the astrolabe is an analog model and computer of the celestial realm, it does not show planetary positions, or, indeed, enable a direct calculation of planetary positions to be made. This being said, if the ecliptic location of a planet is known (through observation, or by calculation), the astrolabe can be used to determine rising, culmination, and setting times. Specialized and highly complex geared devices were eventually constructed, however, to aid in the determination of planetary positions. Amongst the earliest such mechanical devices are the *Albion*, designed and built by Richard of Wallingford[10] circa 1326, the *Astrarium* (astronomical clock) by Giovani De Dondi circa 1364, and the *Equatorie* (literally, equation solver) designed and built by John of Westwych circa 1394 [2].

4.4. The Paradox that is the Antikythera Mechanism

When they come to model heav'ns,
And calculate the stars, how they will wield
The mighty frame; how build, unbuild, contrive,
To save appearances; how gird the sphere,
With centric and excentric, scribbl'd o'er,
Cycle and epicyle, orb in orb.

John Milton, *Paradise Lost*

Wonder piled upon wonder: there is no better way to describe the Antikythera Mechanism (Figure 4.12). It is an absolute gem. Plucked from a long-lost shipwreck by sponge divers in 1900, the mechanism is, to paraphrase Winston Churchill, a mystery contained within an enigma [3–5]. This incredible mechanism is (though perhaps it is more correct to say was) a mechanically geared device constructed

---

[10] Richard of Wallingford also designed and oversaw (until his death in 1336) the construction of the astronomical turret clock installed in St. Albans Abbey starting circa 1327. The exact meaning behind the name *Albion* is unclear.

Figure 4.12. An exploded view of the Antikythera Mechanism as recreated by Freeth *et al.* [8]. To the far left is the front face which shows the zodiacal positions of the Sun, Moon (along with phase), and the planets. To the far right is the back plate which shows a series of calendrical dials capable of predicting the times at which an eclipse will occur. Image courtesy of Tony Freeth and University College London.

with the clear intent of describing a number of calendric, planetary, solar, and lunar cycles. While it is incredible that any fragments of the mechanism survived the watery shipwreck, the additional fact that it was constructed circa 100 BC is even more remarkable. The mechanism is unique, having no surviving prototypes, no surviving copies, and no surviving provenance. Indeed, geared mechanisms of similar sophistication were not to be made in Europe until the first half of the 14th century. All in all, the origins of the Antikythera Mechanism are a mystery — we do not know who built it, or who specifically designed it, and to a large extent we have no clear idea of its intended use. This latter point, however, is a topic of great debate, and like all good mysteries, there is much to tease the imagination and just enough solid evidence to hint at various design possibilities. Perhaps one of the greatest oddities about the Antikythera Mechanism is that there is no mention of it in Ptolemy's *Syntaxis*.

This absence is suggestive of the fact that its existence, even in ancient times (Ptolemy was writing some 200 years after the devices were constructed), was not widely known. It is also suggestive that it was a one-off device, made not for a practicing astronomy, but for a wealthy patron. It was not a device to be used, as such, but it was a prestige piece intended for display.

Since its discovery and recognition as an object of great importance, there have been many attempts to explain the Antikythera Mechanism. The basic problem is that the mechanism is largely in fragments, and it is clear that many pieces are missing, there are, however, some 30 recognized gears (most of which reside in the single large fragment — see Figure 4.13). The late Yale University historian of science Derek de Solar Price published the first comprehensive

Figure 4.13. The 82 fragments that constitute the Antikythera Mechanism. For further details and images, see http://www.xtekxray.com/antikythera_images.htm. See also the web page for the Antikythera Mechanism Research Project at: http://www.xtekxray.com/antikythera_images.htm.

description of the Antikythera Mechanism in 1974, and in a remarkably detailed study suggested a reconstruction of the machine using a total of 31 gears [3]. A recent reconstruction by mathematician Tony Freeth and co-workers at University College London uses 36 gears, of which 7 are hypothetical (with one of the surviving gear fragments not actually being used in the reconstruction).

While the precise details of its construction may never be known, the existence of the Antikythera Mechanism clearly tells us one important, hitherto unrecognized, thing, and that is that there must have been a vibrant community of skilled metal workers, engravers, and artisans, in the ancient world over 2000 years ago. Indeed, the discovery of the Antikythera Mechanism has forced historians to re-evaluate their assumptions about ancient Greek technology. There are a few hints found within the surviving literature that suggest the existence of magnificent astronomical machines, but no useful details or descriptions are provided by the ancient authors. In their recent work, however, Tony Freeth and co-workers [6–8] have upped the ante somewhat, and suggest that Hipparchus may have been involved in the mechanism's design — a suggestion based upon the realization that a mechanical version of his lunar theory is incorporated within the surviving gearing. Other evidence points to Archimedes being involved in the mechanism's design.

Many suggestions have been made concerning the intended user, or users, of the Antikythera Machine, but this issue continues to remain unresolved [4]. Not only this, but the question also arises as to why was it built at all. In spite of its sophistication, it seems reasonably clear that it was not intended to make precise, hour of the day, predictions of astronomical phenomena. Indeed, the deduced gearing ratios, scale markings, and the hand-crank operation of the Mechanism, while very good, would have at best provided only approximate times at which, for example, eclipses would occur, and at what approximate times various planetary alignments might take place. At best, the machine would have only followed what astronomers (astrologers) of the time already knew was likely to be happening in the sky — there are no apparent means, for example, of setting the device to some specific date either into the future or the past.

Second, one could argue, the prediction of eclipse times and other astronomical phenomena was achieved well enough through hand and table-based calculations by scribes. The other important point to bear in mind is that information as to eclipse times, and calendar interpolations was not something of any great importance in ancient Greek society. In addition, the complex construction and intricate workings of the Antikythera Mechanism rule out its use as a visual demonstration piece — showing, as it were, the clockwork of the heavens (a term that would only have significance and resonance from the late 17th century onwards — see Chapter 6). Indeed, the Antikythera Mechanism is a classic example of what John Herschel, in the early 19th century, would call "very childish toys" (see Chapter 8). Meaning by this, that such a device might mimic celestial motion, but it only produced the motion through hidden and mysterious (to the observer) ways. This all suggests that the Antikythera Mechanism was actually an exhibition or trophy piece, probably built for a rich patron, who never actually used it. The Antikythera Mechanism is a treasure for sure, but a calculator built for real astronomical (or astrological) use, by a practicing astronomer, mostly likely not. For all this, the Antikythera device is beautifully designed, crafted, and engineered, but it is an essentially useless device. One of the reconstructed dials, for example, shows the 4-year cycle associated with the Olympic games — a significant social event, yes, but hardly a complex astronomical cycle requiring mechanical evaluation.

4.5. Ptolemy's Theorem — A Mathematical Aside

Claudius Ptolemy was a polymath, interested in many areas of study — not just astronomy. He was a pioneer and a compiler of knowledge in a number of fields, including geography, cartography, optics, and astrology. Mathematics, and specifically geometry, however, was at the core of much that he did, and he must have known Euclid's *Elements* in detail. Indeed, Ptolemy had strong opinions about mathematics, arguing in the *Syntaxis* that mathematics was as close to knowing the divine as the human mind could approach.

Mathematics, he explained, "is concerned with inquiry into things that are *always such*". This latter constancy, the *always such*, is equivalent to the modern concept that a mathematical proof, once proved, is true for all time. It seems appropriate, therefore, that since history (and the cosmos) has denied Ptolemy his planetary model (see Chapters 5 and 6), he should be associated with an *always such* theorem. Ptolemy's theorem, so-called, is a statement about the interrelationship between the side lengths (chords) of any quadrilateral inscribed within a circle. With reference to the labeling shown in Figure 4.14, Ptolemy's theorem states that the product of the diagonals is equal to the sum of the products of adjacent sides:

$$AC \times BD = AB \times CD + AD \times BC \qquad (4.1)$$

As a corollary to Ptolemy's theorem, it is noted that if the quadrilateral ABCD corresponds to an inscribed square, then AC = BD, AD = BC, and AB = CD, and accordingly, we regain Pythagoras's theorem (recall Chapter 2):

$$AC^2 = AB^2 + AD^2 \qquad (4.2)$$

The proof of Ptolemy's theorem is straightforward, but the details need not concern us here. Although seemingly unwieldy,

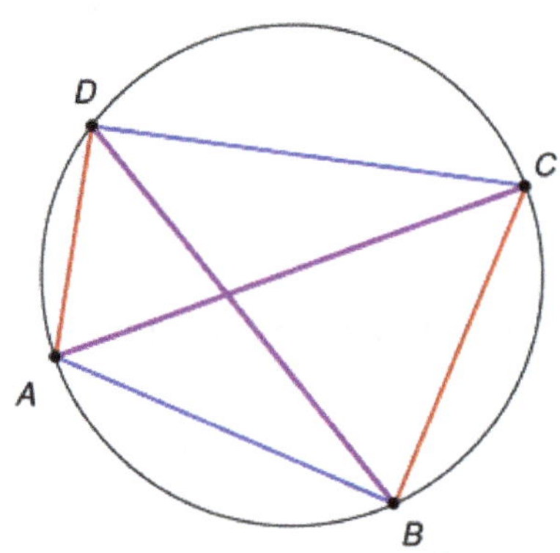

Figure 4.14. Ptolemy's theorem applies to any quadrilateral ABCDA inscribed within a circle.

equation (4.1) is a gem, and it allows for numerous further mathematical constructions. Indeed, it provides an important bridge between geometry and trigonometry. While Ptolemy worked in terms of chord lengths, it is standard practice now (well, starting in the 16th century) to work with the sine, cosine, and tangent functions of an angle. To see how Ptolemy's theorem can be used to establish trigonometric identities, we can re-configure Figure 4.14 so that AC corresponds to the diameter of a unit-length circle — as shown in Figure 4.15.

With this new configuration, Thales theorem indicates that we must be dealing with a rectangle since (by the theorem) $\angle ABC = \angle ADC = 90°$. Furthermore, $\sin \alpha = AD/AC = AD$, since AC has been set equal to 1. Likewise, $\cos \alpha = CD/AC = CD = AB$, since AC is set equal to 1, and $AB = CD$. From equation (4.2) we now find the trigonometric identity: $\cos^2 \alpha + \sin^2 \alpha = 1$, again, since $BD^2 = AC^2 = 1$. Additionally, sum and difference expansions of the sine and cosine functions can be derived with the help of Ptolemy's theorem.

If, in contrast to Figure 4.15, we re-arrange the quadrilateral in Figure 4.14 to the form shown in Figure 4.16, a generalization of Pythagoras's theorem applicable to any triangle, not just right-angle triangles, can be established. Here, in Figure 4.16, we have two congruent triangles ABD and ACB, with DC being parallel to AB. To the diagram, we have added the construction lines DF and CE which are perpendicular to AB. By observation, we now see that

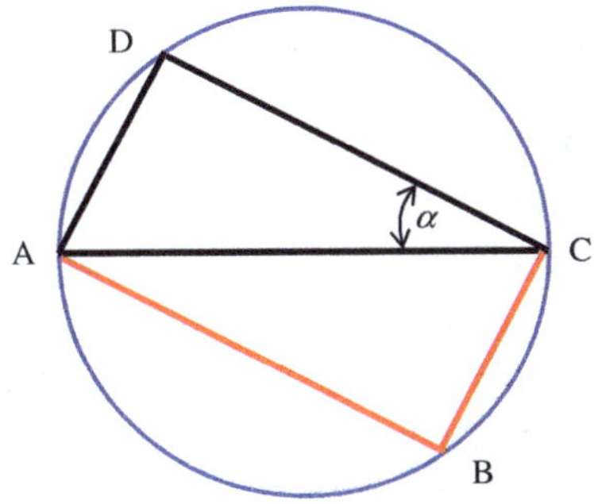

Figure 4.15. Quadrilateral ABCD is configured so that AC lies along the diameter of the circumscribed circle.

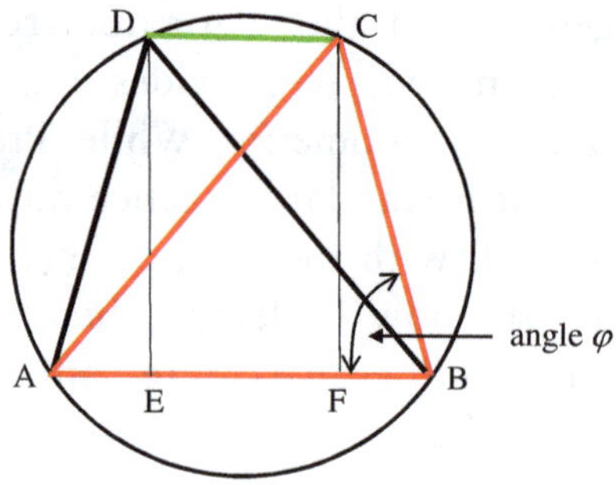

Figure 4.16. Arrangement of two congruent triangles ABC and ABD to form a quadrilateral ABCD.

DC = AB – AE – FB = AB – 2FB since AE = FB. Furthermore, FB = BC cos φ. Applying Ptolemy's theorem gives AC × BD = AC^2 = AD × BC + DC × AB = BC^2 + DC × AB, since AC = BD and AD = BC. Noting now that FB = BC × cos φ, so with a little algebra we obtain the cosine rule that is the generalization of Pythagoras's theorem.

$$AC^2 = AB^2 + BC^2 - 2AB \times BC \cos \varphi \tag{4.3}$$

With equation (4.3), Pythagoras's theorem is recovered when angle φ = 90°. Within equation (4.3) is the measure of all Earthly things, enabling as it does, the art of surveying and the accurate construction of terrestrial maps, but this is another story altogether [9].

4.6. The Mighty *Syntaxis*

Ptolemy's *Syntaxis* was the final word on ancient Greek astronomy — a last, and lasting, hurrah. More than just a compendium on astronomy, however, it promoted the power of mathematics. Indeed, Ptolemy placed mathematics at the very core of all philosophy and advanced the idea that it was only through mathematics that humanity might approach perfection. Like Aristotle before him, Ptolemy divided philosophy into two domains: theoretical and practical. Practical philosophy deals with knowledge concerning action and includes such activities as the arts, the practice of medicine, ethics, and political discourse. Theoretical philosophy, in contrast, is concerned with knowledge for its own sake. Ptolemy, furthermore, divided theoretical philosophy into three distinct domains: natural, mathematical, and theological. Of these, natural philosophy is akin to the modern fields of physics, chemistry, biology, and geology. Mathematical philosophy is concerned with numbers, geometry, theorems, and proofs. Theology (or metaphysics) is concerned with first causes, and the manner in which the divine can be approached through the action of human reasoning. Of these philosophical domains, that of mathematics, according to Ptolemy, was supreme, and what knowledge physics (natural philosophy) provided, beyond that of the conjectural, was solely due to its borrowing of mathematical techniques. Mathematics, Ptolemy believed, provided pure, useful, and certain results. It was through mathematics specifically that the heavens could be ordered and understood, and Ptolemy wrote the *Syntaxis* to convey this order and understanding.

Ptolemy's mathematical, "just so", constructions provide for an extremely good description of planetary motion, and the *Syntaxis* can be viewed as a triumphant piece of human reasoning. That the workings espoused by Ptolemy are not actually built upon a true representation of our planetary system is not actually relevant, or particularly important — the Ptolemaic model works, and it works very well. For all this, however, the *Syntaxis* existed in a state of continual tension both with Aristotelian teachings and Plato's cosmological dictates (as introduced in Chapters 2 and 3). The key sticking points were the epicycles, the equant, and compound motion — none of these essential elements to the Ptolemaic model had counterparts that could be mapped into the Aristotelian cosmos. The tension was real, and the grumblings of Ptolemy's critics and reviewers echoed down the centuries. Ptolemy's mathematical web caught the observed motions of the heavens, but it did so at the price of working outside of the perceived (human-idealized) properties of the cosmos. This underlying tension was always there, right through to the appearance of the Copernican hypothesis in the mid-16th century, but in a practical sense, it could be ignored. Indeed, to the practicing calculator, the Ptolemaic model provided all the mathematical tools necessary for the determination of accurate planetary positions — the rest was a problem for the philosophers.

References

[1] Malin, S. Constructing the first Ptolemaic orrery. *Astronomy and Geophysics*, **58**, 1.41–1.42, 2017.

[2] Falk, S. *The Light Ages: The Surprising Story of Medieval Science*. W. W. Norton and Company, New York, 2020.

[3] Price, D. and de, S. Gears from the Greeks. *Transactions of the American Philosophical Society*, **64**, 1–70, 1975.

[4] Beech, M. The Antikythera Mechanism — Still a mystery. *Journal of the Royal Astronomical Society of Canada*, **101**, 93–94, 2007.

[5] Jones, A. *A Portable Cosmos: Revealing the Antikythera Mechanism, Scientific Wonder of the Ancient World*. Oxford University Press, Oxford, 2017.

[6] Freeth, T., *et al.* Decoding the ancient Greek astronomical calculator known as the Antikythera mechanism. *Nature*, **444**, 587–591, 2006.

[7] Freeth, T., *et al.* Calendars with Olympiad display and eclipse prediction on the Antikythera Mechanism. *Nature*, **454**, 614–617, 2008.

[8] Freeth, T., *et al.* A model of the cosmos in ancient Greek Antikythera Mechanism. *Scientific Reports*, **11**, article 5821, 2021.

[9] Beech, M. *The Pendulum Paradigm: Variations on a Theme and the Measure of Heaven and Earth*. Brown Walker Press, Boca Raton, 2014.

Chapter 5

Siting the Glorious Lamp

Mathematics is written for mathematicians.

— N. Copernicus

5.1. All Change

Something absolutely remarkable happened in the mid-16th century — an alternative planetary model to that described by Ptolemy appeared. This being said, the seed and defining idea that blossomed into the new model had been planted in ancient Greece some 1800 years earlier. To move forwards, Copernicus looked backwards. Indeed, he returned to the ancient authorities, and set the Sun, that glorious lamp, at the center of the planetary system. There was no crisis in astronomy when Copernicus began his task to re-invent the planetary model, all he wanted to do was produce a simple method by which planetary positions could be calculated, and to do this he re-instated the old Platonic ideals of circular orbits and constant orbital speeds. Reinstating these ideals would, he believed, do away with the need for deferents, and epicycles, and such contentious features as Ptolemy's equant. He was only partially successful in this latter respect.

Copernicus did not develop his new planetary model in an intellectual vacuum — this was not the case of an isolated genius having a great epiphany. Rather, Copernicus picked-up an old idea that had been abandoned, and then spent many decades developing the mathematical details needed to make it work. Copernicus was an able slogger, and he knew full well that the system he was developing would not please the authorities. The church authorities would grumble about how it contradicted scripture, the university-based scholars would object to its anti-Aristotelian component, and humanists would grumble that it demoted the central importance of humankind.

To make progress in his new description, Copernicus took up an idea first expounded by Aristarchus circa 250 BC. This being said, there is no actual record of what Aristarchus might have actually written, and it is only through comments made by Archimedes, in his *Sand Reckoner* (recall Chapter 1) that the ancient authority of the idea is preserved. Aristarchus rejected the Aristotelian notion of a central, stationary Earth, and argued that the Sun occupied the central point of the cosmos, returning to a parallel with the Pythagorean concept of a central fire. The Sun so located, dictated that the Earth must orbit around it and additionally spin upon its own axis. Here is an example of a more complicated theory (requiring an orbiting, spinning Earth) being presented as an alternate simpler theory (in which there was just a fixed Earth). Indeed, this is partly why the Sun-centered model never gained any specific traction in the ancient era. Not only was it dynamically more complex (with respect to the requirements of the Earth), it also seemed to violate common sense. Arguments, such as, if the Earth is spinning, then why do we not sense it, held sway, and likewise, the accepted doctrine of the four-elements (recall Chapter 3) required that all earthly material must strive toward a location at the center of the cosmos.

Questioning the underlying structure of Ptolemy's geocentric model did not begin with Copernicus. Rather, it began much earlier with those Arabic scholars who inherited the ancient Greek ideas after the fall of the Roman empire. In particular, the Iranian polymath Abū Raihān al-Birūnī (circa 1000 AD) developed the idea of a heliocentric universe, correctly pointing out that mathematically it made no difference whether the Sun went around the Earth or the Earth around the Sun. Wisely, however al-Birūnī continued to promote the use of the traditional Ptolemaic model, effectively arguing that it was the role of the mathematician to make computations, and that issues such as the fundamental structure of the cosmos were better left to the philosophers and clerics.

As formulated by Copernicus the heliocentric planetary model was a return to pure ideals. All planets orbited the Sun along circular orbits at a constant speed. The steady shift of the constellations during the course of the year was due to the motion of the Earth around the Sun, and the daily nighttime/daytime motion of the Sun was caused by the spin of the Earth upon its central axis. Copernicus introduced a third rotational motion as well — this one acting to keep the Earth's spin-axis pointing in the same direction as it orbited the Sun. In this manner the direction and location of the north celestial pole (NCP) remained fixed with respect to the background stars. Not only did the removal of the Earth from its central location allow Copernicus to return to Platonic ideals of motion, it also satisfied his devout Christian beliefs that the Sun was literally God made visible. Indeed, there was only one location for the home of such an omnipotent operative, and that was at the center of all things.

A long and poorly kept secret, the details of the Copernican system, *De Revolutionibus Orbium Coelestium* (On the Revolution of the Heavenly Spheres), finally appeared in print in 1543. Indeed, Copernicus was in no specific hurry to see the results of his many labors brought before the public and the scrutiny of scholars. Partly goaded, however, and partly abetted by mathematical protégée Georg Joachim Rheticus, the text saw light, with its boiler-plate introduction by Lutheran theologian Andreas Osiander,[1] just as Copernicus himself passed away. By all accounts, the book was an immediate flop, with few copies being brought and/or read. Indeed, *De Revolutionibus* is not an easy book to read, and once a reader gets past the first introductory sections of Book 1 (there were six books, or sections, altogether), they are confronted by a difficult and

[1] Osiander's key aim was to present *De Revolutionibus* purely as the outlines of a mathematical hypothesis with no specific or intended pretentions to representing reality.

mathematically dense tome. Only a handful of scholars were capable of plumbing its depths, but those that did could see that Copernicus was on to something special. For all this, the rumblings of establishment discontent soon began to sound. As early as 1549, Lutheran reformer Philip Melanchthon noted the conflict between the Copernican model and Biblical scripture, and ultimately, 73 years after publication, in 1616, the Catholic church placed *De Revolutionibus* on the Index of Forbidden Books. This latter silencing, came, however, far too late, and the proverbial cat was out of the bag. By this time not only was the heliocentric model well thought of by scholars and astronomers across Europe, it had already attracted the attention of its strongest reformer and its most vociferous supporter — respectively, Johannes Kepler and Galileo Galilei.

It seems entirely appropriate that in addition to the publication of *De Revolutionibus* in 1543, so Andreas Vesalius published his *De Humani Corporis Fabrica* (On the Structure of the Human Body). As Copernicus sought to rearrange the heavens, so Vesalius sought to redefine the study of human anatomy along with the understanding of internal bodily organs. With these two works, both the macrocosm and microcosm, and their each-reflected-in-the-other association, were thrown into question. It was to take several centuries more, but it was the publication of these two revolutionary books that eventually saw the separation of astronomy (made personal by astrology and the date of one's birth) from medicine.[2] It was the beginning of the end for Zodiac Man.

5.2. Practical Geometry

The Ptolemaic model can, in some sense, be thought of as a case of putting the cart before the horse. The point was that the geometry of the model came first, and its attributes were then adjusted until they

[2]The flow of blood, for example, was thought to be influenced, just like the ocean tides, by the Moon, and it was accordingly deemed unwise to perform medicinal blood letting at times close to the Full Moon phase.

came into agreement with the observations. In the Copernican model, the orbits of the planets are determined by actual observations. While the Ptolemaic model worked according to a relative scale, the Copernican approach provided for the actual sizes of planetary orbits. Once one adopts the Sun as the central body, then there can be no quibbling about the size of the gaps between the planetary bodies. Indeed, the scale is set by a new unit of measure — the astronomical unit. This new unit is defined as the distance between the Earth and the Sun. Furthermore, and importantly, the tricky (from a Ptolemaic model perspective) circumstances underlying retrograde motion are fully accounted for by measurable time intervals.

As with Ptolemy's *Almagest*, Euclidean geometry was at the heart of the Copernican model, and to determine the size of a planet's orbit one simply had to measure (or infer) angles in the sky. With Earth being the third planet out from the Sun, the other planets were divided into inferior and superior bodies. The orbital radii for the inferior planets, Mercury and Venus, were derived according to a measure of their greatest elongation — that is their greatest observed angular separation of the planet from the Sun (Figure 5.1). Central to this method is the mathematical realization, that at the moment of greatest elongation the line of sight between the Earth observer and the planet must be tangential to the orbit of the planet about the Sun. This establishes the construction of a right-angled triangle, and in modern notation, the orbital radius of the planet will be given by the sine function

$$R_P = R_E \sin\varphi \qquad (5.1)$$

Equation (5.1) indicates that the orbital radius of the planet is determined relative to that of the Earth's orbit about the Sun ($R_E = 1$ astronomical unit) through the sine of the measured angle of greatest elongation. Copernicus, of course, knew the values for the greatest elongation of Mercury and Venus — indeed, he took them from Ptolemy's *Almagest*.

To determine the orbital radius of a superior planet, knowledge of its orbital period P about the Sun is required. This is

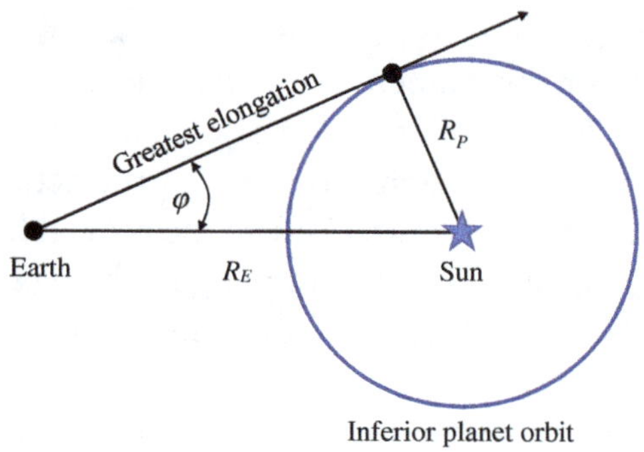

Figure 5.1. To determine the orbital radius R_P of an inferior planet, a terrestrial observer need only determine its greatest elongation (angle φ) away from the Sun.

something that cannot be directly measured from the Earth since the Earth itself is in motion about the Sun. What can be measured from the Earth is the synodic period S which corresponds to the time interval between repeat moments of opposition, when the Sun, Earth, and Planet are in line with each other. A relationship between the synodic period and the orbital period can be determined in a straightforward manner. Given two planets with orbital periods P_1 and P_2, with $P_1 < P_2$ indicating that planet 2 is superior to planet 1, so, the rate of increase Δ, as determined by an imagined solar observer, in the angular separation between the planets is

$$\Delta = 360/P_1 - 360/P_2 \text{ (degrees per year)} \tag{5.2}$$

Given that the synodic period S of planet 2 with respect to planet 1 is the time to repeat opposition alignments, S corresponds to the time required for the planets to accumulate a separation of 360°. That is, $\Delta S = 360°$. Substituting now for Δ in equation (5.2) gives the following result:

$$\frac{1}{S} = \frac{1}{P_1} - \frac{1}{P_2} \tag{5.3}$$

Now setting $P_1 = 1$ year, the Earth's orbital period, so equation (5.3) rearranges to give

$$P = \frac{S}{S-1} \tag{5.4}$$

where the 2 subscripts have been dropped, and the period P now applies to the specific superior planet being considered. By way of example, the synodic period of Mars, as measured by a terrestrial observer, is 2.135 years. From equation (5.4), the orbital period of Mars about the Sun is found to be $P = 1.881$ years. For an inferior planet equation (5.4) changes to $P = S/(S + 1)$, since in this situation it is P_2 that corresponds to the Earth's orbital period.

With the orbital period of the planets being known, their angular rates of motion can be determined through the relation $360/P$ (in units of degrees per year). The angle traversed by a planet along its orbit, in time T, is accordingly $360\ (T/P)$. With this knowledge the orbital radius of a superior planet can now be calculated by obtaining a measure of the time interval T_{OQ} for a planet to move from opposition to quadrature (when the Planet–Earth–Sun angle is 90°) — see Figure 5.2).

With T_{OQ} measured, the orbital radius is evaluated as follows:

$$R_P = R_E \cos\varphi \tag{5.5}$$

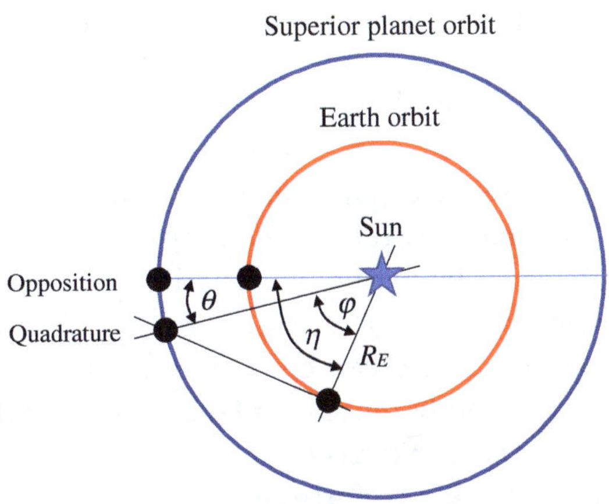

Figure 5.2. To determine the orbital radius R_P of a superior planet. In this case, the time interval T_{OQ} between opposition and quadrature must be measured.

where angle φ is given by the relationship

$$\varphi = \eta - \theta = \left(\frac{360}{1} - \frac{360}{P} \right) T_{OQ} \tag{5.6}$$

with both T_{OQ} and P being expressed in years on the righthand side of equation (5.6). Through equations (5.1) and (5.5), the scale of the planetary system can be determined. The observational data available to Copernicus was sufficiently accurate that his deduced orbital radii are hardly different from those presented in textbooks in the modern era.

While the geometry inherent to the Copernican system allowed for the unambiguous determination of the sizes of planetary orbits, its key feature was the straightforward explanation of retrograde motion (recall Figure 4.2). It was to describe this particular phenomenon that Ptolemy (following Apollonius of Perga) introduced epicycles into his planetary theory (recall Figure 4.1). Copernicus was able to do away with epicycles (but see later) to describe this periodic change in the direction of a planet's motion, measured, and recall, with respect to the background stars, by seeing that it was really a result of the motion of the Earth about the Sun (see Figure 5.3). Indeed, in the Copernican model, retrograde motion is a reasoned phenomenon — a natural consequence of its structure. In contrast, in the Ptolemaic model, it is an imposed phenomenon that needs explanation. For superior planets, retrograde motion always occurs at the time of opposition, and this is the time at which the Earth catches up, draws level with, and then pulls ahead of the more distant planet. In this manner, retrograde loops occur at intervals corresponding to the synodic period *S*. Variations in the appearance of a retrograde loop on the sky (being literally a loop, or a zig-zag shape — see Figure 4.2) are caused by the fact that the planets do not orbit the Sun in exactly the same plane. Indeed, if the planets all orbit the Sun in the plane of the ecliptic, then retrograde loops would trace out a straight, back-and-forth, line on the sky.

The geometry underlying the Copernican system is essentially no different from that contained within the Ptolemaic model. It is practical

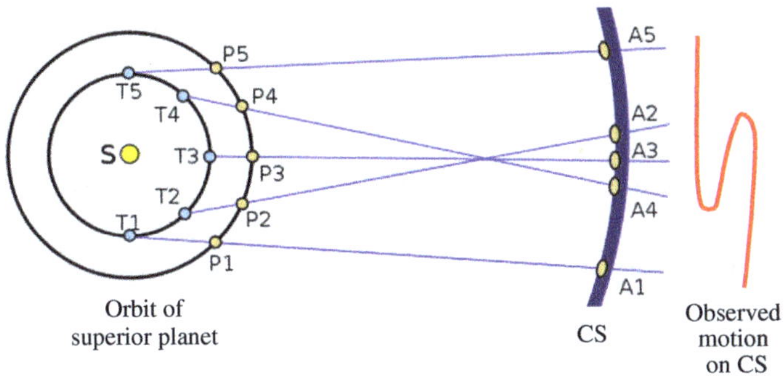

Figure 5.3. Explanation of retrograde loops in the Heliocentric model. S is the Sun, and the orbit of a superior planet P is shown relative to that of the Earth. As time progresses, the Earth catches up with (T1 and T2), draws level with (T3 — the time of opposition), and then draws ahead of (T4 and T5) the more slowly moving superior planet. The position of the superior planet as projected onto the celestial sphere (CS) and seen from the Earth is illustrated by the red curve (at far right). The retrograde loop has *structure* (a zig-zag as drawn) due to the inclination of the superior planet's orbit to the ecliptic.

and workman-like. For all this, the rearrangement of centers and circles allows for a consistent ordering principle, and, in principle, provides for the more straightforward calculation of planetary positions and behaviors (i.e., retrograde motion). The qualifier "in principle" is used in the previous sentence since, in fact, Copernicus did not actually do away with epicycles — they were simply a less obvious part of his model. Indeed, Copernicus needed about 30 circles and epicycles to make his six-planet (and one Moon) system work.[3] From a post-Kepler viewpoint (see Chapter 5) such additions are to be expected since Copernicus insisted upon perfectly circular orbits and uniform motion. Furthermore, it is because Copernicus needed to introduce epicycles, it is legitimate to question the system's literal truth: Does it or does it not represent the true distribution of the planets, or is it just another mathematical model that saves the appearance? This is a question we shall come back to later. First, an aside.

[3] In his *Commentariolus* (published and disseminated to friends in 1514), Copernicus stated that 34 circles were required to explain the entire "ballet of the planets". In his *De Revolutionibus*, Copernicus reduced the number of circles to 28.

5.3. An Aside — The Tychonic Universe

By placing the Earth in orbit about the Sun, Copernicus presented himself with a problem. Why, it could be asked, did the stars not show any indication of a parallax shift? The situation is outlined in Figure 5.4. Since Copernicus adopted, without question, the idea that the stars are all located at the same distance from the Sun (that is on the celestial sphere), there should be a periodic 6-month variation in the angular separation between any two stars. With respect to Figure 5.4, the larger the Earth's orbit about the Sun, the greater the difference $\Delta = \alpha - \beta$. For two closely situated stars, the variation $\alpha/\beta = (1 + x)/(1 - x)$ where $x = R_E/R_{CS}$, with R_E being the Earth's orbital radius about the Sun and R_{CS} being the (imagined) radius of the celestial sphere. If the Earth was located at the center of the cosmos, then there would be no periodic variation in the angular separation of the two stars — this is so since with $R_E = 0$, so $x = 0$, and $\alpha = \beta$. A very definite problem arises, however, when it is assumed, as indeed ancient philosophers did, that the celestial sphere is located

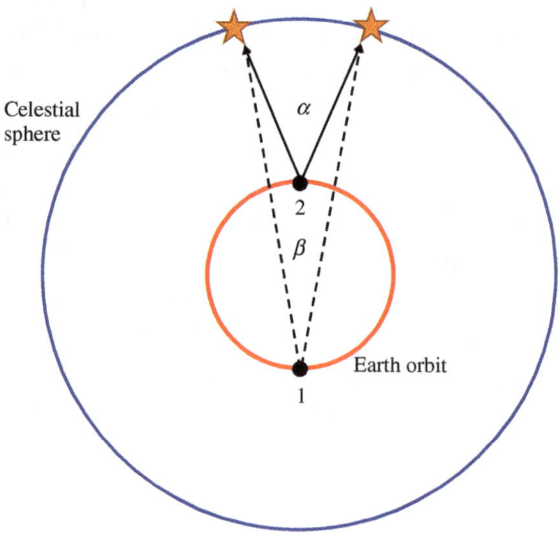

Figure 5.4. Parallax effect related to the motion of the Earth about the Sun. The larger the Earth's orbit, the greater the difference between angles α and β.

just beyond the orbit of Saturn. By modern standards, this makes for a decidedly claustrophobic universe. Since Copernicus had deduced that the orbital radius for Saturn was about 9.5 astronomical units, the size of the universe, R_{CS}, would not be much bigger. If we assume, for example, that $R_{CS} = 20$ astronomical units (making $x = 0.05$), and that angle $\beta = 1.0°$ (equivalent to the angular width of one's index finger held out at arm's length), then, $\alpha = 1.105°$, indicating a 10% variation, or a difference of $\Delta = 6$ arcminutes. To Tycho Brahe, working in the late 1500s, such a six-month angular variation would be readily observable — problematically, however, no such variation was found.

Tycho Brahe was the pre-eminent observer in the 16th century. He pushed the available materials and technologies to their limits, constructing and refining remarkably accurate observing instruments. Certainly, his instruments could measure angles to within 1 arcminute accuracy, and accordingly, his failure to find any measurable stellar parallax was a serious problem. In the modern era, we are used to the idea that the stars are at vast distances away and therefore expect that $x \approx 0$. To Tycho, and many of his contemporaries, however, the notion that the universe might contain vast regions of empty space seemed ridiculous, for indeed, why would the Creator make such an open and wasteful cosmos? Not only could Tycho find no indication of any stellar parallax, but he also stuck to the Aristotelian notion that the Earth was large, and "lazy", and must accordingly be at rest. Tycho's feet were firmly planted upon solid ground. Reluctant to fully dismiss the Copernican model, and equally as determined to move away from the Ptolemaic model, Tycho came up with a hybrid scenario in which the Earth was taken as stationery and orbited by the Moon and Sun. All of the planets, Mercury through to Saturn, however, orbited about the Sun (Figure 5.5). Remarkably, for so it would at first seem, this system can be made to predict accurate planetary positions. Indeed, Tycho, in the last few years of his life, employed Johannes Kepler to work out the mathematical details of his model. Like the Ptolemaic model, the Tychonic system provides a perfectly reasonable (indeed, reasoned) mathematical method by which planetary motion can be explained.

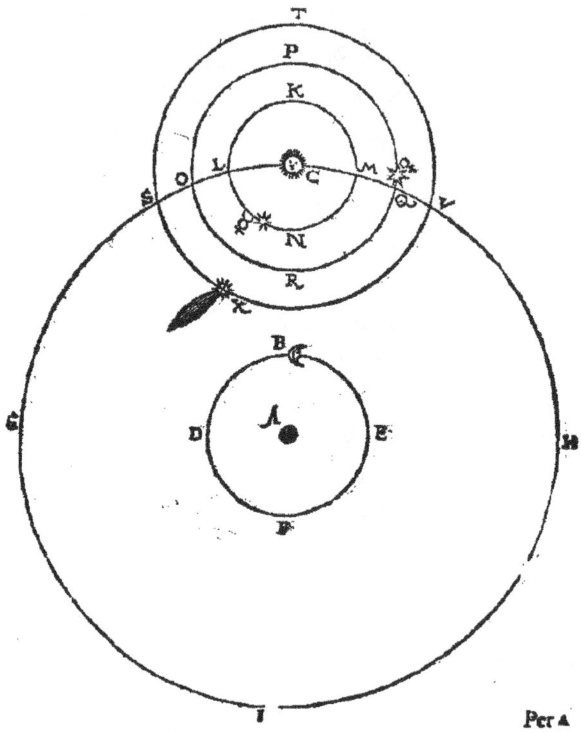

Figure 5.5. The Tychonic system with the central Earth at A being orbited by the Moon (BDEF) and the Sun C (orbit CGHI). In orbit about the Sun are the various planets (only the orbits of Mercury (KLMN) and Venus (OPQR) are shown). Also shown is the orbit (STVX) for the Great Comet of 1577. Diagram from Tycho's *De Mundi Aetherei Recentioribus Phaenomenis Liber Secundus*, originally published in 1588.

Philosophically, there is nothing wrong with this approach. It only becomes problematic if one further asks if the model describes the actual structure of the planetary system. Indeed, the Tychonic system provides a third method by which observed planetary motions can be explained. Mathematically, they are all equivalent systems; philosophically, however, they are worlds apart. Copernicus objected to the Ptolemaic system because of its geocentric nature and its extensive use of epicycles. Tycho objected to the Copernican system first because it didn't agree with observations (to the limit that parallax observations could then be made), and second, because it was inconsistent with Aristotelian concepts of matter.[4] Third, on essentially theological grounds, Tycho further objected to the Copernican model because it implied, if one was to really accept the no-parallax detection limit, that the universe must be extremely large and mostly empty. And, a large and mostly empty universe was something that Tycho could not countenance.

History has not been kind to the Tychonic model, and if it is mentioned at all in modern astronomy texts it is usually as an astronomical sidenote — a fruitless *cul-de-sac*. For all this, it did garner some favor with many scholars, and it was the preferred model for many decades after *De Revolutionibus* was placed on the list of forbidden books. Ultimately, however, the model fell into disfavor — not because its prediction capabilities were any less accurate than those available from the Copernican (or Ptolemaic) model, but because the Copernican model was seen as being slightly less contrived, and esthetically more pleasing in structure. For all this, the Tychonic model was a mathematical model that worked; a result that further underscores the flexibility of adopting a mathematical approach. Indeed, just because sets of equations can be written down and produce results that agree with the available observations, does not mean that the model is a correct representation of reality — to repeat Eddington's quip, the mathematics isn't there until we put it there.

The 1570s proved to be a marvelous decade for Tycho Brahe. On, or close to, 2 November of 1572 a new star (supernova) was discovered in the constellation of Cassiopeia, and Tycho studied it extensively. Just 5 years later, on 2 November (again), 1577, a "terrible and marvelous" comet was discovered, and once again Tycho and his famulus of observers studied its motion intensively. Importantly, these two events brought into question the correctness of received ancient wisdom. To the ancient Greeks, the heavens were perfect and unchanging, and accordingly, the appearance of a new star was

[4] More specifically, it opened up the issue with respect to the Aristotelian notion that nature abhors a vacuum.

impossible. Tycho's observations revealed that the new star had no independent motion (and no measurable parallax), and this implied it was not some atmospheric phenomenon, but a *bona fide* star. In contrast, the comet of 1577 had a measurable parallax, but the measurements placed it farther from the Earth than the Moon's orbit (see Figure 5.5). Once again, against ancient wisdom, the comet was not an atmospheric phenomenon, but a changeable celestial one. For classical Greek astronomy, the late 16th century was the beginning of the end. Not only was the Ptolemaic model being abandoned, but observations were showing that the heavens were not unchanging, but allowed the appearance of ephemeral phenomena.

5.4. Real or Saved

Copernicus did not oversee the actual publication of *De Revolutionibus* — rather he trusted its publication to Rheticus who was then at the University in Wittenberg. Unbeknownst to Copernicus (and Rheticus) Andreas Osiander who oversaw the text's final production, thought it prudent to add a preface in which it was clearly stated that the work was nothing more than a mathematical contrivance. More to the point, the preface also indicated that the text was not intended as a work concerning physical truth. At the time of its publication, this distinction, mathematical versus real and actual, was prudent and justified,[5] but it did undermine the possible reasons why anyone might want to read through its complicated mathematical content.

Copernicus is a little vague concerning the physical truth of his model, writing only that, "the harmony of the whole world teaches us their truth, if only — as they say — we would look at the thing with both eyes". If the proof of the pudding is in the eating, so the proof of a planetary model must be in its ability to predict planetary positions. Such predictions for Copernican theory were formalized

[5] Justified, that is, within the strongly religious settings of the time. Anyone could write a mathematical text, but only Church Authorities could decide what was and what was not true when it came to the universe.

under the Prutenic (that is, Prussian) Tables. Produced by Erasmus Reinhold, these tables were first published in 1551, and they were intended to update, and indeed, replace, the Alfonsine Tables, which had been computed in Toledo in 1252, and were based upon Ptolemaic model predictions. Detailed computer analysis has revealed that the Prutenic Tables could give results that were as much as 5°[6] in error when compared to actual planetary positions. This, in fact, is no better than the positional errors inherent to the Ptolemaic model. So, after all this, why did the Copernican model triumph over its rivals? Certainly, the Copernican model was no easier to use, with respect to calculating planetary positions, than the Ptolemaic system, and it did not rid the universe of epicycles. What it did do, however, was make for an inherently simpler minds-eye view of the planetary system, and it allowed specific features to emerge naturally (e.g., retrograde loops) without mathematical juggling. It offered more to its users than simply saving the appearance — it implied reality, but it did not openly insist upon it. It is clear from the outset that the Copernican model had strong support amongst many university-based scholars, but it was primarily the support that it received from Kepler and Galileo that changed that cloistered support into open acceptance — and eventually changed the minds-eye perspective into a physical "this is reality" stance.

Galileo Galilei is most famously remembered for his role in championing the reality of the Copernican model. Railing against the blind acceptance of ancient wisdom, Galileo knew full well that there were technical problems with the Copernican model (as originally formulated), but he did nothing to try to fix them — this was something that Kepler did (see Chapter 6). For Galileo, the Copernican model was, as much as anything, a convenient alternative model to that presented by Ptolemy. Galileo could not prove that the Copernican model was a true description of the planetary system, even though he (along with other scholars) was convinced that it was. But conviction on its own, of course, is not a valid argument with respect to establishing truth or reality. Indeed, Galileo did not fall

[6] This is equivalent to about the width of one's fist held out at arms length.

foul of the Church authorities because of his open support of the Copernican model, he attracted their ire because he taught that it was true — and this was something that he (or anyone else at the time) plainly could not do.[7]

Like Tycho Brahe before him, Galileo found good reason to question the wisdom of the ancients. His use of the newly invented telescope to discover craters and mountains on the Moon, the phases of Venus, the Moons of Jupiter, Sunspots, and a whole miasma of new faint stars, all pointed toward a much more dynamic, strange, complex, and surprising cosmos. Not only did Galileo find fault with the received wisdom concerning the heavens, but he also found fault with Aristotelian dynamics on Earth, and in this latter respect he did do something about improving the situation. Indeed, Galileo's experiments concerning rolling, falling, and swinging bodies introduced a whole new way of thinking about dynamics and paved the way for what was to become the Newtonian revolution (see Chapter 6). Indeed, following swiftly in the wake of the Renaissance, the scientific revolution, of the 17th and 18th centuries, oversaw the development of an entirely new way of thinking about nature and the cosmos. Indeed, a traction of new ideas and methodologies, that the ancient Greek and Arabic scholars were entirely unable to achieve, was too came about, with practitioners (pre-scientists[8]) becoming more and more skilled in developing and also distinguishing between rival theories about nature.

5.5. The Copernican Principle

The idea that there is nothing special about our (that is humanities) location in space and our view of the cosmos is a philosophical

[7]Galileo also had a knack for annoying people in power, which is not a bad thing in its own way, but it is a skill that should be used sparingly. Rather than being a martyr to science, Galileo was really a martyr to his ego.

[8]The title scientist was not actually introduced until the 19th century, when in 1833, William Whewell first coined the term upon the prompting of Samuel Taylor Coleridge. It is interesting to note that the term science fiction, in terms of literature genre, was coined soon after this time by William Wilson in 1851.

stance that has long been adopted, but it is one which bears very little real weight. Indeed, we are highly special observers, being the only known observers in the entire universe, who live on a highly special planet, being the only known planet to harbor life, in a highly special orbit,[9] about a relatively rare type of star.[10] The idea of our mediocrity encapsulated within the Copernican Principle is entirely based upon wishful thinking, and more than this, it limits our imagination. Historically, the Copernican Principle (also called the Principle of Mediocrity) has essentially been a statement about tempering our collective ego, which has supposedly taken a bruising each time the Earth has been displaced away from the center of things. In the words of psychoanalyst Sigmund Freud, "the ego is not the master of its own house". Yes, the Earth is not located at the center of the planetary system, nor is it located at the center of the galaxy, or, for that matter, at the center of the local group of galaxies, or at the center of the universe, but this does not undo the special conditions that surround our existence as observers. Indeed, Darwinian evolution undermines the central importance of humanity, but it also affirms the extraordinary set of circumstances that must have come about for us, that is, humans, to evolve. There is nothing ordinary about our existence even if our view of the universe isn't central. We may not be privileged observers but this does not in turn mean that our understanding of the universe isn't privileged. Specifically, there is nothing within the Copernican Principle that says our mathematical interpretation of the universe is, in fact, universal. The mathematical tools that have been developed to explain the reality that we believe we see, can make no specific claim to uniqueness, and other, equally non-privileged observers, could have entirely different methodologies to describe the reality that is out there. For the moment, we shall let this issue rest but will come back to it in Chapter 11.

---

[9] Since its formation some 4.56 billion years ago, the Earth's orbit has been permanently located within the Sun's habitability zone.

[10] Only about 8% of all the stars in the galaxy are Sun like stars. This still indicates, however, that there are of order 24 billion Sun-like stars within the Milky Way galaxy.

Chapter 6

Kepler's Harmony, Newtonian Imperialism, and Chaos

Fantasy imposes order on the universe. Or, at least, it superimposes order on the universe. And it is a human order.

— Terry Pratchett

Mathematical harmony, of a sort, was brought to the planetary realm by Copernicus in the mid-16th century. By re-ordering the planetary sequence, making the Sun a star and the Moon an actual moon, Copernicus constructed a system that had structural harmony. Johannes Kepler later appreciated the mathematical harmony introduced by Copernicus, but he pushed the questioning deeper. Why are there only six planets, Kepler asked, and why do the gaps between the planets have the specific values that are observed? To resolve these questions Kepler turned to the ancient authorities for guidance, and he looked for solutions in terms of numerical sequences and geometry.

6.1. Kepler's Epiphany

Kepler's path of reasoning was set on its course on 9 July 1595, when, in the midst of a lecture, so the story goes, he noticed an association between the orbital radii deduced for Jupiter and Saturn and the radii corresponding to the inscribed and circumscribed circles of an equilateral triangle. For any regular polygon having N sides, the ratio of the circumscribed radius to that of the inscribed radius is

$$\frac{r_{circ}}{r_{insc}} = \sec\left(\frac{360}{2N}\right) \tag{6.1}$$

For the triangle $N = 3$, and accordingly $r_{circ}/r_{insc} = \sec(60°) = 2.0$. Using modern-day values, we have $a_{circ}/a_{insc} = 9.52/5.20 = 1.83 \approx r_{Saturn}/r_{Jupiter}$. The concordance between the predicted radii, based upon the inscribed and circumscribed circles to a triangle, and the observed radii (semi-major axis) for Saturn and Jupiter, is far from perfect, but for Kepler, it was enough — enough to suggest that some underlying geometrically ordered sequence might be at play. If a triangle acted as the generating basis for the orbits of Saturn and Jupiter, then perhaps a square would describe the orbital spacing between Jupiter and Mars, and then a pentagram the orbital spacing between Mars and Earth, and so on. The generating sequence for determining planetary orbits and orbital spacing was thus envisioned to be the set of ordered polygons: ▲,■,⬟,⬢,⬣ with the polygon-side sequence being 3, 4, 5, 6, 7. Table 6.1 reveals the planetary orbits and orbital spacing ratios that result from a regular polygon sequence (see also Figure 6.1).

Table 6.1. Kepler's regular polygon sequence for planetary orbits and orbital spacing. N is the number of sides; C indicates a circumscribed circle and I indicates an inscribed circle. The orbits are normalized so that the Earth's orbital radius is 1. The last 2 columns provide actual orbital semi-major axis and orbit spacing data.

N	Polygon	Circle	Radius	r_{circ}/r_{insc}	Planet	Radius	r_{circ}/r_{insc}
1	Triangle	C	3.4961	—	Saturn	9.5212	—
2	Triangle	I	1.7481	2.00	Jupiter	5.2026	1.830
3	Square	I	1.2361	$\sqrt{2} = 1.414$	Mars	1.5236	3.415
4	Pentagon	I	1.0000	$\sec(\pi/5) = 1.236$	Earth	1.0000	1.524
5	Hexagon	I	0.8660	$2/\sqrt{3} = 1.155$	Venus	0.7233	1.383
6	Heptagon	I	0.7803	$\sec(\pi/7) = 1.110$	Mercury	0.3871	1.869

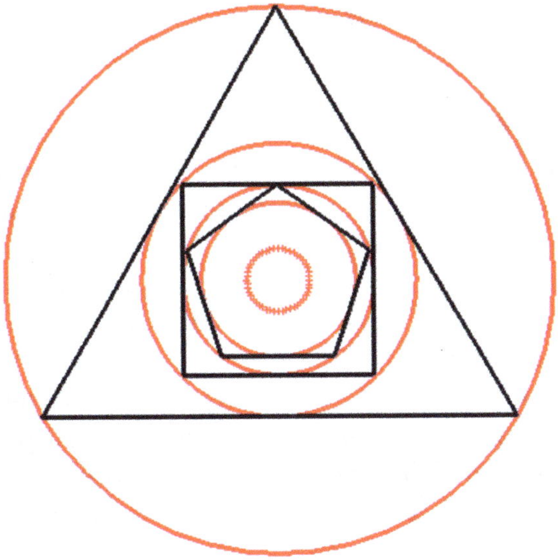

Figure 6.1. The first three (triangle, square, and pentagon) of the nested polygons. The central circle indicates the limit of regression as the number of polygon sides $N \to \infty$ (the radius of this circle corresponds to the so-called Kepler–Poinsot limit).

Table 6.1 reveals the predicted orbital spacing, and Kepler's (presumed) disappointment, with respect to the polygon-sequence numbers; the nested polygon scheme provides only a poor description of the orbital radii and the spacings between the planets. If the Earth's orbital radius is normalized to 1 AU, then the location and spacing for Mars and Venus are found to be about 20% smaller and larger respectively than their actual values; Saturn and Jupiter, however, have orbital radii that are about a factor of 3 too small, while that for Mercury is a factor of 2 too big. Part of the problem (if one wishes to think of it that way) with the nested polygon sequence is that the more one extends the sequence ($N \to \infty$) so $r_{circ}/r_{insc} \to 1$, and the radius limit approaches the so-called inscribed polygon limit, which with $r_4 = 1$ (for the Earth's orbit), gives $r_{N\to\infty} = 0.4018$. We find, as indeed, Kepler found, the nested regular polygon sequence simply does not work — the deduced orbital radii and orbital spacing — the gaps — cannot be brought into any reasonable agreement with the observations.

While the idea that planetary orbits and orbital spacing might be arranged according to an ordered sequence of nested polygons provides some sense of a design principle, it turns out that polygonal ordering can be achieved, and achieved with some accuracy. Let us here complete the argument, although out of temporal sequence with history, and see how far one can take Kepler's polygon-generating idea. To make the system work we simply need to relax the order polygonal sequence constraint. Table 6.2 and Figures 6.2 and 6.3 show the consequences of relaxing the ordered sequence requirement, and allowing for any regular polygon in the fitting arrangement. The least successful fitting polygons correspond to the square between Uranus and Neptune (10.85%), and the triangle between Saturn and Jupiter (9.47%). The best-fitting polygons correspond to the triangle between Saturn and Uranus (1.04%), followed by that of the square between Earth and Mars (1.82%).

Table 6.2. Polygon fitting to planetary orbits. Column 1 indicates the planetary pairs, while column 2 indicates the inscribed polygon. Column 3 corresponds to the ratio of the orbital radii for the outer and inner planets in each pairing. The last column indicates the percentage difference between the perimeters of the inscribed and circumscribed polygons for each planetary pair. A difference of zero percent would indicate the perfect agreement between the perimeter lengths of the inscribed (to the outer orbit) and circumscribed (to the inner orbit) polygons.

Planet	Polygon	r_{outer}/r_{inner}	Difference (%)
Mercury–Venus	Triangle	1.846	8.33
Venus–Earth	Square	1.389	4.00
Earth–Mars	Square	1.500	1.82
Mars–Jupiter	None	3.467	—
Jupiter–Saturn	Triangle	1.827	9.47
Saturn–Uranus	Triangle	2.021	1.04
Uranus–Neptune	Square	1.568	10.85

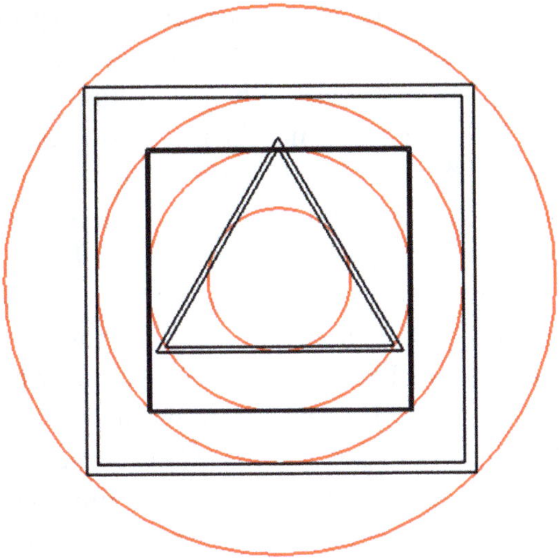

Figure 6.2. The best-fit circumscribed and inscribed polygons between the orbits of Mercury and Mars. The square between Venus and Earth provides the best-fit polygon between planetary orbits in the inner solar system.

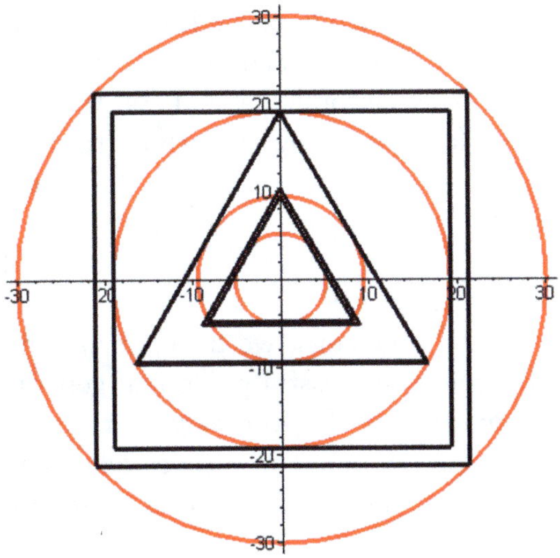

Figure 6.3. The best-fit circumscribed and inscribed polygons between the orbits of Jupiter and Neptune. The triangle between Saturn and Uranus provides the best-fit polygon between planetary orbits in the outer solar system. The scale is expressed in astronomical units.

Table 6.2 indicates that while no regular polygon can provide for a good fit between the orbits of Mars and Jupiter, the other planetary pairs can be matched with just triangles and squares — indeed, out to the orbit of Neptune, the generating polygon sequence is ▲,■,■, *, ▲,▲,■ with the * symbol accounting for the gap between Mars and Jupiter. On a purely visual basis, one might now ask if it is reasonable to posit the existence of two new *planets* between Mars and Jupiter such that symbolically * might be written as ■, ▲ with which arrangement the triangle and square reflection sequence is maintained. The radius of a circumscribed circle to a square containing the orbit of Mars as its inscribed circle is r_c(Mars) = 2.12. The radius of the inscribed circle to a triangle for which Jupiter is the circumscribed circle is r_{ins}(Jupiter) = 2.60. These two radii are not in especially good agreement with the *existence* of 1 Ceres, with r(Ceres) = 2.769, but they do approximately correspond to the inner edge and approximate middle location of the main belt asteroid region.[1] Our symbolic * gap between Mars and Jupiter has now expanded to become ■, ▲. Being even more daring, however, we note that our two hypothetical planets (asteroids) form a near-perfect match to the inscribed and circumscribed circles of a pentagon. This gives our symbolic * gap as ■, ⬟, ▲, and the entire polygonal sequence for generating the orbits from Mercury out to Neptune, including two hypothetical planets (asteroids), is ▲,■,■, ■, ⬟, ▲, ▲,▲,■. The polygonal sequence has a somewhat odd, but not unappealing symmetry to its form, with the polygons changing from squares to triangles about the central pentagon figure.

Inspired by Kepler's musings, we have let our imagination run wild, and our apparently successful polygonal generating sequence, complete with its *prediction* of two new *planets* between Mars and Jupiter is really a sham. We have in essence generated a Fermi elephant, and the moral of this exercise serves to remind us that just

[1] Given the multitude of asteroid orbits, however, there are always going to be some objects that occupy the predicted orbits — asteroids 13 Egeria and 8 Flora, for example, fit the bill, and while both these asteroids are large, 200 and 128 km across respectively, there is nothing specifically special about their appearance.

because we can do something with numbers and geometry, does not mean that the end result has any actual (that is physical) meaning or contact with reality.

6.2. Moser's Problem and Wittgenstein's Paradox

Rule following, as philosopher Ludwig Wittgenstein made much play upon, is not as simple as it seems. Likewise, guessing, or more technically inferring, the next number in a finite list of numbers is fraught with potential problems. Moser's circle problem is one such example where the inference of an apparently simple rule eventually lets one down. Described by Canadian mathematician Leo Moser in 1949, the problem asks how many regions R is a circle divided into when n points on its circumference are joined together in all possible ways (Figure 6.4). This is something that one can draw out on a piece of paper. When $n = 1$, there are no chords and the circle has but $R = 1$ interior region. When $n = 2$, there is one chord, and the circle is divided into $R = 2$ regions. When $n = 3$, there are 3 chords and the circle is divided into $R = 4$ regions. At this stage we might begin to sense that the number of regions R is doubling each time we add a new point on the circumference, varying according to the rule $2^{(n-1)}$. When $n = 4$, there are 6 chords and the circle is divided into $R = 8$ regions. Our inferred rule, at this stage, is looking promising. Moving on, when $n = 5$, there are 10 chords, and the circle is divided into $R = 16$ regions — again, in exact step with our inferred formula. We now start to feel decidedly confident about our inference, predicting that when the next point is added to the circumference ($n = 6$), the number of regions that the circle will be divided into will be $R = 2^5 = 32$. The acid test upon drawing this out, however, reveals a problem — the number of regions that the circle is divided into is actually $R = 31$. Adding another point to the circumference results in the interior of the circle being divided into $R = 57$ regions. Indeed, the sequence of divisions runs as 1, 2, 4, 8, 16, 31, 57, 99, 163, 256, ... Our initial inference, that the divisions will increase as $2^{(n-1)}$, which seemed so good for the first 5 cases, fails for the 6th and all other cases thereafter.

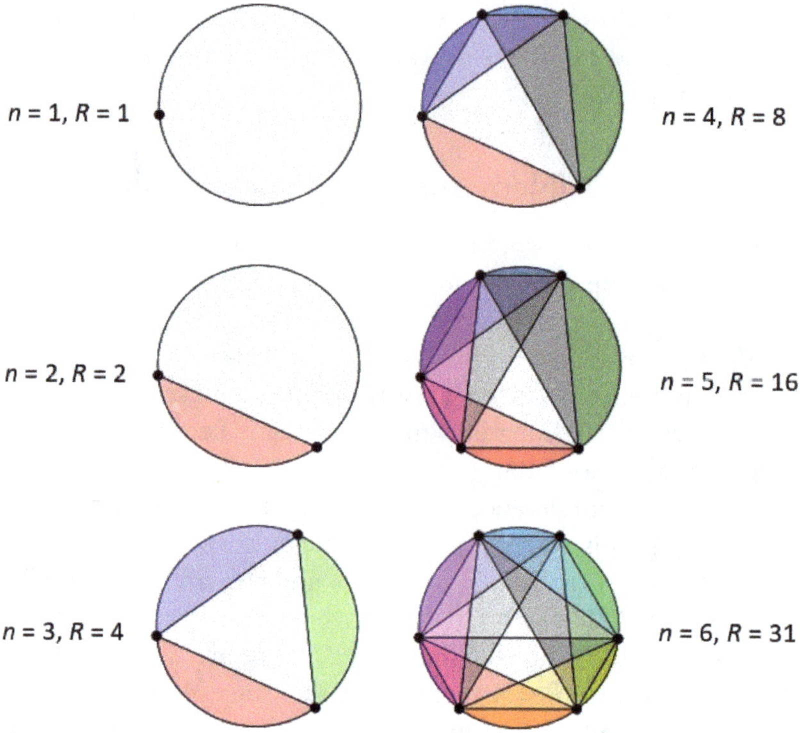

Figure 6.4. Moser's circle problem: How many regions R is a circle divided into when n points on its circumference are joined together in all possible ways?

The deduction and subsequent use of a rule, based upon a known but finite sequence of numbers is something that Wittgenstein realized is fraught with danger. Indeed, Wittgenstein's rule following paradox encapsulates this idea. Writing in his *Philosophical Investigations* (published posthumously in 1953) Wittgenstein argued that even simple rules can be *read* and *used* in non-standard, but entirely consistent (to the user) ways. In proposition 201 he mussed, that there is a paradox since no course of action can be determined by a rule, because every course of action can be made to accord with the rule. In other words, it is not possible to assume that when someone apparently grasps how to use a rule they are interpreting it in the right way. This idea is beautifully exploited by Argentinian mathematician Guillermo Martinez in the novel *The Oxford Murders* (first published in 2005). A subplot in this story considers the tragic outcome of an exam marker who tries to understand why some students provided wrong answers to simple questions — what is it, the examiner wondered, that causes the apparent misinterpretation. Martinez writes, "the series 2, 4, 8, can be followed with the number 16, but also with the number 10, or 2007. You can always find a justification, a rule, that lets you use any number as the fourth term in the series". Wittgenstein went even further with his discussion on rule following, arguing that there can be no final arbiter on how rules should be used. This being said, it is usually the case that the *community* is the arbiter, although the ultimate stance with respect to the use of a given rule, when asked "why so?", is that "because we said so".

When it comes to continuing a finite sequence of numbers, or of shapes — as in our orbit-defining polygon sequence introduced above — Moser's problem indicates that the inference of some simple rule can be misleading. Likewise, assigning physical meaning to a deduced rule does not automatically carry with it strict authority, since any rule can be justified, as exemplified by Wittgenstein's rule-following paradox.

6.3. Taking it up a Notch

Returning now to the temporal flow of events in history, Kepler, upon rejecting the polygonal ordering principle turned to the consideration of 3-D solids. Under this new scheme, with its extra dimension to move in, he found increased order and apparent predictability. Specifically, Kepler employed regular-faced polyhedra in the construction of his cosmological model, and this restricted the shapes that were available to those of the Platonic Solids (recall Chapter 3) — corresponding to the tetrahedron, hexahedron, octahedron, dodecahedron and the icosahedron (Figure 6.5). What caught Kepler's attention with respect to this new nesting sequence was that it had a finite range — there are five, and only five, Platonic polyhedra, and accordingly, there can only be six planets (the then known number)

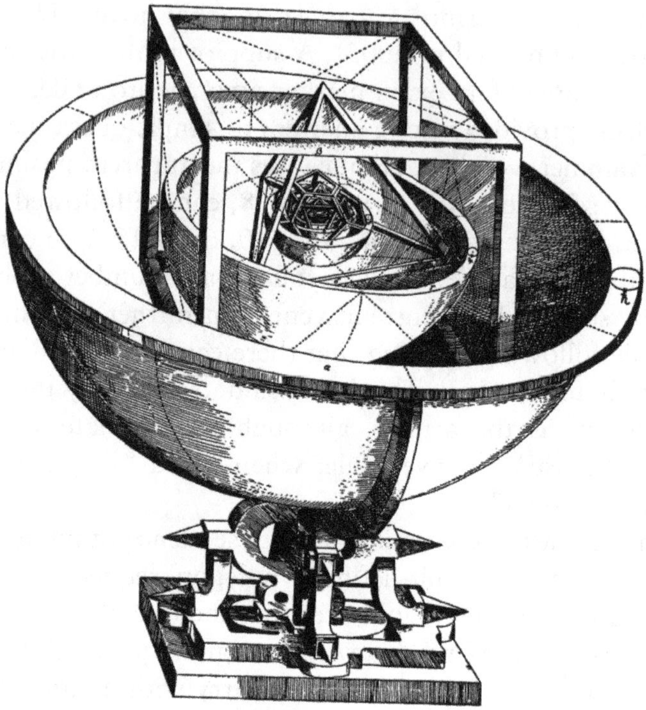

Figure 6.5. Kepler's nested polyhedral sequence for the ordering of planetary radii from Mercury out to Saturn.

set in orbit about their associated inscribed and circumscribed spheres. This apparent explanation for the number of planets would hold true until 1781, a time some 150 years after Kepler's death, in which year William Herschel discovered the planet Uranus. Table 6.3 reveals the sequence of planetary orbits and orbital spacing as invoked by Platonic polyhedra nesting. The problematic *gap* between Mars and Jupiter that was encountered in the nested polygon sequence is nicely bridged in the nested polyhedra sequence through the large circumscribed to inscribed sphere ratio associated with the tetrahedron. Accordingly, and indeed, necessarily so, the nested Platonic polyhedra sequence makes no specific predictions concerning additional planets within the solar system — other than that there shouldn't be any.

Table 6.3. Kepler's Platonic polyhedra sequence for planetary orbits and orbital spacing. Columns 1 through 3 indicate the planet, the orbit-generating polyhedron, and the number of polygonal faces *N* that generates the polyhedron. Setting Earth's orbital radius to be 1.00 AU, the orbital radii *r* for the rest of the planets are given in column 4. Columns 5 and 6 indicate the orbital spacing ratio for the polyhedral sequence and the actual sequence of planets, respectively.

Planet	Polyhedra	N	r(AU) (model)	r_{circ}/r_{insc} (model)	r_{circ}/r_{insc} (actual)
Saturn	—	—	6.54	—	—
Jupiter	Hexahedron	6	3.78	$\sqrt{3} = 1.73$	1.830
Mars	Tetrahedron	4	1.26	3.00	3.415
Earth	Dodecahedron	12	1.00	$\sqrt{(3(5 - 2\sqrt{5}))} = 1.26$	1.525
Venus	Icosahedron	20	0.79	$\sqrt{(3(5 - 2\sqrt{5}))} = 1.26$	1.383
Mercury	Octahedron	8	0.46	$\sqrt{3} = 1.73$	1.869

The correspondence between the model and the actual planetary radii is certainly not exact, nor especially convincing beyond the orbit of Mars (which is already 17% too small) — but the model is remarkably close with respect to the orbit spacing ratio; close enough, in fact, that the mind somehow wants the model to work. And, of course, therein resides the internal tension — the tension between reality, as invoked by human measurement, and a *pleasing and simple* explanation of reality, as desired by the human mind. Kepler was never able to fully let his nested polyhedra sequence model go, and for him it became an *idée fixe*. Realizing, and then gracefully accepting, however, that (correct) measurement must always triumph over theoretical nicety is another philosophical journey altogether, and we do not pursue the topic here (but see Chapter 11). But, for all this, are Kepler's geometrical ideas truly sterile? Is there really no room for his geometrically musings to be true, somewhere, amongst the billions of stars and exoplanetary systems in our galaxy? True, neither of Kepler's schemes, the polygonal and polyhedral, work for the solar system, but this does not mean that they cannot work in some other exoplanetary system (this possibility will be examined more closely in Section 6.7).

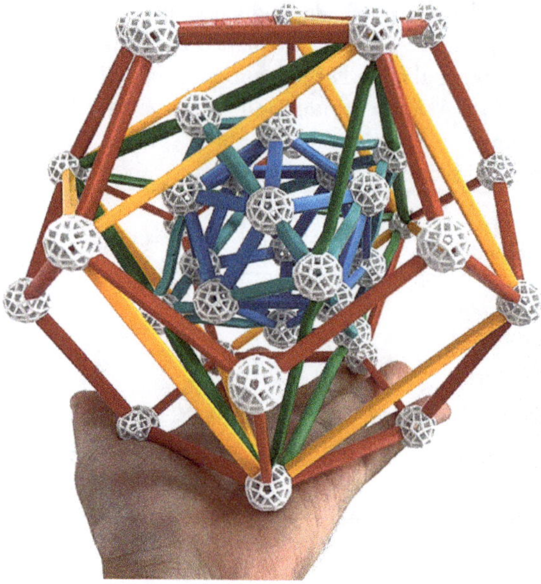

Figure 6.6. John Conway's *Zometool* model of Kepler's Kosmos. Image courtesy of Research Outreach (https://researchoutreach.org).

6.3.1. *Conway's Keplerian Kosmos*

British mathematician John Conway had a great fascination for toys and games — he was also a world-leading authority on the mathematics of symmetry groups [1]. Inspired by the modeling system *Zometool*, however, Conway was able to create a nesting of the Platonic solids that used the smallest possible number of nodes and struts — this model was later named Kepler's Kosmos (Figure 6.6). One of the interesting features of Conway's Kosmos construction is that the various polyhedral, while nicely nested together, actually belong to three different symmetry groups. Such groups describe the manner in which a given shape can be rotated and still preserve its original profile. The cube, for example, has 24 rotation-preserving symmetries, while the tetrahedron has 12 rotation-preserving symmetries.

In Conway's construction, the nesting of the Platonic figures runs from dodecahedron to hexahedron, to tetrahedron, to octahedron, and finally to icosahedron. This is a different ordering to that invoked by Kepler but is arguably more interesting (because of its minimal nodes plus struts construction). For all this, however, the inscribed spheres relating to the Kosmos order provide a very poor representation of the relative spacing of planetary orbits. Scaling the radius of the inscribed sphere to the dodecahedron to be 9.5 AU, corresponding to Saturn's orbit, so the implied orbital radius for Jupiter (the inscribed sphere to a cube) is 6.9 AU (giving $r_{Saturn}/r_{Jupiter}$ = 1.376, which is about 25% in error — see Table 6.3). Here, we have used the formula for the radius of the inscribed sphere to a dodecahedron:

$$r_{Dodecahedron} = a \frac{\Phi^2}{2\sqrt{3-\Phi}} \tag{6.2}$$

where a is the edge length of the pentagonal facets and φ is the golden ratio $\Phi = (1+\sqrt{5})/2$. Setting a = 1 so, $r_{Dodecahedron}$ = 1.1135. The radius of the inscribed sphere to a cube is simply r_{cube} = $k/2$, where k is the side length of the cube. In Conway's ordering k corresponds to the diagonal of the pentagon facets which generate the dodecahedron, giving, $k = a\Phi$, and accordingly, with a = 1, so r_{cube} = 0.809.

While Conway's Keplerian Kosmos nesting dramatically fails to account for the orbital spacing of the planets (of course it was never intended to), chemist Di Sun (Shandong University, China) and co-workers have found that a similar (but not identical) mixing of symmetry properties exists in a synthesized nanocluster made of 90 silver atoms (Ag_{90}) [2]. This particular nanocluster has an inner shell with the silver atoms arranged according to the vertices of an octahedron, a second shell in which the silver atoms are arranged according to the vertices of a truncated octahedron, and an outer shell where the silver atoms are arranged at the vertices of a (wonderfully called) rhombicosidodecahedron[2] [3]. With perhaps a little more

[2] This Archimedean solid is composed of 20 triangles, 30 squares, and 12 pentagons and has a total of 60 vertices and 120 edges. It was additionally a solid described and named by Kepler in his *Harmonices Mundi* (published in 1619).

human synthesizing skill and development, one day, the microcosm equivalent of Kepler's Kosmos might yet be realized.

6.4. The War with Mars

Before moving on from Kepler's model for planetary ordering, it is useful (and appropriate) at this stage to reflect upon what Kepler very definitely got right about the planets. Indeed, Kepler's three laws of planetary motion changed the whole way in which planetary orbits were described — they introduced a new picture and a new narrative. The laws came about because Kepler felt that a difference of just 8 arc minutes between the observed and the predicted position of Mars, was unacceptable. This highlights a change in Kepler's approach to the planets and contrasts the contradictory figures of Kepler the mystic against Kepler the scientist [4]. In the former case, we find a Kepler ready to impose Platonic order on the planets according to a sequence of embedded polyhedra (in spite of all its inaccuracies). In the latter case, we find a Kepler ready to doggedly quibble over an angular difference of about 1/10th of a degree.[3]

Kepler was exactly right to quibble about the small difference — this was his mathematical side coming to the fore. Indeed, if the Copernican model is intended to be truly a correct representation of reality, then it should provide precise planetary positions — or at least positions to better than arc minute accuracy, and preferably much less. Determined to reduce the apparent mismatch between observations and prediction, Kepler set about studying the orbit of the planet Mars. He had, care of Tycho Brahe, a good selection of positional data for this planet, and he used these to map out the planets' orbit about the Sun. Geometry was at the heart of Kepler's approach, and he used the innovative idea of triangulation to fix a set of positions for Mars. To do this he compared observations (or interpolated between observations) of the position of Mars at intervals corresponding to its sidereal period of 1.883 years (Figure 6.7).

[3] Your index finger held out at arms length subtends an angle of about 1-degree on the sky.

In this manner, Mars would be at the same location in its orbit about the Sun, but the viewing angle from Earth will be different — the shift in the Earth's position (which can be calculated) can then be used to enable the triangulation and placement of Mars in space, and more specifically at a specific point in its orbit about the Sun. In this manner, Kepler was able to find the positions of Mars at multiple locations in its orbit about the Sun (these are the points M1 to M5 in Figure 6.7). Having derived such a set of positions for Mars, Kepler then set about reverse engineering the orbit — that is, he asked what closed curve best fits the Mars location points, and where is the Sun located with respect to the closed curve. He soon found that a circle centered on the Sun did not provide a good fit to the positional data. Eventually, after much computation, over a

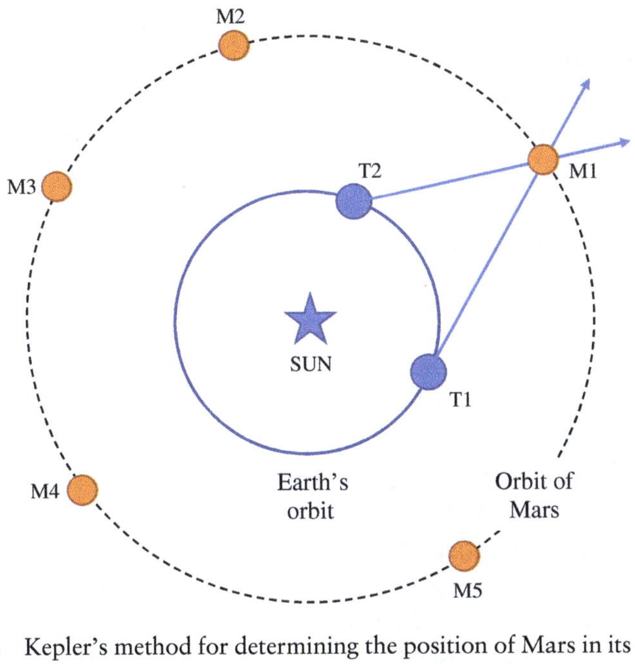

Figure 6.7. Kepler's method for determining the position of Mars in its orbit about the Sun. The time interval between observations T2 and T1 corresponds to the sidereal period of Mars and the intersection point of the sightlines from Earth at times T1 and T2 fixes the location of Mars at M1. Other sets of observations enable the positioning of Mars, in its orbit about the Sun, at locations M2 to M5.

timespan that he called his War with Mars, Kepler found that the best-fit curve was that of an ellipse with the Sun located at one of the focal points — this is indeed the result stated in his first law of planetary motion (see below). Furthermore, the displacement of the Sun away from the center of the ellipse (along the major axis — see Chapter 10) enabled the eccentricity e of the orbit to be determined. The eccentricity measures the deviation of the orbit from that of a perfect circle (for which $e = 0$), and Kepler's data indicated the orbit of Mars had an eccentricity of $e = 0.094$. This eccentricity, it turns out, is about 5½ times larger than that of Earth's orbit about the Sun. A comparison of planetary orbit properties need not concern us just now, and we will defer this topic until Chapter 10.

Kepler won the war with Mars by sheer dogged determination. It was a monumental result, and with his new insights, he was able to do away with all the *fudges* that Copernicus had introduced in order to make his original model work. Having found the shape appropriate to planetary orbits, Kepler pressed on to find additional rules for describing planetary motion (see the following section). He also, eventually, set about producing a new set of planetary tables — the Rudolphine Tables — which saw publication in 1627. The main purpose of these tables was to provide information for the accurate prediction of planetary positions, and its first success was in correctly predicting a transit of Mercury on 7 November 1631. Further refinements of Kepler's calculations by Jeremiah Horrocks (see Chapter 7) further enabled the correct prediction of a transit of Venus on 4th December, 1639.

With Kepler's searching analysis of planetary motion, the Copernican model was brought into exquisite agreement with observation. By the close of the 1630s, the new tables and mathematical methods available to astronomers allowed them to accurately predict planetary positions past, present, and future. From this time onward the Ptolemaic model, following a millennium and a half long dominance, fell from grace. The revolution was a quiet one, however, in the sense that all that had really changed was the geometry behind the calculations. There was no crisis of prediction that oversaw the demise of the Ptolemaic system. Rather, it was a change in the minds-eye viewpoint. The Copernican model, in spite of Kepler's refinements, was still a theoretical construct, and it was far from clear that it represented reality any more so than the Ptolemaic system. Certainly, the Copernican model was mathematically refined, and less arbitrary in its design (that is, it required fewer tweaks) than the Ptolemaic system, but there was no absolute dictate that it must represent reality, and fully describe the true nature of the planetary system. Copernicus made no strong claims that his planetary model represented reality — this job fell upon the capable shoulders of Kepler and Galileo. Most astronomers, however, from the mid-17th century onwards, were convinced of the reality of the Copernican model, taking it beyond the Ptolemaic (that is, ancient Greek) concept of simply saving the appearance. Importantly, therefore, by accepting the Copernican model as an actual representation of reality, astronomers were better able to construct a self-consistent minds-eye picture of the planetary system that was in agreement with the new sets of observations being made available with the newly invented telescope.[4] For all this, however, there was still no answer to Kepler's question concerning the number of planets, and it was still unclear just what physics (or laws) kept the planets in their orbits and controlled their spacings.

6.5. Kepler's Laws

Kepler introduced what are now known as his three planetary laws of motion in two books published 10 years apart. The first two laws appeared in his 1609 publication, *Astronomy Nova*, with the third appearing in his 1619 publication *Harmonices Mundi*. In modern form, these laws are as follows:

K1 — Planets orbit the Sun along elliptical orbits with the Sun being located at one of the focal points.

[4] The telescope first exploded onto the scene in 1608, and it was soon being used to map out the heavens and observe the Moon and the planets (and their moons).

K2 — The planet–Sun line sweeps out equal areas in equal intervals of time.
K3 — The cube of the orbital semi-major axis a is proportional to the square of the orbital period P. That is, $P^2 = Ka^3$, where K is a constant.

The meaning behind such terms as focal point, and semi-major axis will be explained in Chapter 10, and for the moment we shall concentrate on the meaning and implications of the laws. In terms of historical impact, **K1** and **K2** are the game changers. Effectively, these two laws indicate that, in spite of Plato's insistence, planetary orbits are not circular and the planets do not move at constant speed along their orbits. Indeed, **K2** requires that a planet moves at its fastest speed when at perihelion (its closest point to the Sun), and at its slowest speed when at aphelion (its greatest distance from the Sun). These two rules broke, for all future time, the chains of ancient Greek astronomy, and they brought into question the very essence of *prisca sapientia*. **K1** tells us about the shape of planetary orbits, they are ellipses, and accordingly, they have a very specific geometrical form and mathematical description (see Chapter 10 later). **K2** informs us about planetary dynamics in the sense that it dictates exactly where a planet must be in its orbit at successive intervals of time ΔT. That is, it is an area–time rule, such that if in a given time interval ΔT, the Sun–planet line sweeps out an area ΔA within the ellipse, then the Sun–planet line must sweep out the same area ΔA within the ellipse over all subsequent time intervals ΔT. Note, this law does not say where a planet will be in its orbit at some specific time T following perihelion passage.[5] The third law, **K3**, links together one of the geometrical properties of the elliptical orbit (effectively its largest diameter) and the orbital period — it is a dimension-time rule.

All revolutions require their moment of resonance. The times must be right for a revolution to proceed, else they will fizzle. With Kepler's laws, the time for revolution was right, and his laws were rapidly adopted, and incorporated into astronomy. Copernicus had

[5] This is another problem all together — in fact, it is generally known as Kepler's Problem.

set the stage for change, but it was Kepler who saw the changes through to completion. Other astronomers had little problem with accepting and accommodating the mathematical implications of Kepler's laws — albeit, that **K3** was seen as problematic. The problem with **K3**, of course, was why are cubes and squares of quantities required. In the famous words of physicist Isidor Rabbi,[6] "who ordered that". At the time, 1619, that **K3** was first articulated there was no known reason why such powers should apply, and the law was initially treated, in spite of the actual data, with some suspicion. Here was a law that required a clear explanation, and neither Kepler nor his compatriots could provide an answer. An explanation was eventually provided, however, by Isaac Newton (see Section 6.8).

6.5.1. *Just too good to be true*

Kepler's three laws of planetary motion are generally taken to be fundamental laws. Certainly, in the modern era, they are taught in every astronomy course, at all colleges and universities, and they are given special status in every introductory astronomy book. Indeed, they are held up as a triumph of human intellectual reasoning, and as an example in which observations and mathematical analysis have been combined to reveal the true nature of reality. The laws are so ingrained that we invariably forget that they are not, in fact, absolute truths. The planets do not actually obey Kepler's three laws of motion, rather the laws apply in an average sense over short intervals of time. Indeed, Kepler's laws are only exact and immutably true in two-body systems — that is a system in which the gravitational interaction is between two objects only (see Section 6.8). The solar system is a multibody system, and all planetary orbits evolve over time — their orbits are not exactly the same elliptical shape from one perihelion passage to the next, and their orbital periods, semi-major axis, and eccentricities vary with time. In the 2-body world, the mathematics describing Kepler's planetary laws is precise and

[6] Rabbi is said to have made this comment in 1936 after hearing of the discovery of the muon, the heavier relative of the electron.

absolute. In the real solar system (and universe), however, the 2-body dictate is never fully realized,[7] and the mathematics that underlies Kepler's laws is only an approximation to actuality.

6.6. Exoplanetary Systems

The discovery of the first exoplanet orbiting around another star filled a major gap in astronomical knowledge. Long reasoned that such exoplanets should exist, it was essentially a case of time and technological advancement catching up with expectation. Similar, in many ways, to the situation with respect to the fabled planet between Mars and Jupiter in the solar system, astronomers reasoned on the basis of the Principle of Mediocrity (for all its narrowness — recall Chapter 5) that our planetary system could not be the only one in the galaxy. With this philosophical drive, along with the growing body of knowledge of just how planetary systems form, astronomers set out to find new planetary systems, with the full expectation that they would be found. Indeed, not finding any exoplanets would have initiated a philosophical crisis. The first exoplanet to be found as part of a deliberate survey was 51 Pegasi b — a Jupiter-sized planet with an orbital period of just 4.23 days about the Sun-like star 51 Pegasi.[8] The planet was detected spectroscopically by (2019 Noble Prize winning) astronomers Didier Queloz and Michel Mayor [5]. Since this first discovery, in October 1995, over 5000 additional exoplanets have been identified — and the number keeps growing by the day. Indeed, many of these exoplanetary systems have been discovered through the *Kepler Mission* — a dedicated exoplanet search mission run by NASA. The various ground and space-based surveys

---

[7] Even within binary star systems, Kepler's laws will not be exactly followed for all time since gravitational perturbations, from close-passing stars, molecular clouds, and indeed, the rest of the galaxy, will always apply. Furthermore, as stars age, they lose mass, and if they are close enough, they will interact tidally.

[8] Technically speaking, the first exoplanets to be found were those in orbit around the neutron star and pulsar PSR 1257+12 on January 9, 1992. The planets in this system were found serendipitously, and it is not clear if they formed from supernova ejecta material, or are the remnant, solid cores of shredded Jupiter-like planets.

have revealed many different system architectures, with close-in, single hot Jupiter systems, extremely wide, cold-Jupiter systems, multiple-planet systems, and highly compact, multiple terrestrial planet systems being found. Once again expectation leads us onwards, and while neither the polygonal nor the polyhedron architectures developed by Kepler apply to our solar system, there is absolutely no reason to suppose that they won't be found to apply in some exoplanetary configurations. This expectation, however, is predicated on the requirement that the physical processes that result in planet formation allow, at least on occasion, for totally random spacings between planetary orbits to come about. On this basis, sooner or later, Kepler's specific orderings will appear.

The current haul of exoplanetary data[9] indicates that there are 2 systems containing 8 planets (the solar system and Kepler-90[10]); there are 3 systems containing 7 planets (HD 10180, HR 8832, & TRAPPIST-1) — of these, however, only the TRAPPIST-1 system contains 7 actually confirmed planets (rather than a total of 7 including several suspected, but still unconfirmed, planets). The number of known 6 and 5 exoplanetary systems come in at 5 and 15, respectively. The comparison to be made below is the closeness with which the ratio of observed orbital radii (working inward from the outermost planet to the innermost one) agree with expectations against Kepler's polygonal sequence given by

$$R_N = r_{\text{circ}}/r_{\text{insc}} = 1/\cos(\pi/N), \quad \text{where } N = 3, 4, 5, \ldots \tag{6.3}$$

In equation (6.3), we use the circumscribed and inscribed circles for a triangle ($N = 3$) for the outermost two planets and then move inwards with subsequent inscribed circles being generated by a square ($N = 4$), pentagram ($N = 5$), and so on. An error term is evaluated at each step with $E(N) = (R - R_N)/R_N$, where R is the actual orbital ratio between successive planets. Accordingly, the error term

---

[9] Since the discovery of new exoplanetary systems are published continuously, this number (when you read this) is likely out of data, but it was correct as of mid-2022.

[10] The Kepler here refers to the NASA *Kepler mission* which operated from 2009 to 2018.

Table 6.4. Comparison between orbital ratios for Kepler's polygonal sequence (R_N), the classical solar system planets, and the planets in the Kepler-11 exoplanetary system.

N	R_N	Solar System	R_{obs}	Kepler-11	R_{obs}
3	2.00	Saturn–Jupiter	1.83	f–g	1.86
4	1.41	Jupiter–Mars	3.41	g–e	1.28
5	1.23	Mars–Earth	1.52	e–d	1.26
6	1.15	Earth–Venus	1.38	d–c	1.45
7	1.11	Venus–Mercury	1.87	c–b	1.18

is zero if a perfect match is obtained — the larger (positive or negative) the error term so the poorer the comparison is deemed to be. A total error is then evaluated as $E_T = \Sigma\sqrt{E(N)^2}$, where the sum extends over the total number of steps (that is planet pairs) in the particular sequence. Systematically working through the list of the 24 known systems with 5 or more exoplanets, it is found that only one system, Kepler-11, provides a relatively good fit to Kepler's ideal polygon sequence (see Table 6.4) and Figure 6.8.

In terms of the 5 known 6-planet systems, the classical solar system (planets Mercury to Saturn) provides (in terms of a total error of E_T = 0.29) the best fit to Kepler's nested polyhedron sequence (see Table 6.5) — only the exoplanet system Kepler-62 (a 5-planet system — see Figure 6.9) provides for an otherwise reasonably good fit to the polyhedral sequence (omitting in this case the inscribed sphere to the octagon), with a total error of E_T = 0.39. Living according to the dictates of Pascal's wager, however, and not giving up the chance to make a prediction (in spite of my earlier dismal of such possibilities), if another, inner-most planet exists within the Kepler-62 system (Figure 6.10), then its orbit, predicated on the radius of the inscribed sphere to an octagon, will be 0.032 AU (with a period of 2.5 days).

In terms of nested 2-D polygon sequences, the TRAPPIST-1 (T1) system provides for an interesting case study since its various planets are distributed in step with a near-constant 3:2 orbital resonance scheme (see Chapter 10) [6]. Indeed, the orbital semi-major axis ratios for T1c/T1b, T1d/T1c, T1e/T1d, T1f/T1e, and T1h/T1g

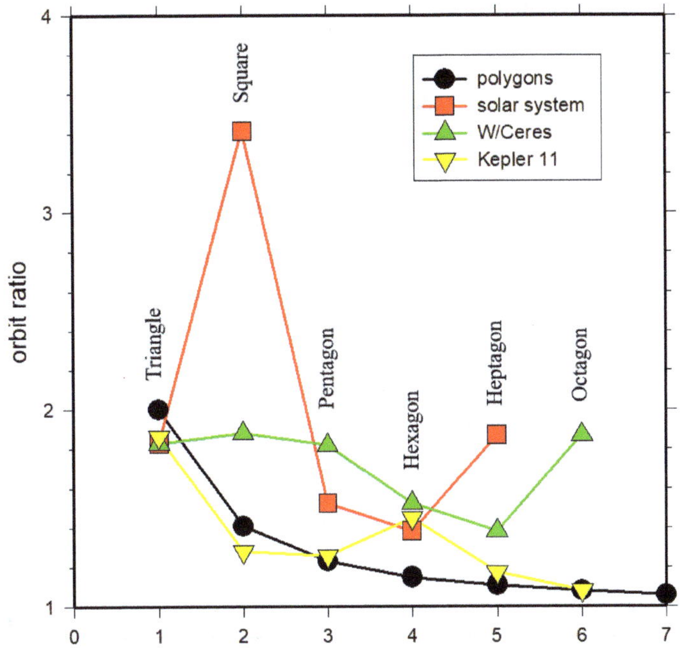

Figure 6.8. Comparison between successive orbital ratios for Kepler's polygonal sequence (black line), the classical solar system planets (red line), and the planets in the Kepler-11 exoplanetary system (yellow line). Also shown are the orbital ratio data points (labeled W/Ceres — green line) for our solar system when asteroid (now minor planet) Ceres is taken to be a legitimate intermediary in the gap between Jupiter and Mars.

Table 6.5. Comparison between orbital ratio for Kepler's polyhedron sequence, the classical solar system, and the planets in the Kepler-62 exoplanetary system. The rows are here arranged according to the number of faces (column 2) associated with the Platonic polyhedral. Column 3 indicates the ratio of the radii to the circumscribed and inscribed spheres. The columns labeled "Obs." indicate the observed orbital ratios.

	Faces	r_{circ}/r_{insc}	Solar system	Obs.	Kepler-62	Obs.
Tetrahedron	4	3.00	Jupiter–Mars	3.41	e–d	3.56
Hexahedron	6	1.73	Saturn–Jupiter	1.83	f–e	1.68
Octahedron	8	1.73	Venus–Mercury	1.87	—	—
Dodecahedron	12	1.26	Mars–Earth	1.52	d–c	1.29
Icosahedron	20	1.26	Earth–Venus	1.38	b–c	1.68

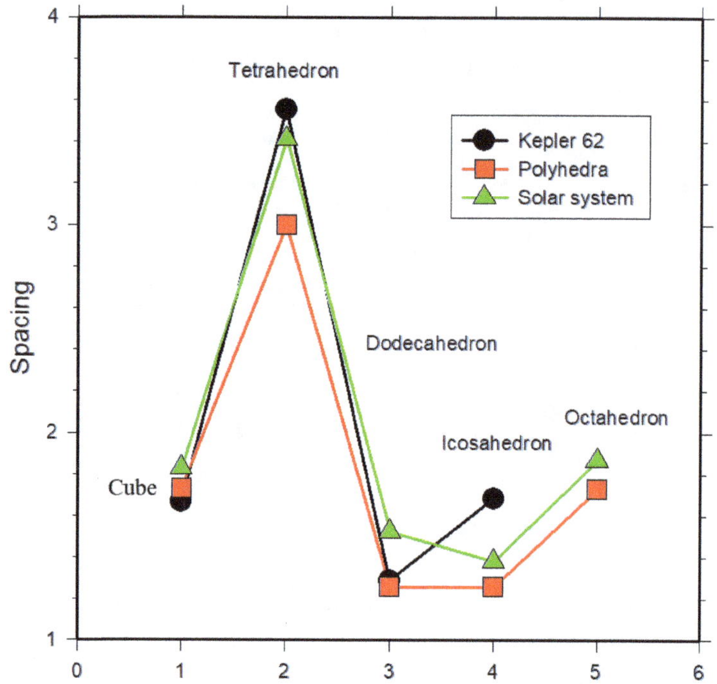

Figure 6.9. Comparison between orbital ratios for Kepler's polyhedron sequence (red line), the classical Solar System (green line), and the (known) planets in the Kepler-62 exoplanetary system (black line — see also Figure 6.10).

average to 1.36 ± 0.04, with only T1g/T1f showing a slightly smaller value than this average, coming in at 1.22. These ratios are very close to those expected for the inscribed to circumscribed circles from a square ($R_4 = 1.41$ — see column 2 of Table 6.3) and the inscribed to circumscribed circles of a pentagram ($R_5 = 1.23$). Accordingly, the orbital spacing of the T1 exoplanets (from b to h) is well described (see Figure 6.11) by the inscribed and circumscribed circles to a polygon sequence consisting of 4 nested squares followed by a pentagram (between the orbits of T1f and T1g) and finally the circumscribed circle to a square for the orbit of planet T1h (with an impressive total sequence error of just $E_T = 0.015$). The sequence for the T1 system would presumably not have appealed to Kepler's aesthetic outlook even though polygons provide a very good framework for

Figure 6.10. Comparative diagram of the planetary arrangement in Kepler-62 and the solar system. The region labeled as the habitability zone indicates the restricted domain in which an Earth-sized planet with an atmosphere could support liquid water on its surface — and thereby possibly host extraterrestrial life. Image courtesy of NASA.

the orbital spacings — this being so since it is not an ordered or even constant sequence of nested polygons. Rather, its arrangement appears to provide a nice example of Wittgenstein's rule-following paradox (recall Section 6.2), and it is impossible to establish an unambiguous rule to describe the next outcome (or action) in response to a given finite sequence of past outcomes. In the T1 case, what this boils down to is the fact that there is no possibility of predicting what the *i*th circumscribed polygon will be, even when the polygonal profiles for all of the previously inscribed polygons are known. The geometry emulates the orbital spacings of the planets in the T1 system, but it does not explain them.

Not all planetary systems with components displaying remarkable orbital structure, such as that seen in TRAPPIST-1, can be described in terms of inscribed and circumscribed circles to polygons

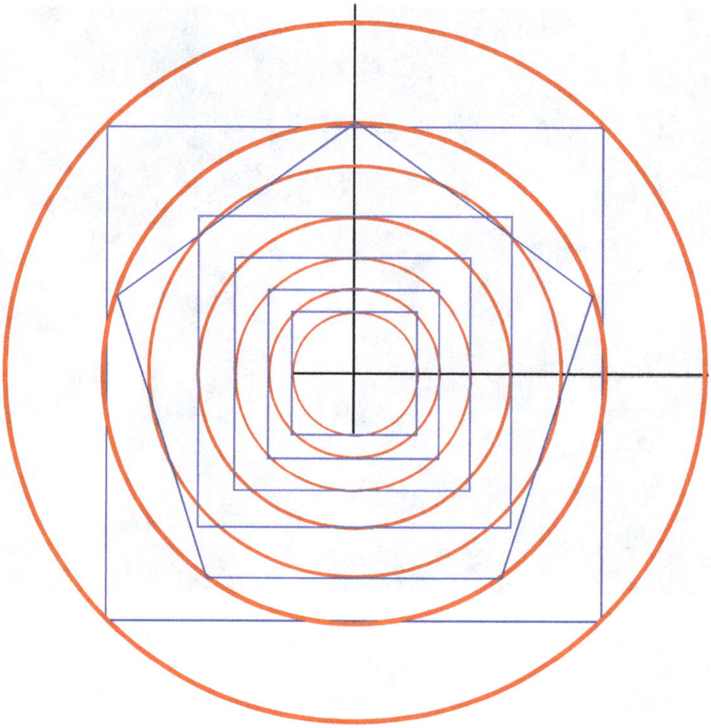

Figure 6.11. The polygonal sequence for the orbital radii of the exoplanets within the TRAPPIST-1 system.

or Platonic polyhedral. The six-planet system TOI-178 discovered by the *Characterising Exoplanets Satellite* (CHEOPS), for example, is remarkable for its harmonic (orbital period) structure, with the outermost 5 planets exhibiting a whole series of resonances [7]. The outermost planet g makes 3 orbits about the parent star for every 4 orbits of planet f. Planet f in turn makes 2 orbits about the parent star for every 3 of planet e, planet e makes 2 orbits for every 3 of planet d, and planet d makes 1 orbit for every 2 of planet c. Planet b, the innermost planet appears to break the integer resonance link, making 1.7 orbits about the parent star for every 1 orbit of planet c. Accordingly, the planets c through g provide for a remarkable chain of resonances that runs as 2:4:6:9:12. How, and at what stage of system formation, such a chain of resonances might come about is unclear. While the orbital periods for the planets in TOI-178 provide an impressive run of resonances, only the ratio of orbital radii for planets b and c closely match that for the inscribed and circumscribed circles to a square, while those of planets g and f match those for a pentagon. None of the other planetary pairs provide a good fit for the inscribed/circumscribed circles of any regular polygon. Geometry and nested polygonal sequences are not, after all, the be-all and end-all of interesting orbital structures. In contrast, the remarkable 6-planet system K2-138 is constructed according to a chain of near 3:2 resonances [8]. For this system, each successive planetary pair can be set according to the radii of the inscribed and circumscribed spheres to a nested sequence of dodecahedra (or as a nested sequence of icosahedra — recall Table 6.5).

The existence of systems such as TRAPPIST-1 (Figure 6.11) and K2-138 leads us to the inevitable conclusion that, in principle, there is likely to be some exoplanetary system, somewhere, with planetary spacings that will correspond to any inscribed-circumscribed polygon sequence that one might care to construct — even the polygon sequence described in Table 6.2 [9]. Indeed, it is likely that there is some exoplanetary system, somewhere, that will satisfy the orbital spacings corresponding to (almost) any arbitrary monotonic increasing arithmetic sequence. Somewhere, there is a planetary system in which the spacings are predicated upon the Fibonacci sequence (1, 2, 3, 5, 8, 13, ...), the triangular number sequence (1, 3, 6, 10, 15, 21, ...), the prime number sequence (1, 2, 3, 5, 7, 11, ...), the demiurge sequence from Plato's *Timaeus* 1, 2, 3, 4, 9, 16, 27, ... (recall Chapter 3), or the numerical sequence 0, 3, 6, 12, 24, 48, ... which underpins the Titius–Bode rule (see Chapter 9).

6.7. Newtonian Imperialism

If Kepler's mathematical dexterity and genius dominated the first part of 17th-century astronomy so that of Isaac Newton dominated the later part. And, while Kepler has been likened to a mystical realist, Newton was a hardcase pragmatist. The work that set science upon its new pathway was *Philosophiae Naturalis Principia Mathematica*, published in 1687. As anchor points in history go, where everything changes, and new beginnings prevailed, the appearance of the *Principia* counts as a cracker. Its appearance introduced a whole new approach to the "doing" of science, and it explained numerous long-standing problem issues, including that of accounting for Kepler's three planetary laws. Furthermore, Newton placed mathematics at the very heart of his analysis. Indeed, after the appearance of the *Principia*, the only acceptable way to draw firm conclusions about the workings of nature and nature's laws became through the straitjacket of mathematical analysis.

Building upon ideas introduced by Galileo, Newton presented three laws of motion in the *Principia*, and he developed the equation describing the centrifugal force felt by an object of mass m orbiting a more massive object of mass M along a circular orbit of radius R — this being $F_{cen} = mV^2/R$, where V the orbital velocity. Furthermore, Newton realized, that in order for the smaller mass object to move in a circular orbit about the central mass M, then some centrally acting force F must continuously act between the two bodies. Changing the run of history somewhat, if we take that central force to vary as the radius to some power n, then $F = Km/R^n$, where K is a constant. Equating F and F_{cen}, we obtain the result that $V^2 = KR^{(1-n)}$. In the case of a circular orbit, the velocity will be constant, and related to the size of the orbit and the orbital period P, with $V = 2\pi R/P$. Substituting for V now yields a relationship between the orbital period and the size of the orbit, with

$$P^2 = \frac{4\pi^2}{K} R^{(n+1)} \tag{6.4}$$

With equation (6.4) established, a match with Kepler's third law (**K3**) is obtained when $n = 2$ — that is, to explain **K3**, the centrally acting force must obey an inverse square law.[11] This centrally acting force, of course, is gravitational attraction. While here developed

[11] Newton assumed an inverse square law from the outset of his analysis, taking his (unacknowledged) lead from Robert Hooke.

assuming a circular orbit, the more general case of elliptical orbits provides us with the same result. Importantly for future astronomy, it is possible to show that in the general case the constant $K = G(M + m)$, where G is the universal gravitational constant. This formulation is invaluable to astronomers since it enables the determination of component masses to be made from a study of observable orbital properties.

An explanation of Kepler's first and second law was also provided for by Newton in the *Principia*, and it seemed as if a complete explanation of planetary orbits had been found. Newton explained Kepler's three empirical laws in terms of a new universal force, the force of gravity. Newton could not say what gravity was[12] but he could, and assuredly did, describe mathematically how gravity worked and that it acted as an inverse square law. Indeed, Newton championed the idea that purely mathematical descriptions were all that one needed in order to account for the physical world. This is not the same, however, as saying that mathematics is reality.

The *Principia* was (and still is) a work of monumental importance. Upon its publication, it delighted and amazed a stunned audience of practitioners, and it enabled the opening up of numerous new fields of study. Indeed, an indulgent sense of Newtonian imperialism came about, with the exactitude, extension, and success of Newton's mathematical methods bringing about a whole new cosmological viewpoint. The imperialism of Newton's methodology resulted in a dominant mechanistic outlook, with the workings of the universe being likened to the actions and behaviors of the cogs in a complex watch. The universe behaved like some precision mechanical system, ticking solidly and reliably, requiring, according to Newton, just the occasional re-set and/or winding by the omnipotent watchmaker. Even the presence of a benefactor deity was done away with by later practitioners, with, for example, Pierre-Simon Laplace championing the idea of causal determinism. Indeed, Laplace

[12] Indeed, concerning what gravity *is*, Newton famously wrote, *hypothesis non fingo* (I fain no hypothesis). It was Einstein, in 1915, who explained gravity as a manifestation of the curvature of spacetime.

replaced God with a demon and argued that should that demon know, at any one instant, the position and momentum of all the atoms in the universe, then the future behavior of the universe could, in principle, be determined precisely by Newtonian mechanics. This opened up deep-rooted philosophical problems with respect to the concept of free will, but it turns out, as we shall see below, the Devil (as ever) is in the details, and even Laplace's demon would be stymied by the fiendish machinations of nature.

Success breads arrogance, and as problem after problem fell to the onslaught of Newton's methodologies, confidence in the new physics, and specifically Newton's formulation for the force of gravitational attraction, grew and grew, ultimately becoming absolute. Newtonian methods worked, and they gave the right answers when applied to real-world problems. The dominance of Newtonian imperialism is perfectly illustrated in the discovery of the planet Neptune. In this case, it was the apparently wayward motion of the newly discovered planet Uranus that caught the attention of astronomers. Indeed, after the discovery of Uranus by William Hershel in 1871 it was soon realized that its observed motion was in disagreement with the predictions. Accordingly, astronomers had two choices, either the laws and formulae introduced by Newton only applied over a limited domain, or there was another planet beyond Uranus that was perturbing its path. Showing absolute faith in Newtonian methods, Urbain Le Verrier in France, and John Couch Adams (Figure 6.12) in England, set about using the difference between the observed and predicted positions for Uranus to determine the location of the new perturbing planet. The mathematics presented a formidable challenge, but both Le Verrier and Adams were able to make predictions for the new planet's location. History informs us that Le Verrier was successful in prompting Johann Gale at the Berlin Observatory to make a search for the new planet, and sure enough (almost exactly as predicted) planet Neptune swam into view on 23 September 1846. In many ways the discovery of Neptune was a false triumph — yes, it was newly discovered, yes, it was discovered through an immense mathematical effort, but Le Verrier and Adams had absolutely no

Figure 6.12. John Couch Adams (with a faraway, and concentrated look in his eyes). From the comfort of his study in Cambridge, Adams (and Le Verrier working independently in Paris), by mathematical reasoning alone, was able to deduce the location of a new planet beyond Uranus. That such a calculation made sense as a mental exercise was entirely due to an unwavering faith in Newtonian imperialism.

doubts that the planet must exist — such was the dominance of Newtonian imperialism.

The discovery of Neptune illustrates the power of the human intellect, showing that the structure of the planetary system can be deep-probed and revealed by the mind and mathematics. With the discovery of Neptune, astronomers could rightly feel that they had a full understanding of planetary dynamics. Confidence, even swagger, was justified, and indeed, true to history and life in general, a fall from grace was just around the metaphorical corner. Buoyed by his success with Neptune, it was entirely reasonable that, when the perihelion advancement of planet Mercury was found to be at odds with prediction, Le Verrier, in 1859, should argue that the small deviation, amounting to about 40 arc seconds per century, must be due to the existence of another planet orbiting interior to Mercury. Thus was born planet Vulcan. The hypothetical planet had an orbital radius of 0.14 AU and a mass 1/17th that of Mercury. Such an object would have a diameter of about 1900 km and could account, so Le Verrier argued, for the deduced difference between the observations and the results required by Newton's laws. With a size half that of Earth's Moon, Vulcan would have been a readily accessible object for astronomers to find, and identify, and indeed, reports of successful detections were claimed. For all this, and in spite of the many researchers looking for Vulcan, no clear consensus emerged. Indeed, the lack of clear detections prompted some researchers to question the strict obeyance of gravities inverse square law. American astronomer Asaph Hall, for example, noted in an 1894 publication that changing the radial dependency of the gravitational attraction term from $n = -2$ to $n = -2.00000016$ could account for Mercury's anomalous perihelion advancement [10]. Such a small numerical change seems hardly problematic, but the philosophical consequences of such a change would have been revolutionary. If the inverse square law could be *tinkered* with every time an anomaly appeared, then the whole foundation of Newtonian imperialism was put into question.

History tells us that planet Vulcan was never found, and indeed, its very existence was negated by the emergence of a new theory of gravity. Indeed, the first great triumph of Einstein's general theory of relativity, which first saw publication in 1915, was its ability to successfully account for the anomalous motion of Mercury. Einstein wrote of his time of triumph, that his heart shuddered in his chest, and that it was as if something had snapped within him. Once more, human intellect and mathematical analysis had triumphed over the planetary system, and as American astronomer Olin J. Eggen succinctly noted in 1953, "the mathematical necessity for Vulcan was established by a French astronomer [Le Verrier] at his desk in Paris [in 1859]. The necessity was removed by a German mathematician [Einstein] at his desk in Berlin, 500 miles distant and 56 years later [in 1915]" [11]. There are cracks in the foundations of Newtonian imperialism after all, but (and it is an important but) they are cracks pertaining to domain. Just as Einstein's theory of general relativity has magnificently passed all of the modern tests that have been set against it, so too, in the correct domain of applicability, has Newton's theory magnificently passed close scrutiny. The two theories are related mathematically, with Newtonian theory emerging from general relativity in the weak limit, but philosophically they are universes apart.

In recent decades the reach of Newtonian imperialism has been extended to the outer regions of the solar system, beyond Neptune, with the introduction of a new hypothetical planet — so-called, planet 9. Here we have a similar story to that which inspired the discovery of Neptune, although in this case the existence of the new planet is predicated upon the basis of an unlikely distribution[13] of orbits exhibited by (presently) less than a dozen distant Kuiper Belt Objects. On the basis that this odd distribution of orbits is a real phenomenon (and it may not be), Michael Brown and Konstantin Batygin (California Institute of Technology, Pasadena) have run many hundreds of computer simulations to tease out the orbit, and current location of the required perturbing object (planet 9) [12]. To date, no planet 9 has been found, but time will tell how all this plays out (see Chapter 10).

Just as planet Vulcan was written out of existence by the pen of Einstein, through his development of general relativity, Newtonian imperialism may find its domain foreshortened in the very low acceleration domain that pervades the deep outer solar system, and beyond. This outer limit is set according to the domain in which modified Newtonian dynamics (MOND) holds sway. There are many versions of MOND, but the key idea, as developed by Mordecai

[13]Not all researchers actually agree that the distribution is unusual but argue that it is merely the consequence of various observational selection effects.

Milgrom (Weizmann Institute, Israel) calls for a modification of Newton's second law[14] once accelerations drop below levels of order $a_0 = 10^{-10}$ m^2/s [13]. For the Sun, this low acceleration domain begins once the heliocentric distance exceeds about 8000 AU. The MOND concept was initially introduced in order to account for the near-constant rotation velocities observed for stars in the disks of spiral galaxies, but it remains unclear if this phenomenon is to be explained by dark matter or by modified Newtonian dynamics[15] — or something entirely different from these two.

6.7.1. *The rise of the pure mathematician*

In addition to the pursuit of natural philosophy, Newton also left his mark upon mathematics, not least in the development of his method of fluxions, the forerunner of modern calculus. For all this, however, Newtonian imperialism also set the trajectory for mathematics, especially so in England, for the next several centuries. Prior to the Newtonian shift, and throughout the Middle Ages, mathematics, within the university setting, was essentially catered for under the guise of the *Quadrivium*. These are the combined topics of arithmetic, geometry, astronomy, and music, with these topics being combined to cover practical issues such as navigation, astrology, and cosmography. Under the guise of Newtonian imperialism, however, mathematics was very much seen as subservient to the physical sciences, and the role of the mathematician was to develop techniques that had utility and practical application. Indeed, the idea of studying mathematics for its own sake was something to be assiduously avoided. Writing in a 1755 serialized commentary, *Man. A Paper for Ennobling the Species*, Peter Shaw noted that it would, "be absurd to

[14] Newton's second law of motion asserts that the acceleration a produced by a force F acting upon a mass m is related by the equation $F = ma$. Milgrom argues that in the very low acceleration domain, this equation should be modified to yield $F = m(a^2/a_0)$.

[15] Dark matter in its own right also implies new physics, but to date, no experimental constraints (or even laboratory detections) exist with respect to what dark matter might possibly be.

spend time in the bare study of logic, or mathematics; which are only means of our rising higher, and preparing us for the nobler sciences". To this Shaw further noted, "it cannot be necessary, for the promoting of human happiness, that the world should abound with mere logicians, or mere mathematicians". Harsh words indeed, even as they continue to hold resonance for all arts and classics majors enrolled at university to this very day. Accordingly, the idea of being a professional pure[16] mathematician simply didn't exist until relatively recent times. At Cambridge University, Newton's *alma mater* and home for many years, for example, pure mathematics only slowly emerged as a subject in its own right. Indeed, its first mathematics chair, founded in 1663, and held by the Lucasian Professor of Mathematics, was typically given to a practicing applied mathematician. Newton was Lucasian Professor, so too was astronomer George Biddell Airy, and computer pioneer Charles Babbage. Electrical engineer and physicists George Gabriel Stokes held the professorship, so too did Paul Dirac, and most recently the chair was held by Steven Hawking. Chemist Isaac Milner held the chair from 1798 to 1820, and during this time did not give a single lecture on mathematics. This being said, Cambridge University emerged, in the 19th century, as the foremost university for developing mathematical talent — especially under the guise of its highly competitive, and publicly ranked, wrangler system.[17] In spite of graduating many extremely talented mathematicians, however, there was still little opportunity for the continuance of a pure mathematical career, and most wranglers moved into other fields of scientific research, and/or general employment. This all began to change, however, in the late 19th century, with, for the first time, the appearance of professional mathematicians,

[16] The distinction between pure and applied mathematics essentially emerged in the mid-19th century. At Cambridge University, the first pure mathematics chair, the Sadleirian Professorship, was only established in 1860.

[17] Those obtaining the equivalent of first-class honors were called wranglers, with the highest scoring candidate being called the senior wrangler. Candidates obtaining second- and third-class degrees were refereed to as senior optimes and junior optimes, and those with a pass degree were known as poll men (and until the late 19th century, they were all men).

studying mathematics for the sake of studying mathematics. Mathematics, from this time moving forward, could be studied in a manner free from the fetters of utility or application, and graduates in mathematics could begin to find jobs that supported their pure mathematical interests.[18] The pure mathematician, as a professional label and title, long-stifled in its appearance by Newtonian (utilitarian) imperialism, emerged as the Victorian era (characterized by its zeal for growth and utility) came to a close. In this manner, it is no great surprise that the strong push by mathematicians (working worldwide) to establish the fundamental foundations of such topics as number theory, and mathematical logic (see Chapter 10) only took hold toward the close of the 19th century. Furthermore, it was in the early 20th century, that pure mathematicians, as they struggled to justify their profession, introduced the notion that pure mathematics (in contrast to applied mathematics) was directly concerned with the production and contemplation of beauty. Bertrand Russell, writing in 1907, argued this point, indicating that, "Mathematics, rightly viewed, possesses not only truth but supreme beauty — a beauty cold and austere". Godfrey Hardy, foremost pure mathematician of his era, re-iterated this point in his famed *A Mathematician's Apology* (first published in 1940), where he argued, "beauty is the first test; there is no permanent place in the world for ugly mathematics". The problem, of course, is that beauty is highly ambiguous, and not a rigidly defined scientific quantity (see Chapter 10).

6.8. Uncertainty Regained

Newtonian imperialism, and the all-seeing eye of Laplace's demon, appear, at first glance, to impose rigidity and certainty upon the universe. The equations of classical dynamics, and the boundary conditions required for their solution, appear to offer absolute knowledge, with the future being built-on and causally determined

[18] In many ways, this was a case of Cambridge (and by default England) catching up with the rest of Europe, which for many earlier decades had both encouraged and celebrated its pure mathematicians.

by the conditions that prevailed at the very first moment of creation. This, of course, is an appalling notion, and fortunately, neither our everyday lives nor the universe runs according to the trammeled straightjacket of mathematical formulation.[19] The very idea that the future is fully determined by running the appropriate sets of equations forward in time is totally repulsive, and fortunately, it is an ill-founded, pipe-dream of a concept. Laplace's demon, it turns out, is an impotent imposter. What saves us, and provides for free will and a non-predictable future, are the boundary conditions, specifically the sensitivity of boundary conditions that result in chaos. The equations may be mindless, rigid, and totally dictatorial, but the exact boundary conditions are unknowable. This unknowability is not just a case of the lack of data (or computational power), it is an unknowability gap that can never (ever) be overcome. Ten years before the close of the 20th century, Cambridge mathematician Sir Hermann Bondi lambasted the idea of Laplacian determinism and argued that it was a doctrine that should never have entered into physics [14]. For all this, Bondi also offered a reason why determinism became an integral part of Newtonian empiricism. The reason, effectively, was the acceptance, and blind confidence in mathematical proof. Indeed, the existence theorem established for second-order differential equations offered the false belief that unique solutions to the equations of all dynamical systems must exist. Not only must a solution exist, but the orbit it defined must be uniquely defined by the boundary conditions. This mathematical solution-existence proof is a spectacular result, but it fails to account for a vital real-world condition. The key point is that the existence of a solution theorem does not say that a small change in the initial boundary conditions must produce a small change in the solution. Here is the doorway through which chaos is able to creep or to mix metaphors; this is where the tail can begin wagging the dog. Stated another way, the downfall of determinism was latent within its own foundations.

---

[19] The caveat here, of course, is that our lives are more and more run by computer algorithms of one form or another, but this is only because we allow ourselves to be so herded like mindless beings.

The emergence of chaos, where it is no longer possible to make predictions about future system behavior breaks the stranglehold of Newtonian imperialism — it also acts to bedevil Laplace's demon. Seemingly insignificant changes in the initial boundary conditions can ripple and multiply over time to create divergent dynamical results. Indeed, the emergence of chaos is nature's way of saying that knowing the mathematical equations is not enough to fully capture future system behavior. In spite of the Newtonian promise of determinism, it transpires that we will never be able to mathematically predict, ahead of time, the full future behavior of the solar system — likewise, we will never be able to fully describe the exact manner in which it formed.

Chaos emerges from classic, deterministic equations and, as Hermann Bondi pointed out, it destroys the Newtonian image of a clockwork planetary system. Certainly, in one and two-body gravitational systems exact analytic solutions can be found. Once one moves beyond two gravitationally interacting bodies, however, into the domain of N-body dynamics ($N \geq 3$), then not only are there no analytic solutions to the equations, but system instability is the rule rather than the exception. Newton toyed with the analysis of many-body systems, but abandoned his attempts with the comment that it effectively gave him a headache, suggesting that it, "exceed[s], if I am not mistaken, the force of any human mind". Some two centuries after Newton had given up on many-body systems, French mathematician Henri Poincaré felt that it was time to give the problem another airing. Inspired by the possible catch of a prestigious prize that was being offered by King Oscar II of Norway and Sweden, Poincaré tackled the question of solar system stability. Stripping the problem down to its barest form, he looked at the equations governing the (so-called, restricted) 3-body problem where two of the objects are taken as being massive and substantive with the third being effectively a massless point. In this manner, the negligible mass particle has no gravitational effect on the two more massive bodies. By greatly simplifying the problem, Poincaré (and the prize judges) felt that he had solved all the niggly technical details and he accordingly set about describing the trajectory of the massless particle, arguing that, in principle, an infinite number of periodic solutions were available. Unfortunately, after being awarded the Prize, Poincaré realized his analysis had a problem — indeed, he found that even the smallest of changes in the initial conditions could produce orbits with wildly different characteristics. What Poincaré had discovered, in fact, was not a general solution to the (restricted) 3-body problem,[20] but that chaos rules.

Realizing that no straightforward, or readily findable, solutions to the general N-body problem were possible, Poincaré along with fellow French mathematician Paul Painlevé asked in the early 1890s just how bad can the dynamics of multiple point-mass systems, interacting via gravitation attraction only, become. Collisions between such objects would represent one such critical situation since the equations say nothing about the outcome of such interactions.[21] Additionally, in 1895, Painlevé conjectured that for $N \geq 4$ systems, non-collision singularities were possible in which one or more of the bodies could disappear to infinity in a finite period of time.[22] The eventual proof of Painlevé's conjecture was established in 2014 by Jinxin Xue (Tsinghua University, Beijing). Specifically, Xue showed that non-collision singularities were possible in $N = 4$ systems — the situation for $N \geq 5$ systems having been proven by Zhihong Xia (Georgia Institute of Technology) in 1988.

The discovery of deterministic chaos in the late 19th century ripped the very foundation from beneath the edifice of Newtonian imperialism — it is fundamentally impossible to fully describe the dynamical behavior of a many-body system into the deep future. The importancy of Laplace's demon is not from want of computing power, however, and nor is it due to not knowing the exact mathematical

---

[20]There are, in fact, some highly symmetric, highly artificial solutions to the 3-body problem [15].

[21]Recall, the Poincare–Painleve question concerns gravitational interactions only between point-mass objects with no physical structure. As described in Chapter 10, however, the effects of collisions, resulting in fragmentation, accretion, and mixing, must be included when considering solar system and planetary body formation.

[22]Recall that this is a classical (rather than a relativistic) problem, and accordingly infinite speeds are allowed to occur.

equations to follow — it is due to a fatal inability to fully know from where to start (that is, to know the actual boundary conditions). The equations that underpin Newtonian dynamics may well be exact and solvable, but they, and their solutions, do not fully capture the machinations of reality. Once again, we are forced to conclude that there are in this universe knowledge gaps that no amount of mathematical reasoning can bridge and that mathematics is limited in what it can say about the workings of the universe. This is not to say that mathematics can't tell us an awful lot.

References

[1] Roberts, S. *Genius at Play: The Curious Mind of John Horton Conway*. Bloomsbury, New York, 2016.

[2] Su, Y. M., *et al.* A Keplerian Ag_{90} nest of Platonic and Archimedean polyhedral in different symmetry groups. *Nature Communications*, **11**, article 3318, 2020.

[3] Cromwell, P. *Polyhedra*. Cambridge University Press, Cambridge, 1997.

[4] Voelket, J. *Johannes Kepler and the New Astronomy*. Oxford University Press, Oxford, 1999.

[5] Mayor, M. and Queloz, D. A Jupiter mass companion to a solar-mass star. *Nature*, **378**, 355–359, 1995.

[6] Gillon, M., *et al.* Temperate Earth-sized planets transiting a nearby ultracool dwarf star. *Nature*, **533**, 221–224, 2016.

[7] Leleu, A., *et al.* Six transiting planets and a chain of Laplace resonances in TOI-178. *Astronomy and Astrophysics*, **649**, article A26, 2021.

[8] Christiansen, J., *et al.* The K2-138 system: A near-resonant chain of five sub-Neptune planets discovered by citizen scientists. *The Astronomical Journal*, **155**, article 57, 2018.

[9] Beech, M. Looking for Kepler's ideal polygon law in Kepler Mission data. *The Observatory*, **139**, 105–110, 2019.

[10] Hall, A. A suggestion in the theory of Mercury. *The Astronomical Journal*, **14**, 49–51, 1894.

[11] Eggen, O. J. Vulcan. *Astronomical Society of the Pacific*, Leaflet No. 287, 1953.

[12] Batygin, K., *et al.* The planet 9 hypothesis. *Physics Reports*, **805**, 1–53, 2019.
[13] Milgrom, M. MOND theory. *Canadian Journal of Physics*, **93**, 107–118, 2015.
[14] Bondi, H. The prehistory of chaos. *Bulletin of the Institute of Mathematics and Its Applications*, **25**, 107–108, 1989.
[15] Stewart, I. Cosmic tennis blasts particle to infinity. *New Scientist*, issue 1841, October 2, 1992.

Chapter 7

The Laws of Horrocks and Wright

Kepler never relented on his *idee fix* of finding a geometrical scheme to explain the spacing of planetary orbits — mathematically, it was far too good an idea to let go of. Indeed, mathematical nicety provides for a strong organizing principle when investigating new possibilities, and other astronomers soon followed Kepler's lead. In this chapter, two imposed, but now discarded, rules will be considered. The first rule is that introduced by Jeremiah Horrocks, who developed a scheme for describing planetary sizes, and the second is that by Thomas Wright, who tried to impose a universal constraint upon the characteristics of cometary orbits. Interestingly, both impositions were made on the basis of observed and measured quantities. Where the schemes went wrong, however, was not, as such, a result of incorrect measures but in the assumption that small data samples are representative of the properties of a larger population.

7.1. The Equal Angles of Horrocks

British astronomer Jeremiah Horrocks (1618–1641) believed in the exactness of heavenly design, and in the ability of human beings to make sense of that design. Furthermore, he believed that exact and precise predictions could be made about heavenly motion. His inspiration was Johannes Kepler, and like Kepler, Horrocks's intent was to tease out from the observations those mathematical laws that described the grand design of the universe. It was Horrocks who first

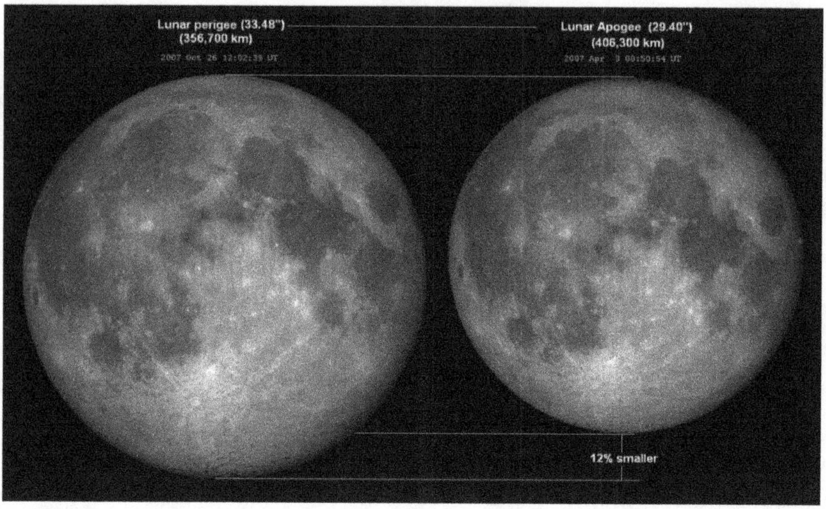

Figure 7.1. Angular size of the full Moon when located at perigee (left) and when located at apogee (right). Image courtesy of Wikimedia Commons: https://commons.wikimedia.org/wiki/File:Lunar_perigee_apogee.png.

demonstrated that, in keeping with Kepler's laws, the Moon's orbit was elliptical. That this must be the case was, in fact, a result that had long been staring humanity in the eye — indeed, since the time that humans first began observing the Moon. The problem, however, is that the evidence is hidden below the threshold of human perception. Figure 7.1 indicates the issue at hand, and while both images in the picture correspond to a full Moon, it is evident that the two images have different sizes. Indeed, there is a difference in the angular size amounting to about 4 minutes of arc (a 12% variation in size). This variation is far too small for the un-aided human eye to convincingly decern, but with the advent of telescope technology circa 1610 onwards such changes could be observed and quantified — if one had the mindset to do so, and Horrocks did have such a mindset. While Figure 7.1 shows the maximum variation possible at the same full Moon phase, the angular size of the Moon actually varies by the same angular amount approximately every two weeks.[1] That is, each time it completes a full-Moon to new-Moon cycle.

[1]Technically at intervals corresponding to half of the anomalistic month or once every 13.7773 days.

Having established (by repeated observations) that the angular size of the Moon really does vary in a repeated manner, the question then becomes what is it that is causing the variation. Could the Moon be pulsating, periodically expanding, and contracting in size? In principle, this is a viable hypothesis, and it would be a necessary result if the Moon's orbit were actually a perfect circle. Horrocks realized, however, that a more reasonable hypothesis is that the Moon has a constant physical size, but that its distance from the Earth is periodically changing. Equation (7.1) indicates the straightforward geometrical relationship between the angular size θ of the Moon (a quantity that can be measured from the Earth), the diameter of the Moon, and the Moon's distance from the Earth — vis

$$\tan\frac{\theta}{2} = \frac{\frac{1}{2}\text{Diameter}}{\text{Distance}} \qquad (7.1)$$

In order to establish that the Moon's orbit is indeed elliptical, a set of assumptions need to be put in place, and a series of measurements over many Moon phase cycles need to be made. First, as already indicated, the assumption that the Moon is not changing in its physical size has to be agreed upon. This is something that must be stated if the measurements showing a periodic variation of the Moon's angular size are to be interpreted as a change in physical distance. It seems like a reasonable assumption, but it is not a result that can be easily tested. In principle this assumption of constant physical size must also apply to the Earth — if the Earth was periodically changing in size, then the distance between the observer (on the Earth's surface) and the Moon (in say a fixed circular orbit) will also vary periodically. Being pedantic, one could actually argue that the Earth's size is varying (locally, not globally) in a periodic manner due to the Moon's tidal influence — but this effect is far too small to account for the variation seen in the Moon's angular size.[2] Agreeing

[2]We are familiar with the rise and fall of ocean tides, but the landscape also heaves and relaxes, by a few millimeters every day, due to the gravitational attraction of the Moon. In the case of the collider experiments carried out at CERN's Large Hadron Collider, such tidal flexing of the landscape must be included in the final data analysis.

that the Earth and Moon really do have constant physical sizes,[3] then any set of observations showing a periodic variation in the Moon's angular diameter can be safely interpreted as indicating a change in the Earth–Moon distance.

The largest θ_{max} and smallest θ_{min} measurements of the Moon's angular size will occur when it is at its closest (perigee) and greatest (apogee) distance from the Earth. Using Figure (7.1) as our guide, with $\theta_{max} = 33.48$ minutes of arc and $\theta_{min} = 29.40$ minutes of arc, and taking equation (7.1) along with the equations for the extreme distances from a given focal point in an elliptical orbit (see Chapter 11). Then,

$$\frac{q}{Q} = \frac{a(1-e)}{a(1+e)} = \frac{\tan\theta_{min}/2}{\tan\theta_{max}/2} \tag{7.2}$$

And, from (7.2), upon inserting values, we derive the eccentricity of the Moon's orbit to be about $e = 0.065$. A more refined analysis gives the actual eccentricity of the Moon's orbit to be 0.0549, indicating that the images in Figure 7.1 were not actually obtained at the exact positions of perigee and apogee — so, while the methodology is sound, we are about 18% to big in our example case. Note, while the eccentricity can be determined from equation (7.2), we do not have the means of finding the semi-major axis *a*, it cancels in the equation, and indeed, this term must be evaluated by a different set of observations.

There is a saying that, "the deeper you dig, the muddier you get", and the Moon's orbit and its motion against the background constellations have proved notoriously difficult to describe. Importantly, it turns out that, the Moon's motion must be considered as a three-body problem composed of the Earth, Moon, and Sun, with the Moon being sufficiently close to the Earth that tidal interactions are important. All these interactions result in the emergence of a complex orbital dance, with the Moon's orbit gradually

[3] Alfred Wilks Drayson famously argued in 1859 that the Earth was actually expanding as it aged. To this end, he predicted that all transatlantic telegraph cables would ultimately break through over-stretching — a prediction that came true, of course, but not for the reasons that Drayson advocated.

changing in its orientation, eccentricity, and semi-major axis. Horrocks was well aware of some of these details, and they were later a major headache to those 18th-century observers and theoreticians who set out to tame the Moon's celestial motion in order to enable the determination of longitude by ships at sea.

Purely by chance, the average angular diameter of the Moon (approximately 0.5 of a degree in the sky[4]) is almost exactly the same as the average angular diameter of the Sun. Essentially the Sun is about 200 times farther away from the Earth than the Moon, but the Moon is about 200 times smaller than the Sun. Accordingly, the Moon's shadow during a total solar eclipse only just reaches the Earth's surface, and the full eclipse condition is only visible from a relatively narrow strip, no more than 267-km wide, on the Earth's surface. Similar exacting conditions apply to viewing other objects (e.g., planets Mercury and Venus) when they pass between the Sun and the Earth. Not only must the planet be seen against the Sun's disk in the sky, but the planet must also be in the ecliptic at the same time. To calculate the times at which such transit-allowing alignments might take place requires a precise knowledge of the Earth's orbit as well as that of the planet. Kepler predicted that transits of Mercury and Venus would take place in 1631, but erred in predicting that a transit of Venus would also take place in 1639. It is at this point that Horrocks enters the picture, and based upon his own refined calculations from Kepler's 1630 *Rudolphine Tables*, Horrocks discovered that a transit of Venus was to occur about 3 p.m. on Sunday the 4th December, 1639. Only two people in the entire world knew of and then saw this Cytherean transit, and they were Horrocks (Figure 7.2) and his friend William Crabtree. While news of the transit soon became known across Europe, Horrocks's reputation as a gifted astronomer was only established posthumously with the publication of his manuscript[5] *Venus in Sole Visa* in 1662 [1]. Not only did this publication reveal Horrocks's great achievement in

---

[4] As a handy guide to this scale, the width of your index finger held at arms length is about 1° on the sky.

[5] The manuscript was published privately by renowned Polish astronomer Johannes Hevelius, but it was actually written circa 1640.

Figure 7.2. Jeremiah Horrocks observing the transit of Venus (on a projected solar image) in December 1639. Oil painting by artists Eyra Crowe (1891). Image courtesy of The Walker Art Gallery Liverpool.

unraveling the mathematical complexities of solar system dynamics, so as to make a successful prediction, but it also contained his observation that the angular diameter of Venus, measured as it slowly crossed the Sun's disk, was 76 arc seconds (Crabtree measured it to be 63 arc seconds, which is actually a better value with respect to present-day numbers). It is from these latter observations that Horrocks determined his planetary-size law.

Writing in his *Epitome of Copernican Astronomy* (1618–1622), Kepler speculated upon the observed brightness of the planets and correctly reasoned that brightness should depend upon distance and physical size. Kepler argued: "Nothing is more in concord with nature than that the order of magnitudes should be the same as the order of the spheres", and if this is to hold true then the sizes of the planets must increase with distance from the Sun. The implications of Kepler's argument are clear, and he is suggesting that there should be a steady increase in physical size from Mercury (the smallest and closest planet to the Sun) to Saturn (the largest and most distant planet from the Sun). By combining angular size measurements with deduced orbital distances, Horrocks correctly reasoned that Kepler's assertion concerning the growth of planet size with distance could be tested.

The 1631 transit of Mercury had already set the stage, and Horrocks knew that Pierre Cassendi, observing in Paris, had estimated that the angular diameter of Mercury was about 20 arc seconds (this is actually about a factor of 2 to big). To convert this Earth-based measurement to that which an imagined observer at the center of the Sun would record, straightforward geometry reveals

$$\vartheta_{ms} = \vartheta_{me}(1 - D_m)/D_m \tag{7.3}$$

where D_m is the orbital radius of Mercury in astronomical units, ϑ_{ms} is the angular size of Mercury from the Sun, and ϑ_{me} is the angular size of Mercury as measured from the Earth. Taking the orbital radius of Mercury to be 0.4 AU, so equation (7.3) indicates that, from the Sun, Mercury has an angular diameter of about 30 arc seconds (the uncertainty, of course, being largely in the uncertainty in Cassendi's angular diameter measurement). If we take an average of the angular diameter measurements of Venus, as made by Horrocks and Crabtree, then $\vartheta_{ve} = 69.5$ arc seconds, and for the imagined Sun observer, this gives for Venus, an angular size of $\vartheta_{vs} = 29.8$ arc seconds, taking $D_v = 0.7$ AU. Here is the result that caught Horrocks's eye — as seen by an imagined observer at the Sun's center, the angular sizes of Mercury and Venus are almost identical. This observation was not just too good a coincidence to overlook, and it was exactly what Kepler had argued for in his *Epitome of Copernican Astronomy*. Horrocks was now ready to formulate his planet-size law [2].

HPSL: All planets (with the exception of Mars) have the same angular size when seen from the Sun, with this size being 28 arc seconds.

A number of comments and questions immediately follow from this statement. First, what is going on with Mars? This exception to the law immediately makes one suspicious of its validity, especially as no reason for it is given. Second, why an angular size of 28 arc seconds? This latter point is more readily resolved since Horrocks allowed for the eccentricity of the orbits of Earth and Venus in reducing the angular size that it would have from the Sun, and then he took this as the definitive value. But, for all this, why 28 arc seconds — what is so special about this angular size? To which, of course, the answer is that there is absolutely nothing special[6] about 28 arc seconds or, indeed, any other number that it might have chanced to be.

Figure 7.3 is a comparison of the actual angular sizes of the planets, as seen from the Sun, against the angular size predicted by Horrocks's law. Here the point is not so much what the actual numbers are, but to note that Mercury, Venus, Earth, and Saturn do have approximately the same angular size (about 17.5 arc seconds) for a solar observer. The anomalies are Mars (being much smaller, at 6.2 arc seconds) and Jupiter (being much larger, at 37.9 arc seconds). Taking an average of the heliocentric angular diameters (excluding Mars), the observations indicate an angular size law of 23″ ± 9. If instead of taking the angular diameter of the planetary body of Saturn, but use that of the rings (with a diameter of 280,000 km), then the average heliocentric angular size is 27″ ± 9, which is almost identical to the angular size suggested by Horrocks. Pushing the data, a little more, and moving beyond a simple average to allow for a linear increase in planet size with orbital radius, a least-squares fit to the data for a solar observer (excluding Mars and taking the ring diameter for Saturn) yields the relation

$$\vartheta\,(\text{arc seconds}) = 18.5 + 2.6R(AU) \tag{7.4}$$

---

[6]There is the story told by G. H. Hardy, however, about a visit he made to an ailing Srinivasa Ramanujan. Apparently, Hardy had remarked that his taxicab number was 1729, which seemed like a rather dull number. Ramanujan responded, however, that rather than being dull, it was the smallest number expressible as the sum of two cubes in two different ways. As far as 28 goes, it is the number of dominoes in a standard double-six set, as well as the first triangular number expressible as the sum of two cubes. It is also (approximately) the number of days in a lunar phase cycle.

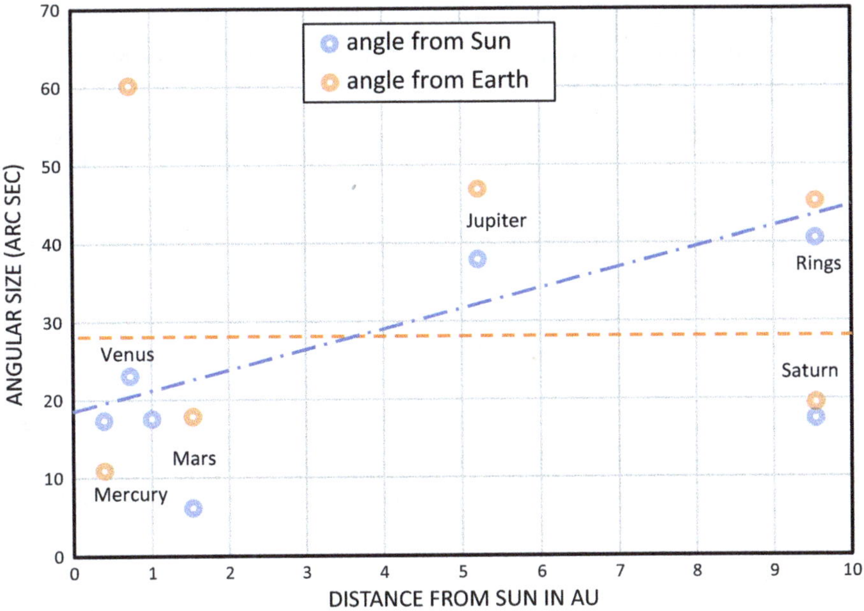

Figure 7.3. The angular diameter of the planets out to Saturn (including that for Saturn's rings) as seen from the Earth (orange circles) and from the Sun (blue circles). The horizontal dashed red line corresponds to Horrocks's planetary law, while the dash-dot (blue) line corresponds to the least square linear fit to the data (as given by equation (7.4)).

The linear law is not particularly good, but neither is it horribly bad. In a very broad sense, one could attempt to argue that Horrocks's constant angular diameter law, or a slightly improved least-square linear model, hints at some ordering principle for planet size in the solar system. Given that, *a priori*, there is absolutely no reason for there to be any correlation between angular size and distance, this is seemingly a remarkable result. But, of course, we have twisted and molded the data in each derivation to fit a perceived ideal, and we have completely failed to account for why the law cannot accommodate the angular size of Mars, and, for that matter, why it is the rings, rather than the body of Saturn, that provides for the best linear model fit. In short, Horrocks's law is entirely bogus, and so too is the linear increase with distance rule. They are both failed attempts to impose aesthetic order on the nature of things. There is, we must conclude, no correlation between planetary size and distance from the Sun.

While Horrocks worked according to measurements of actual angular sizes, Kepler expressed his ideas in terms of planetary brightness (or magnitudes). This is a slightly more complicated term to work with, and indeed it is not at all obvious that it is correct. Specifically, the order of maximum brightness for Jupiter and Saturn is seemingly reversed. Recall, Kepler had argued that the brightness should correlate with size and distance, and accordingly Saturn should be brighter than Jupiter.

The brightness B of a planet will depend upon its reflectance characteristics or albedo A, its cross-section area $S = \pi R^2$, where R is the planet's radius, its distance from the Sun r, and its distance from the Earth Δ, with

$$B = B_0(1 - A)\frac{S}{r^2\Delta^2} \tag{7.5}$$

where B_0 is a constant. Accordingly, as indicated by equation (7.5) the brightness decreases in step with increasing distance — to the order of the distance r squared — but increases with increasing size (through S). Jupiter and Saturn will attain their maximum brightness from the Earth when they are in opposition, at which time (in astronomical units) $\Delta = r - 1$. Taking the heliocentric distances of Jupiter and Saturn to be 5.2 and 9.5 AU, respectively, then for Saturn to be at least as bright as Jupiter when at opposition, equation (7.5) indicates

$$\frac{B_{\text{Saturn}}}{B_{\text{Jupiter}}} = 1 = \left(\frac{R_{\text{Saturn}}}{R_{\text{Jupiter}}}\right)^2 \times (0.07315) \tag{7.6}$$

In equation (7.6), we have assumed that the reflectance characteristics of Jupiter and Saturn are the same. Accordingly, however, for Saturn to be at least as bright as Jupiter it will need to be some 3.7 times bigger than Jupiter. This, as we now know, is far from the case — even if we include Saturn's ring system. If Saturn was, say,

4 times larger than Jupiter, then its angular size as seen from the Sun would be about $\theta = 81$ arc seconds. At this angular size, Saturn would not actually plot in Figure 7.3 given the scale adopted, and it would additionally be some 3 times larger in size than Horrocks's constant angular diameter law would predict.

There are no physical reasons why observed planetary size should vary in any systematic manner with increasing heliocentric distance. The size distribution can in principle do one of four things, vary at random, increase with distance, remain constant irrespective of distance, or, decrease with distance. Of these options, Kepler and Horrocks were banking on the last three being the case, since, to their reasoning, such would be the hallmark of a rational demiurge or designer at work. To their reasoning, no self-respecting deity would create a universe without some specific plan of action. In the same manner that Kepler reasoned that the distribution of planetary orbits was predicated upon the unique mathematical characteristics imposed by the Platonic solids, so the sizes of these same planets must also be described by some mathematical plan. From a modern perspective, of course, this is nothing more than a human-imposed ideal, based upon whished-for design.

When introducing his planetary size law, Horrocks added the useful (but confidence-deflating) boiler-plate condition, that, "I do not put forward this conjecture as a certain demonstration, but as a probability". In other words, Horrocks was essentially saying that the law might be true, but equally, it could well be wrong. Horrocks was clearly a wise man. Improved observational measurements made and collected in the half-century following Horrocks's death showed that his law was indeed false. Likewise, Kepler was wrong in his assertion that planetary size should increase with distance from the Sun. For all this, however, as indicated earlier, the planets do actually increase in physical size with heliocentric distance, just not in the manner argued for by Horrocks and/or Kepler. The Jovian planets (Jupiter to Neptune) are indeed larger, much larger, and more massive than the terrestrial planets (Mercury to Mars), but this relates directly to the region in which they formed within the primordial solar nebula. Intriguingly, however, a study by Lauren Weiss

(University of Montreal) and co-workers of exoplanetary system architectures revealed that there is a strong correlation between neighboring planet radii, with $R_{J+1}/R_J = 1.29 \pm 0.63, J = 1, 2, 3, \ldots$ [3]. This rather weak correlation, however, even if true is not the same as a Horrocks-like law, and indeed, it may simply be a result of detection bias, since small planets are less easily detected than large ones. More data will no-doubt resolve the situation, but Wiess *et al.* argue that "there is a correlation between planet size and planet spacing: smaller planets tend to have smaller orbital period ratios".

7.2. Wright's Constant Area Law

Thomas Wright of Durham (Figure 7.4) typically enters into the dialog on the history of astronomy via his 1742 musings concerning the spatial distribution of stars, and the structure of the Milky Way. These thoughts, contained within the text, *Clavis Cœlestis, being the explication of a diagram entitled Synopsis of the Universe or, the Visible World Epitomized*, would probably be less memorable today if it were not for an incorrect reading of its content by philosopher Emmanuel Kant. Indeed, it was Kant who converted Wright's ideas into the philosophically more useful nebula hypothesis for the formation of planetary systems. A fascinating character, Wright became well known in his lifetime as a skilled draughtsman, dial-maker, mathematician, astronomer, philosopher, architect, and landscape gardener; he published a large number of works, on many topics, but it is one particular suggestion, in his 1750 work *The Universe and The Stars, being an original theory on the visible creation founded on the laws of nature*, that concerns us here. Specifically, writing about the properties of cometary orbits, Wright comments, "I am strongly of the opinion, that the Comets in general, through all their respective orbits, describe one common area, that is to say, all their orbits with regard to the magnitude of their proper [that is orbital] planes, are mathematically equal to one another". This is an extraordinary claim, which from a modern perspective is easily negated: the orbital area, for example, of comet 1P/Halley is equal to 253.6 astronomical units squared, while those of 2P/Encke and 3D/Biela are 8.1 and 25.6 astronomical units

Figure 7.4. Thomas Wright of Durham (1711–1786). Portrait and print by Thomas Frye, dated April 18th, 1737. Image courtesy of The National Portrait Gallery, London.

squared, respectively. To step back to the 18th century, however, we find that the reasoning behind Wright's idea builds upon a reasonable-enough solution to a practical problem. The problem for us, however, is that just because some mathematical imposition seems reasonable, this does not mean that nature actually works that way.

In the year 1742, Wright published an extraordinary map of the heavens. Impressive in both physical size and scale, the map displayed multiple insets, and was printed on four 3-feet by 2-feet panels; the complete map, when assembled, was 6-feet (1.83 m) long and 4-feet (1.22 m) wide. The map revealed, to scale, the orbits of the planets and it additionally displayed the orbits of three comets: C/1661 C1, C/1680 V1, and 1P/1682 Q1. These comets were those originally highlighted by Edmund Halley, in his 1705 *Synopsis of the Astronomy of Comets*, as being periodic in nature. The comet of 1682 is Halley's Comet, while the other two are now cataloged as being lost, long-period comets. The problem faced by Wright in the construction of his map, however, was how to determine the greatest distances by which his three "periodic" comets strayed from the Sun.

As a detailed textual companion to his scale map of the solar system, Wright produced in 1742, *Clavis Cœlestis*, and in this work, he reviews the contemporary knowledge of those, "surprising bodies in the system call'd comets". Indeed, at this time Wright appears to have been particularly interested in comets; specifically observing and reducing the orbital elements for the comets of 1742 (C/1742 C1) and 1744 (C/1743 X1). While these latter two comets appeared at a time too late for representation on his map of the solar system, he did depict the full orbits of Halley's three "periodic" comets. To place these comets in his diagram, Wright estimated their aphelion distances from their deduced orbital period. Taking Halley's Comet, for example, Wright had a derived orbital period of $P = 75$ years, and this yields via Kepler's 3rd law an orbital semi-major axis $a = 17.78$ AU. Taking the *Orbis Magnus* of the Earth (the modern-day astronomical unit) to be 81 million (English) miles, Wright then calculates the greatest distance of the comet from the Sun as $Q = (2a - q) \times 8.1 \times 10^7$ miles, where the perihelion distance q is taken from Halley's 1705 *Synopsis*. Of the three comets considered by Wright, that of 1680 had the greatest deduced heliocentric distance — being some 11,200,000,000 miles [based upon an adopted orbital period of 575 years, and by using Halley's 1705 *Synopsis* derived value for the perihelion distance: $q = (612.5/100{,}000) \times 8.1 \times 10^7$ miles]. With this calculation methodology in place, Wright was able to complete his scheme and thereby reveal in his map the full extent of the solar system; showing that the comets roamed the depths of space out to distances at least 14 times further from the Sun than the then outermost known planet Saturn — in modern terms, this would take then out to distances of some 130 AU away from the Sun.

The suggestion that all comets might have the same orbital area appears as a singular idea in Wright's *The Universe and The Stars*, there being no mention of such a rule within *Clavis Cœlestis* nor in his analysis of the comets of 1742 and/or 1744. Likewise, no such similar claim or statement has been found in the writings of any of Wright's contemporaries. There is no president for the idea to be found in Halley's 1705 *Synopsis*, nor in William Whiston's *Astronomical Principles of Religion, Natural and Reveal'd: in nine parts* (first published in 1717; 2nd edition, 1725). A survey through Roger Long's *Astronomy in Five Books*, published in two volumes between 1742 and 1764, reveals but short notice being given to comets and no mention of an equal area relationship. Likewise, Richard Turner in his *A View of the Heavens* (published in 1783), while clearly setting down the characteristics of cometary orbits, makes no mention of any common area rule. Certainly, there is no theoretical requirement for such an equal area rule to exist — a conclusion that can, in fact, be drawn directly from Isaac Newton's *Principia* (published in 1687), where it is shown that a centrally acting force and the conservation of angular momentum suffice to explain all planetary dynamics — and specifically Kepler's three laws of planetary motion. The rule of equal area, which no other philosopher appears to have endorsed, seems to build upon the idea that comets are somehow different from the planets; their orbital motion being constrained not only by Kepler's laws but by an additional rule, presumably creator-ordained and built-in *ab initio* to the planetary system. Certainly, the physical make-up of comets was entirely unknown at the time that Wright was developing his ideas, and it was then generally thought that comets were similar to planets in both size and structure. For all this, however, Wright's justification for the equal area rule is hardly convincing, and it requires the reader to accept the equality of areas deduced for the comets of 1680 and 1682. Essentially Wright takes the area of each comet's orbit to be $A = \pi a(qQ)^{½}$, where q is the perihelion distance (as deduced from direct observations), and where, appropriate to an elliptical orbit, the aphelion distance is $Q = 2a - q$. Accordingly, Wright found that the ratio of their orbital areas was $A_{1682}/A_{1680} \approx 17/13$. Wright then notes that if the perihelion distance to the comet of 1680 was just 1.6 times larger than that given by Halley in his 1705 *Synopsis*, an equality of the areas could be realized. This factor of 1.6 change in the perihelion distance, Wright argues is a mere 1/35,000th of the comet's major axis, "which is an error every Astronomer will readily grant is very easily made". While clearly working on less than secure ground, and injecting just a little bluster, Wright finally argues, "if we can with any show of reason, and a probability on our side, bring the areas of these two extreme comets, as I may call them, to an equality, sure we may conclude it is a subject highly worthy to be more considered and enquired into".

We hear no more of the equal area rule for cometary orbits after Wright's 1750 speculations, and it is completely done away with in his unpublished manuscript *Second or Singular Thoughts upon the Theory of the Universe* written sometime circa 1771. Indeed, in this latter work, Wright essentially abandons all of the ideas expressed in his (1742) *Clavis Cœlestis* and within his (1750) *An Original Theory, or New Hypothesis of the Universe*. The Sun and solar system are now deemed to be contained within a solid spherical shell; the stars are linked to light emanations associated with volcano-like structures embedded upon the inner surface of the shell, and comets are identified with the material ejected from these same starlight-volcanoes. Indeed, in his *Second or Singular Thoughts*, Wright argues that there is no such object as a periodic comet: "they have no common connection with our system so as to be looked upon as regular parts of it but as all of them totally independent and subject to other laws". What these "other laws" might be Wright leaves as an unarticulated blank. There is nothing wrong, of course, with an author changing their opinion with respect to an earlier argument, and Wright is indeed struggling with a topic, essentially that of cosmology and the large-scale distribution of matter, that continues to occupy (and quite often frustrate) the minds of astronomers to this very day. Intriguingly, however, for all of Wright's vacillation, there is, it transpires, some small value in taking his original idea concerning the area of cometary orbits a little further, and indeed, reacting to his 1750 suggestion that the topic is worthy of further inquiry.

The area A of an ellipse is given by the product of its semi-major a and semi-minor axis b such that $A = \pi ab$. The semi-minor axis is further related to the semi-major axis and the eccentricity e as $b = a(1 - e^2)^{1/2}$, giving $A = \pi a^2(1 - e^2)^{1/2}$, and this, as seen earlier, can be related to the perihelion and aphelion distances: $A = \pi a(qQ)^{1/2}$. To take this result a little further algebraically, however, we introduce the Tisserand constant[7]:

$$T = (a_j/a) + 2[(a_j/a)(1 - e^2)]^{1/2} \cos(i) \tag{7.7}$$

where a_j is the semi-major axis of Jupiter's orbit and i is the orbital inclination. Using this relationship and Kepler's 3rd law, we now find

$$\left(\frac{P(\text{yr})}{A(\text{au}^2)}\right) = \frac{0.28 \cos(i)}{T - (a_j/a)} \tag{7.8}$$

where the right-hand side of equation (7.8) may be taken as a (near) constant term parameterized according to the values chosen for T, (a_j/a) and $\cos(i)$ — at this stage we note that we still have the semi-major axis term a on the right-hand side of equation (7.8), but in what follows we take the ratio (a_j/a) to be either unity or zero, corresponding to the situations where $a \sim a_j$ and $a >> a_j$. Additionally, we adopt the approximation that the orbits are predominantly confined to the ecliptic plane and take $\cos(i) = 1.0$. The construction of equation (7.8) builds upon a number of idealizations [4]: the Tisserand constant is only strictly constant in the case of the restricted 3-body problem under the constraint of circular orbits; likewise, Kepler's 3rd law is only strictly true for the 2-body problem (recall Chapter 6). Indeed, in the 2-body situation, such as that exhibited (approximately) by the individual planets, we have $P(\text{yr}) = [A/\pi]^{3/4}$, where the area A is expressed in astronomical units squared. By introducing the Tisserand constant, equation (7.8), just

[7]Introduced by Felix Tisserand, this parameter, even in the face of drastic orbital alteration via gravitational interactions, remains approximately constant. It is a means, therefore, of identifying where specific objects in the solar system originally came from — even if their initial orbits have undergone significant evolution.

like Kepler's 2nd and 3rd laws, generates a relationship between the geometrical properties of an orbit (through the orbital area) and the orbital dynamics (through the orbital period and the Tisserand constant). Additionally, by using the Tisserand constant an allowance is made for orbital evolution via gravitational interactions, and this, of course, is of great importance with respect to regulating the characteristics of cometary reservoirs and families.

Figures 7.5 and 7.6 show plots of orbital area A versus orbital period P for the known periodic comets, and through these diagrams the existence of several distinct cometary groups, is revealed. Indeed, the orbital area versus orbital period diagram nicely brings out the existence of the Jupiter family of comets (Figure 7.5), the Halley-type comets, and the Centaur/Chiron-type objects (Figure 7.6). In Figure 7.5, there are a great number of comets, having orbital periods between ~3 and ~10 years with orbital areas numerically set between those corresponding to the planets Mars and Jupiter. Indeed, these comets are otherwise described as having Tisserand parameters between 2 and 3, and an orbital semi-major axis comparable to that of Jupiter ($a_j/a \approx 1$), that is, they are the Jupiter family of comets. For those comets with $5 < P(\text{yr}) < 10$, we find, from Figure 7.5 data to a reasonably good order, that $P(\text{yr}) \approx 3.0 + 0.1A$. For comets with $10 < P(\text{yr}) < 20$, there is a large spread in the associated period-orbital area data points, although the trend is for the data points to fall toward the line parameterized by $T = 3$ and $a_j/a = 1$, giving $P(\text{yr}) \approx 0.14A$. In Figure 7.6, two additional groups of periodic objects are revealed: those which have $T = 2$ with $a_j/a = 1$ and those having $T = 3$ and $a_j/a = 0$. The former of these groups corresponds to the Halley-type comets, while the latter corresponds to Centaur/Chiron-type objects. For the Halley-type objects, we find for $20 < P(\text{yr}) < 200$ that $P(\text{yr}) \approx 0.28A$, while for the Centaur/Chiron-type objects, $P(\text{yr}) \approx 0.09A$.

Rising to the challenge of a "considered enquiry" some 2½ centuries after Thomas Wright speculated, in his 1750 work *The Universe and The Stars*, that all cometary orbits have the same area, we find, to no modern-day surprise, that his "opinion" was entirely wrong. This, of course, is no indictment on Wright's other

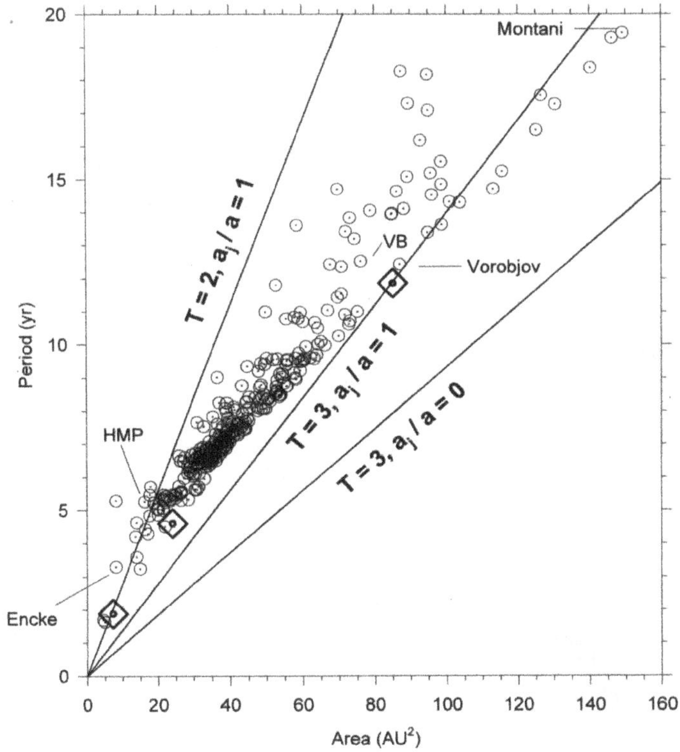

Figures 7.5. Orbital period ($P < 20$ years) versus orbital area for the known periodic comets. The three labeled lines correspond to various parameter combinations for equation (7.8), and the three dotted-diamond points indicate the orbital period and orbital area of Mars, Ceres, and Jupiter. Several data points are identified for reference purposes only and correspond to comets 2P/Encke, 46P/Honda–Mrkos–Pajdušáková (HMP), 276P/Vorobjov, 314P/Montani, and 53P/van Biesbroeck (VB).

contributions to astronomy. What we do find as a result of this analysis, however, is that the orbital-area versus orbital-period diagram for the periodic comets usefully reveals, in a visual manner, the existence of at least three distinct cometary groups or families. This visual clarity stands apart, it is suggested, in contrast to alternative diagrams in which orbital eccentricity is plotted against the semi-major axis, or the Tisserand constant is plotted against the inverse semi-major axis (the latter quantity being a measure of orbital

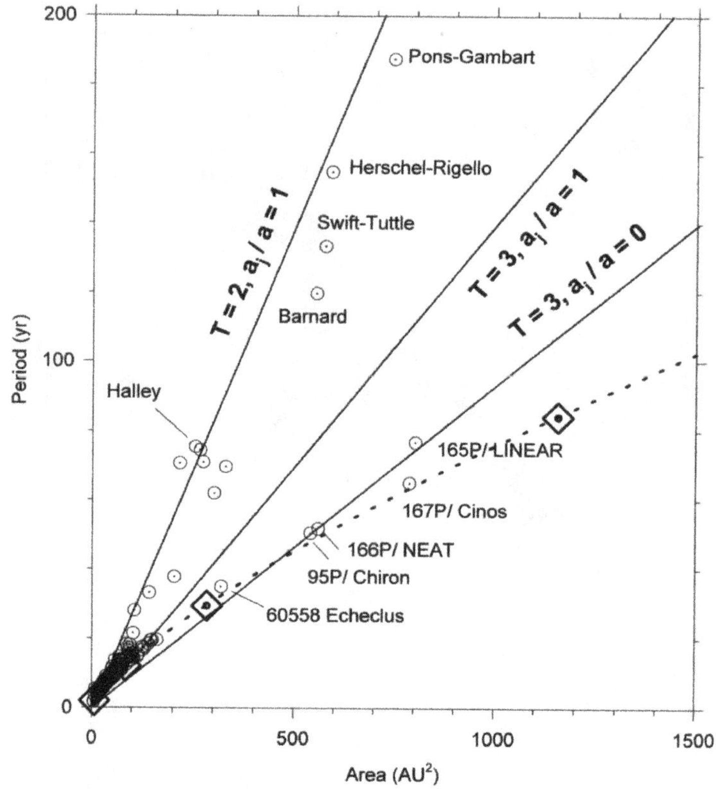

Figure 7.6. Orbital period ($P < 200$ years) versus orbital area for known periodic comets. The three labeled lines correspond to various parameter combinations for equation (7.8), and the four dotted-diamond points indicate the orbital period and orbital area of the planets Mars, Jupiter, Saturn, and Uranus. The dashed line corresponds to the period–area relationship in the restricted 2-body, circular-orbit situation: $P(\text{yr}) = 0.424A^{3/4}$. Various data points are labeled, for reference purposes only, along with the Halley-family branch and the Chiron/Centaur branch.

energy). While these typical diagrams contain the very same information as the orbital-area versus orbital-period diagram, the comet families are less readily evident to the eye.

The classification of cometary families has long been a topic of great discussion, and various schemes have been proposed over the years; some related to the orbital period only, others invoking various dynamical parameters. Figures 7.5 and 7.6 essentially provide a graphical representation of the taxonomic schemes outlined by Hal Levison in 1996 [5] and by Johnathan Horner *et al.* in 2003 [6]. The classification scheme proposed by Levison uses the Tisserand constant as the major discriminate, and indeed, the Tisserand constant is a measure of Jupiter's gravitational influence upon a comet's orbital evolution. Accordingly, the Jupiter family of comets is defined as those objects constrained to move in the ecliptic with $2 < T < 3$, and such bodies will typically have $a_j/a \approx 1$. Those objects with $T < 2$ constitute the Halley-type group of comets. Those bodies with $T > 3$ and with $a_j/a < 1$ fall into the Centaur/Chiron-type family of objects. For this latter group, as emphasized in the taxonomic scheme of Horner *et al.*, the orbital evolution is determined not only by the gravitational influence of Jupiter but by close encounters with the planets Saturn and Uranus. Main-belt asteroids, in contrast, are wholly confined interior to the orbit of Jupiter and have $T > 3$ and $a_j/a > 1$ — see Figure 7.5 for the location of Ceres. Additionally, it is seen that the active asteroids (or main-belt comets), defined as having $T > 3.08$ with $a_j/a > 1$, and which show evidence for mass loss in the form of a coma or tail(s), are not specifically distinguished in Figure 7.5, but rather join smoothly into the small area $[A(\text{au}^2) < 32]$, small orbital period $[P(\text{yr}) < 6]$ domain of the Jupiter family comets.

The simple period-orbital area expressions derived from Figures 7.5 and 7.6 do not provide any new and/or specifically profound description of cometary orbits, but they do bring out a clear and readily observed distinction between the Halley-type objects and the Chiron/Centaur-type objects. Indeed, the distinction between the Halley-type objects and the Chiron/Centaur objects in the period-orbital area diagram is reflective of the fact that the former objects tend to have a relatively large semi-major axis (and hence, by Kepler's 3rd law, a long orbital period) and a high eccentricity (which for a given value of the semi-major axis tends to reduce the associated orbital area). The latter objects, in contrast, have a restricted range in the semi-major axis and allow for only a relatively small variation in orbital eccentricity. The eccentricity constraint in the case of the Chiron/Centaur objects is that if it becomes too large, then a close gravitational encounter with either Saturn or Uranus must come about, and accordingly a dramatic orbital change is likely to occur (see Chapter 10). Likewise, the Jupiter-family comets are restricted to a limited range in semi-major axis and eccentricity, requiring that the aphelion distance be close, but not too close, to the orbit of Jupiter, and that the perihelion distance be interior to the orbit of Mars.

Wright's equal-area rule for cometary orbits is perhaps an aesthetically pleasing idea, but it has no anchor point in reality. At times of great innovation, new discovery, and speculation, however, such oversimplifications (experimental ideas) are not uncommon. Such rules often have some apparent basis in the available observational data, and while ultimately found to be erroneous, they can, and indeed sometimes do, provide a continued stimulus to the construction of new and unanticipated results. The moral that has been underlined in this chapter is that even wrong ideas can sometimes bring out interesting and useful results — or more specifically, sometimes a right result can be deduced from a wrong set of arguments.

Not only did Wright make grand speculations about the area of cometary orbits, he made the speculations solely on the basis of considering just two cometary orbits (granted, this was all the data available to him). Likewise, Horrocks made his argument for the equality of planetary angular size, as viewed from the Sun, upon the deduced measure of just two planets (Mercury and Venus). It is true that the first three planets are remarkably similar in angular size when viewed from the Sun, but Mars immediately breaks the trend — and it does so in a dramatic way. It was one thing to exclude Mars from his planetary size law, but Horrocks provided no rationale for its exclusion.

Writing some 200 years after Wright's death, in 1949, renowned astrophysicists Fred Hoyle and R. A. Lyttleton picked up on the point of imposition when solving for the equations of stellar structure. Their comments resonate still [7]:

"So far no one has succeeded in obtaining stellar models wholly free from both mathematical and physical approximations. Some form of simplification has to be introduced, and the decision that invariably has to be made is whether to maintain a closely correct physical picture and make mathematical approximations to deal with the resulting equations, or to make considerable simplifications of the physical picture and thereby maintain as strict a mathematical treatment as possible."

In essence, Hoyle and Lyttleton are saying, at least in their second procedural option, that just because the mathematics is precise this does not mean that physical insight has been maintained — in other words, being able to *do* the mathematics is not the same as saying that one understands the physical reality.

7.3. The Law of Small Numbers

Both Horrocks and Wright were guilty of imposing idealized (human, minds-eye) pictures onto reality. They were pretty, tidy, and simple rules, but they were entirely wrong. Present-day researchers would (hopefully) be more cautious in their claims than Horrocks and Wright, and indeed, given the, so-called, law of small numbers (LSN), they would be entirely right to be guarded against making dramatic generalizations from limited data. The LSN was first described in a 1971 research paper by Amos Tversky, and Daniel Kahneman [8]. Published in the *Psychology Bulletin*, these authors essentially caught up with human nature, and the manner in which people tend to interpret survey results. In essential terms, the LSN argues that:

LSN: Random events converge to their average value only in the long run. In the short-term, anything goes.

What the LSN is warning against is the false inductive generalization (the one most often adopted in social media surveys and political polls) that a survey based upon a small sample will be representative of the entire population. Such small-sample surveys are almost invariably wrong. And, this is what Wright fell foul of when he made his equal area proposal for cometary orbits — with just two orbits being considered. Another example of an incorrect astronomical inference being made from too small a survey was that by Charles

Blagden in the late 18th century [9]. Blagden was charged by the Royal Society of London to investigate the appearance of several bright (fireball) meteors that had been observed from across Europe. Based upon the data from just a few events, Blagden concluded that all bright meteors traveled from north to south, and accordingly asserted that they must be electrical phenomena — since they were following the Earth's magnetic field lines. This is not a bad idea,[8] but, of course, fireballs do not actually travel exclusively along north-to-south trajectories — as, in fact, a more thorough search of historic fireball records would have informed Blagden. In more recent times the arguments surrounding the existence of Planet 9 might also fall under the LSN lens. In this case, it is an unexpected parameter clustering in a small collection of orbits relating to extreme trans-Neptunian objects (ETNOs) that are being invoked to justify the possible existence of a new planet. In spite of extensive searches being made for the planet itself, it might at this stage, be more useful to spend greater amounts of time searching out more ETNOs (they must be out there), in order to determine if the apparent clustering anomaly still holds true when the orbital data set has doubled, trebled and even quadrupled in size.

The development of the LSN was a case of mathematicians catching up with the actions of human beings, it runs as a corollary to the law of large numbers (LLN). First articulated by Gerolamo Cardano in the 16th century, and mathematically proved in the 18th century by Jacob Bernoulli, the LLN argues:

LLN: The average of results obtained from a large number of trials will converge to the true arithmetic mean or expected value.

The other way of stating the LLN is that very large samples will indeed be highly representative of the population as a whole. The trick, or art of the game, is to know whether, in conducting a survey, the number of trials made falls within the domain of the LSN or LLN. This latter issue is particularly relevant in the modern era of particle physics, where an apparent signal, suggestive say of a new

[8] Meteors are, in fact, the light phenomenon associated with the heat vaporization of millimeter-sized grains entering the Earth's atmosphere at high speed.

particle being found, is based upon the statistical interpretation of experimental data. There are many examples where hints of a new kind of particle, and/or new physics being evident, appear in an early data set, with the *signal* only to fade and then completely disappear in later data sets when many more experimental trials have been conducted.

7.4. The Law of Anomalous Numbers

Known under a number of names, the law of anomalous numbers (LAN) is also called the first digit law, Benford's law[9] [11], and sometimes the Newcomb–Benford law. It is one of those laws that is counterintuitive, and it takes us, in a useful way, beyond the LSN and LLN. It is one thing to impose a law, rule, or geometrical idea on nature, but it is an unforgivable act for a scientist to deliberately fake their data in order to support some particular claim. Such actions are not uncommon in the history of science, and indeed, Ptolemy has been charged with (what is actually a ridiculous claim) fixing his data tables. Even in the modern era, the sin of scientific data fixing and fraudulent behavior is not unheard of, although it is very rare.[10] Normally it is difficult to determine if the entries within a numerical data set have been tampered with, but the LAN is one way of searching out signs of manipulation as well as strangeness.

The first discussion of the LAN was presented by astronomer Simon Newcomb[11] [12] who specifically noticed that in a table of logarithms, it was the first few pages that showed the most wear and tear. This, he realized, indicates that most of the numbers that were being looked up started with a small digit, like 1 or 2, or 3, and only

---

[9] This label refers to electrical engineer and physicist Frank Benford (University of Michigan).

[10] Wikipedia has a page dedicated to scientific misconduct which has examples from pretty much every branch of science.

[11] Canadian born Newcomb grew to international renown for his observational work at the US Naval Observatory and his tenure at John Hopkins University. It has been suggested by Bradley Schaefer that Newcomb was Arthur Conan Doyle's inspiration for the character Moriarty — the Napoleon of crime!

rarely were numbers beginning with a 7, 8, or 9 being looked for. This is seemingly odd since given that all numbers must start with a first significant digit drawn from the list 1, 2, 3, ... , 7, 8, 9, one might expect that each digit (1 through 9) would appear 11.11% of the time in any given large data set — that is they are uniformly distributed. In fact, the LAN indicates that under many different circumstances, 1 will appear as the leading digit 30.1% of the time, the digit 2 will appear as the leading digit 17.6% of the time, and so on to digit 9 which appears as the leading digit just 4.6% of the time. Benford pushed Newcomb's observation still further and considered the leading digit of some 20,299 numbers, gathered from numerous sources. He noted with respect to the leading digit distribution that, "there is here the peculiar fact that numbers that individually are without relationship are, when considered in large groups, in good agreement with a distribution law [the LAN]" [11]. Figure 7.7 shows the expected variation in the percentage appearance of first digits for the LAN (blue curve), and this trend has been recorded in lists of such diverse quantities as the length of rivers, the first 100 numbers found at random within a newspaper, the first 308 numbers found in an issue of *Readers Digest*, the home address numbers in a list 342 scientists chosen at random, population numbers, the tabulated values of 1389 specific heats, and mathematical constants. Indeed, the leading digits of the numbers derived from the Fibonacci sequence satisfy the LAN distribution, so too do the factorials and the powers of 2. Intriguingly, however, the first digits for square roots and the reciprocals of the natural numbers do not satisfy the LAN. A recent application of the LAN to a large astronomical data set looked at the parallax estimates of 1.3 billion stars in the GAIA[12] catalog [14]. Indeed, the data closely follows the LAN, with stars having a parallax starting with the digit 1 being five times more numerous than stars having a parallax starting with the digit 9. This is actually a comforting result since it tells us that there are no unknown

[12]The GAIA space observatory is operated by the European Space Agency and is designed to measure stellar proper motion and distances (parallax). Launched in 2013, the mission is expected to continue until at least 2025, by which time about 1% of all the stars within our galaxy will have been studied.

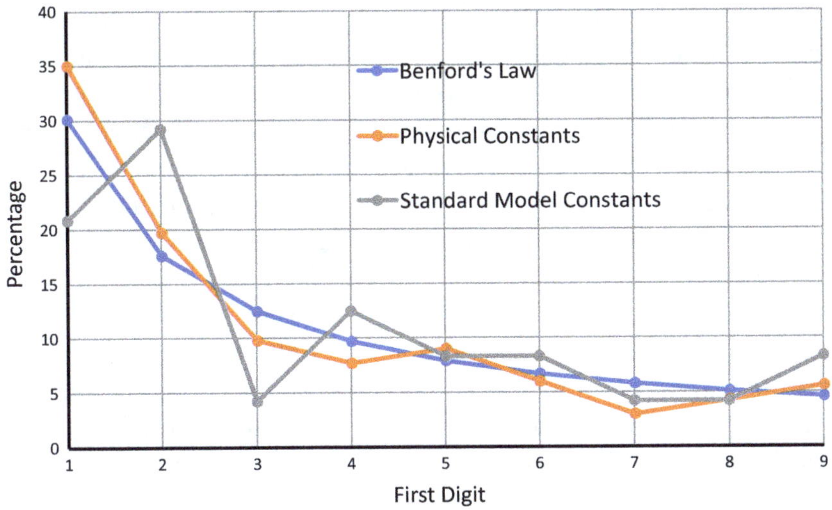

Figure 7.7. Illustration of the LAN (Benford's law). The blue curve shows the variation in first digit occurrence for the LAN, the orange curve is based upon a compilation of 234 physical constants published by NIST. The grey curve indicates the first digit occurrences for 24 of the fundamental constants of nuclear physics and cosmology.

systematic errors in the GAIA analysis procedure skewing the catalog results. In terms of physical constants, Figure 7.7 shows the first digit distribution for two data sets. The orange curve corresponds to a study of the first digits in 234 physical constants extracted from the National Institute of Standards and Technology (NIST) webpage. The grey curve is based upon the leading digits from the 32 fundamental constants[13] of nuclear physics and cosmology — as given, for example, by Tegmark [13]. The latter constants, in principle, should explain the values of all other physical constants, and observables in the entire universe. The NIST physical constants list (orange curve) follows the LAN closely, but interestingly, the fundamental physical constants do not — these data (grey curve) show too

[13] Only 24 of these constants have been measured with any great certainty — the remainder are only constrained experimentally through the derivation of lower or upper bounds.

few occurrences of the digits 1 and 3 and too many occurrences of digit 2 — what this might mean is unclear at the present time.

One of the key requirements for a specific data set or sequence to satisfy the LAN is that the numbers cover a large range, distributed over many orders of magnitude. Other requirements are that there is no specific maximum or minimum number, or that the numbers are increasing sequentially. Importantly, however, where the LAN has found interesting applicability is in detecting fraudulent data sets, for example, in tax return data and in election result returns. The main reason behind this useful ability of the LAN, however, appears, in part, to be a consequence of human nature. Specifically, when someone attempts to construct (fake) random data they tend to distribute the figures uniformly, and importantly, the uniform distribution does not satisfy the LAN — indeed, a uniform distribution will yield an 11.1% occurrence rate for each digit. Diplomatically, Benford concluded his 1938 article [11] with the prescient words, "the theory of anomalous numbers is really the theory of phenomena and events, and the numbers but play the poor part of lifeless symbols for living things".

You can run but you cannot hide. This might be the mantra of the LAN. While Horrocks and Wright attempted to impose specific rules on nature, there was no fraud attached to their undertakings — they were simply neat ideas that nature had no use for. Moving into the present and the future, however, the power of the LAN is that it can potentially highlight problem issues with large numerical data sets. The issue here is not about fraud, as such, but about the ability (through the LAN) to recognize biases or systematic problems in data reduction methods, and in an increasingly data-rich world, this can only be a good thing.

7.5. The Imposition of Horoscopes

In an ironic twist of applying cause and effect, perhaps the most successful imposition on astronomy has been that of astrology. The term ironic is used since in earlier times there was no clear distinction between these two activities, and arguably astronomy, as the study of the heavens, grew out of the desire to make astrological predictions. However, astronomy as a science, and astrology as a human imposition upon the heavens, are now entirely separate preoccupations. Indeed, here we have an example of a situation in which numbers, no matter how used, small, large, measured, faked, or guessed, can resolve the situation. For all our belief in the truth associated with numbers, practitioners of, and believers in, astrology refuses to accept the data presented to them. Indeed, such practitioners rebel against numeration, since there is no statistical/ scientific evidence to support any astrological prediction that has ever been made.

While there is absolutely zero evidence for any causal link between the positions of the planets, Sun and Moon along the zodiac with the events happening here on Earth, or during a specific human's lifetime, people continue to believe in and read their daily horoscopes to this very day. Why is this? Indeed, it is unfathomable, and it speaks volumes about the human mind (the same basic mind that also created mathematics). Indeed, recent surveys indicate that nearly 1/3rd of Americans believe in astrology, and this begs the question of how this state of mind can persist when it is fully contradictory to our scientific understanding of how the world and universe work. It is glib nonsense to argue that we don't know everything about the solar system and the planets, or the workings of the cosmos. It is nonsense to argue thus since we clearly do know more than enough to show that astrology is a complete delusion. And yet ... here we are, in the modern era, with such beliefs being not uncommon. Interestingly, however, what numbers, in the form of statistical analysis, can do for us is find correlations between the belief in astrology and other human traits. This has recently been done, for example, by Ida Andersson (Lund University) and co-workers, Julia Persson and Petri Kajonius [10]. Assessing the data from 264 participants, who claimed a belief in astrology, Andersson *et al.* set about correlating this belief against individual personality traits and cognitive biases as measured by standard testing procedures. Their main conclusion was that a belief in astrology is anti-correlated with intelligence, and positively correlated with narcissistic tendency — the higher the narcissism index for an individual, the stronger their belief in astrology. Remarkably, it is as if the strong converts to astrology find self-satisfaction in the belief that even the stars find them important. Given such a correlation, it would appear that the human imposition of astrology upon the back of astronomy, in spite of the numbers, will persist for a good while yet.

References

[1] Chapman, A. Jeremiah Horrocks, the transit of Venus, and the 'New Astronomy' in early seventeenth-century England. *Quarterly Journal of the Royal Astronomical Society*, **31**, 333–357, 1990.

[2] Hughes, D. W. Horrocks's Bogus Law. *Astronomy and Geophysics*, **46**, 1.14–1.16, 2005.

[3] Weiss, L., *et al.* The California-Kepler Survey. V. Peas in a Pod: Planets in a Kepler multi-planet system are similar in size and regularly spaced. *The Astronomical Journal*, **155**, article 48, 2018.

[4] Beech, M. Second thoughts beyond Kepler: The equal-area rule of cometary orbits by Thomas Wright of Durham (1711–1786). *Observatory*, **137**, 1–8, 2017.

[5] Levison, H. Comet taxonomy. In *Completing the Inventory of the Solar System*, Astronomical Society of the Pacific Conference Proceedings, **107**, T. W. Rettig and J. M. Hahn, Eds., 173–191, 1996.

[6] Horner, J., *et al.* The populations of comet-like bodies in the solar system. *Monthly Notices of the Royal Astronomical Society*, **343**, 1057–1066, 2003.

[7] Hoyle, F. and Lyttleton, R. A. The structure of stars of non-uniform composition. *Monthly Notices of the Royal Astronomical Society*, **109**, 614–630, 1949.

[8] Tversky, A. and Kahneman, D. Belief in the law of small numbers. *Psychology Bulletin*, **76**, 105–110, 1971.

[9] Blagden, C. An account of some late fiery meteors. *Philosophical Transactions of the Royal Society*, **74**, 201–232, 1783.

[10] Andersson, I., Persson, J., and Kajonius, P. Even the stars think I am superior: Personality, intelligence and belief in astrology. *Personality and Individual Differences*, **187**, 111389, 2022.

[11] Benford, F. The law of anomalous numbers. *Proceedings of the American Philosophical Society*, 78, 551–572, 1938.
[12] Newcomb, S. Note on the frequency of use of the different digits in natural numbers. *American Journal of Mathematics*, 4, 39–40, 1881.
[13] Tegmark, M. *Our Mathematical Universe: My Quest for the Ultimate Nature of Reality*. Alfred A. Knopf, New York, 2014.
[14] de Jong, J., de Bruijne, J., and de Ridder, J. Benford's law in the GAIA universe. https://arxiv.org/pdf/2008.12271.pdf.

Chapter 8

Gulliver's Travels, Luck, and Guesswork

Felicitous and inexplicable strokes of inventive talent.
— William Whewell

Many events in this world are predictable, and we can sometimes determine, with very high confidence, what their future behavior will be. Some phenomena, however, sit beyond the boundaries of knowledge, and even beyond the reach of mathematics. Yet other circumstances produce results that are so unexpected, and yet so delightful, that the human mind is tricked into thinking that something profound has happened. The latter, for want of a better phrase, can be called the false-positives of history. Another way to attain a false positive, and to seemingly bridge a gap in one's knowledge, is to guess and trust in the workings of serendipity. Guessing is not usually included among the tools-in-trade of the scientist, but it has, at least historically, played many an interesting hand. No scientist would admit to guessing a result, but they will certainly acknowledge a lucky break. The two are certainly different, since guessing implies desperation, while luck is always welcome. Max Planck famously described the desperate state of affairs that surrounded the discovery of his law describing the characteristics of blackbody radiators. This discovery, which from a classical perspective was an act of desperation, and which initially was purely a mathematical *fix*, resulted in the emergence of quantum mechanics.

There have been many fortuitous observations in the history of astronomy when a new phenomenon or object has sailed into someone's eyepiece. The discovery of planet Uranus, by William Herschel in 1782, was lucky for him, and it was one of the highlights of his illustrious career. Although, if not discovered by Herschel, then someone else would have eventually found the planet (indeed, historically, they had). Sometimes, it is simply a case of being in the right place at the right time that makes, or breaks, a discovery. Novae, comets, and meteorites — all are serendipitous discoveries, and someone has to be the first observer (officially, or unofficially). The discovery of pulsars, and quasars, for example, were inevitable, and a direct result of many research groups beginning to systematically study the heavens with radio telescopes — one team or another had to make the lucky discovery — which is not to deny the hard work and sheer tenacity required to make the observations and then interpret them correctly. Other discoveries are entirely fortuitous, and have nothing to do with expectation and/or foreknowledge — every meteor, new comet, asteroid, Kuiper Belt Object, exoplanet, nova, supernova, and galaxy fits into this category.

Guessing is different from imposing. The anti-Earth of the Pythagoreans, for example, was not a guess. To the Pythagoreans, it was a necessary and essential part of the cosmos. While different from imposing, guessing can lead to apparently successful results. In 467 BC, it is said by Plutarch, (who was writing more than 500 years after the event), that Anaxagoras predicted the fall of a meteorite that fell in Aegospotami. It is probably safe to assume that a meteorite did fall at the place and time indicated, and it is certainly possible that Anaxagoras made a prediction of a fall — but the only link between the prediction and the event actually occurring is luck. Immanuel Kant was lucky in his 1755 suggestion (guess) that he could see "island universes" (what we would now call galaxies) in the process of forming planets.[1] There was no physical basis, data, or definitive observation upon which he could make such a claim, and it was pure luck that his suggestion had anything to do with the formation of planetary systems. Norman Pogson[2], in Madras, India, discovered his comet of 1872 on the basis of a telegram sent out by Ernst Klinkerfues, in Berlin. The telegram announced the onset of the 1872 Andromedid meteor storm over Europe, and the prediction was that the stream's parent comet should be close to the coordinates of the anti-radiant point in the sky. Rather than finding the remnants of comet Biela (the actual parent comet of the Andromedid meteoroid stream), however, Pogson found an entirely new comet that had nothing to do with the Andromedid meteor shower. Klinkerfues's prediction was pure guesswork; Pogson's find was pure good luck. A more recent example of guesswork at play is that of the Drake Equation. This much abused and greatly misinterpreted equation provides an estimate of the number of intelligent lifeforms in the galaxy. It is a product of disjointed expressions that when combined say something, and it is certainly not clear what, about those factors that we think, or guess, will affect the number of extraterrestrial civilizations within our galaxy. In reality, we have absolutely no idea how to determine the number of possible civilizations, and what astronomical, physical, psychological, and societal factors determine the galactic population.

Being lucky is always a good trait to cultivate, and long may it ever be so. And, while guessing is not a recommended scientific strategy in general, it sometimes leads to a lucky association — but it is usually luck that follows the maxim that even a stopped clock is right twice a day. Philosopher William Whewell was more optimistic in his outlook on guesswork, and drew attention to the importance of happy guesses (see the opening quotation to this chapter), writing,

[1]Bertrand Russell in his 1910 review of Kant's *Cosmology* was to write, "Kant's premises are usually false, and his reasoning is usually fallacious. But his theories have a value as a stimulus to the scientific imagination".

[2]This is the same Normal Pogson that introduced the magnitude scheme, still used by astronomers, for stellar brightness.

that some discoveries, "are not improperly described as happy guesses; and that guesses, in these as in other instances, imply various suppositions made, of which someone turns out to be the right one". Indeed, it is sometimes written that science fiction is the forerunner of science-fact, but this is simply not true. Indeed, the argument is entirely based upon just a few cherry-picked results that have chanced to come true (and then often in only some vague manner of similitude). Compared to all the predictions ever made by futurists and the authors of science fiction the ones that are plain wrong far outweigh the few that chance, by pure luck, to come true. To guess is to cast one's thoughts to the turbulent winds of serendipity,[3] and to hope, thereafter, that the goddess Fortuna is watching as events unfold.

8.1. The Mystery of Mars's Moons

One of the more remarkable false positives of astronomy, and a result of pure guesswork, is encountered in one of the most satirical books ever written. The book in question is Jonathan Swift's *Gulliver's Travels* (published in 1726). Indeed, while there are many underlying themes of social-commentary in Swift's book, it is the fact that he described the existence of two Martian moons that is remarkable.[4] To the modern reader this is not odd at all, for indeed, Mars does have two moons — but the point is, when Swift was writing his satire, no such moons were known to exist. The text in question concerns Gulliver's voyage to the small island of Luputa. There he learns that the astronomers have, "discovered two lesser stars, or satellites, which revolve about Mars; whereof the innermost is distant from the center of the primary planet exactly three of its diameters, and the outermost, five; the former revolves in the space of ten hours, and

[3] In the case of Harry Potter, the approach might be to first prepare a draft of *Felix Felicis* (liquid Luck).

[4] Voltaire further referenced two Martian moons in his 1752 sci-fi novella *Micromégas* in which a giant alien from a planet in orbit around the star Sirius visits the Earth and sets out to tour the solar system with a giant companion who happened to live on Saturn.

Table 8.1. Comparison between the orbital radius (expressed in Mars diameters) and orbital period in hours, as predicted by Jonathan Swift and as determined by observations.

	r (Swift)	P (Swift)	r (actual)	P (actual)
Phobos	3	10.0 hours	1.38	7.66 hours
Deimos	5	21.5 hours	3.46	30.35 hours

the latter in twenty-one and a half; so that the squares of their periodical times are very near in the same proportion with the cubes of their distances from the center of Mars". That Swift references Kepler's 3rd law with respect to the orbital characteristics of the moons is not a great surprise, since by the 18th century it was a well-established physical law,[5] rather, it is the apparent similarity of Swift's stated orbital parameters that is noteworthy and surprisingly accurate with respect to modern-day observations. A comparison between the stated and the actual orbital parameters is shown in Table 8.1. In terms of the orbital radii, Swift's values are overestimated by factors of 2.17 and 1.45, which in many respects is not very good, but the point is that Swift could have chosen any numbers for the orbital radii (big or small). A choice of 1 and 4 would have been more impressive, in terms of agreement with the actual observations, but the adopted 3 and 5 are not bad given the large range of choices available. In terms of comparisons that Swift would have had at the time of writing, the Moon's orbital radius is equal to 30.2 Earth diameters, while the 4 Galilean Moons of Jupiter, Io, Europe, Ganymede, and Calisto, come in at 3.0, 4.8. 7.7, and 13.5 Jupiter diameters, respectively. For Saturn and Titan, the orbital radius is 10.5 Saturn diameters. Swift's choice of 3 and 5 therefore mimics (or indeed, were taken from) those numbers appropriate to Io and Europa.

Historically, there has never been a credible claim that planet Mercury might have a companion moon or moons. This is no doubt predominantly due to the difficulty of observing the planet, even with a telescope, since it is only ever visible within the glow of

[5] Effectively, Isaac Newton had established this, and Kepler's other laws in his *Principia*, published in 1687 — recall Chapter 6.

twilight. The planet itself is also relatively small (being just 1/3rd the size of Earth, and not much bigger than the Moon), requiring a sizeable telescope aperture to resolve its disk. The situation for planet Venus, however, is a little more colorful, with telescopic sightings of a companion moon being reported on numerous occasions during the late 18th century. These reports, we now realize, were clearly spurious, and presumably the result of light-scattering artifacts in the telescopes then being employed. The repeated observation of a Cytherean moon, however, by different observers, at different times, using different telescopes, suggests that there was some considerable herding instinct at play.[6] It is not uncommon that once one observer has reported seeing some vague but specific astronomical/physical phenomenon, other practitioners begin to *see* it too. Such historical similarities include the *N*-rays of Prosper-René Blondlot (*discovered* in 1903), the *discovery* of Martian canals, as promoted by Percival Lowell in the early 20th century, the cold fusion experiments of Martin Fleischmann and Stanley Pons (first described in 1989), and the detection of meteor trails (shooting stars) in an assumed lunar atmosphere in the 1950s. For all this, however, no reasonable claims for the detection of a Cytherean moon have been made since the beginning of the 19th century. The only exception to this is the dramatic suggestion made by Immanuel Velikovsky in his 1950 book *Worlds in Collision*. Velikovsky claimed that Venus had, in fact, been ejected from Jupiter, as a comet-like body, some 17 centuries ago. While moving to its present orbital location, Velikovsky maintained, the wayward body dramatically changes the orbit and spin-axis of the Earth, and this, he suggested, might explain the many catastrophe-based mythologies found within the memory and writings of ancient cultures. Under this reasoning, however, Venus was not one of the original planets. While Velikovsky's thesis is completely untenable, it is, when all is said and done, a little surprising that Venus is not accompanied by one or more moons. The main

[6]The herding instinct is very much alive and well to this day in the sense that scientists (almost) invariably pursue research in those areas that are funded — it is not the case that public funding follows individual research interests.

reasons for this latter assertion are its extremely slow and retrograde spin-rotation. Both of these effects are readily explainable in terms of an off-center, protoplanet collisions in the early stages of planet formation. Indeed, Earth's relatively rapid spin and large Moon are generally explained in terms of an off-center, grazing impact between the Earth and a Mars-sized protoplanet some 80 million years after the formation of the solar system began. An alternative to the collision origin of our Moon, however, was articulated by Martyn Bailey in an article published by the journal *Nature* in July of 1969. Indeed, Bailey suggested that the Moon could have been a former planet, originally located between Mercury and Venus. Gravitational interactions between Luna, Bailey's name for the Moon during its planetary childhood, and the inner most planets destabilized its orbit and it was subsequently captured by the Earth. Bailey further argues that the presence and subsequent destabilizing of Luna's orbit resulted in the anomalously high orbital eccentricity observed for Mercury's orbit. This capture scenario for the Moon, while one of low probability, reminds us of the fact that when looking at planet–moon systems there is an inherent dynamism at play and that the moon count for a specific planet can certainly vary as a function of time. For all this, Baily's suggestion for the origin of Earth's Moon is generally not held to be tenable.

The notion that Mars might have two companion moons actually predates Swift's guess by some 110 years, but even this initial seed idea was based upon a mistake — or, to be more generous, its origins lie amid a remarkable set of circumstances concerning the planet Saturn. The first hint that there was something strange about the then outermost planet was noticed by Galileo circa 1616. Unlike his earlier rush to describe and publicize his telescopic observations of the heavens, in the case of Saturn, he uncharacteristically held back. At stake, in fact, was the very notion that telescopes were capable of presenting a true image of the heavenly realm. Indeed, the very real issue of image distortion long plagued early telescopic observations, and this in turn provided ammunition for those critics reluctant to see the universe deep-searched and analyzed — both processes that, in their view, undermined divine wisdom and church authority. Indeed, let us remember that Galileo was not later persecuted by the Church because he argued in favor of the Copernican system, but because he taught that the Copernican system was true without actual proof — and, at that time, the Church authorities were correct: Galileo had no direct and/or indisputable proof to support his assertions. The fact that the Copernican viewpoint was ultimately shown to be correct, is not the point — what Galileo was guilty of was overstepping his authority in matters of the Church concerning truth and reality (this and the fact that he seemed to have an uncanny knack for making enemies in high places[7]). With respect, however, to the telescope, it was argued against Galileo that it was potentially prone to producing distortions, and that accordingly what was seen through the eyepiece could not automatically be relied upon as being true. And this, it is suggested, is why Galileo was uncharacteristically low-key about his observations of Saturn. He did inform some of his friends about his observations, but rather than simply state in an open letter what he had seen, he sent an anagram to Johannes Kepler. The anagram (which was to be deciphered in Latin) reads the following:

SMAISMRMILMEPOETALEUMIBUNENUGTTAUIRAS

Decoded correctly the anagram would read (in English translation): *I have observed the most distant of planets to have triple form.* This was Galileo's claim — the telescope revealed Saturn to be a closely-packed triple planet; a main body and two large moons. That a planet could have moons was no surprise, since in his *Siderius Nuncius* (*The Starry Message*[8]), published in March of 1610, Galileo had described the four moons of Jupiter — and, of course, the Earth has the Moon. Saturn was different again, however, since Galileo thought he was seeing three conjoined planets, rather than a planet

[7]Historian of astronomy Allan Chapman has written that Galileo was a martyr to science who brought his own fuel.

[8]I am using the translation *Starry Message* here since this is what Galileo actually intended the title of his pamphlet to be. It was Johannes Kepler who incorrectly translated and then introduced the term *Starry Messenger*.

and two moons. It is often recorded that Kepler misread the message, but given that there were absolutely no clues as to what the message might refer to, the fact that he found any coherent string of words from within the anagram is truly remarkable, if not miraculous. What is even more remarkable is that he found a sensible, wrong message in the anagram. Kepler's solution (in English translation) was as follows: *Hail, twin companionship, children of Mars* — in other words, Mars had two moons. It is this particular false-positive translation that appears to have launched the idea that Mars must have two moons. That Galileo did not come straight out and claim he had discovered a triple-planet in Saturn relates to the fact that observations taken at later times revealed no such triple-system, but just one spherical planet. Such observations can, of course, be described by some form of system rotation, but now one has to convince a skeptical audience that not only was Saturn a triple-system, but a rotating one as well. It was clearly a case, and Galileo knew it, that the simpler interpretation of the observations was that the telescope was presenting false information to the eye — and this, of course, was something that Galileo was not prepared to countenance. Ultimately larger and better telescopes than those constructed by Galileo were constructed, and in 1647 Polish astronomer Johannes Hevelius published an account of Saturn's appearance, arguing that it was in fact an elliptical planet having two crescent-shaped handles attached to its surface. In 1656, however, Christian Huygens argued that Saturn was in fact a ringed planet, with the equatorial ring, which nowhere touched its surface, having a diameter at least twice that of the primary planet. To establish the priority of his discovery, Huygens announced to his fellow philosophers:

AAAAAA CCCCC D EEEEE G H IIIIIII LLLL MM
NNNNNNNNN OOOO PP Q RR S TTTTT UUUUU

Once again to be deciphered in Latin but read in English translation as follows: *He* [Saturn] *is surrounded by a thin flat ring, which does not touch him anywhere and is inclined to the ecliptic.* There is no account of anyone having solved Huygens anagram — which is perhaps not surprising given that there are some $61! \sim 5 \times 10^{83}$ possible letter combinations — this is more combinations than there are atoms in the observable universe (recall Chapter 2 and the Eddington number).

Even with the refined telescopic observations produced by Huygens it was not immediately accepted that Saturn must be a planet accompanied by a ring system — part of the problem now being, what is it that the rings are made of, and why is it only Saturn that sports such an appendage. It is now known, of course, that all of the Jovian planets have ring systems, and it has also been determined, as a result of measurements gathered by the *Huygens* spacecraft, that the rings of Saturn may be a temporary structure, having formed within the last few hundred million years. Two hundred years after Huygens first resolved the rings about Saturn, however, Cambridge University announced that its Adams Prize topic was to be concerned with the constitution of the rings. The prize was awarded to James Clerk Maxwell (he was also the only person, in fact, to submit an essay) in 1858. It was in this remarkable, mathematically dense, essay that Maxwell demonstrated that the only stable configuration for the rings was that composed of myriad small satellites (or moonlets) — solid or liquid rings, he demonstrated, would be unstable to fragmentation and disruption. While the problem of Saturn's ring system was largely resolved with the publication of Maxwell's study in 1859, which is not to say that planetary ring systems are understood, there were still no detected moons in orbit about Mars — indeed, their discovery was still set 18 years into the future.

8.1.1. *What choice Swift?*

Writing in 1908, astronomer Robert Ball argued that when it came to guessing the number of Martian moons, Swift had a logical choice of it being either 2 or 3. The reasoning (rather than logic) is that the Earth has one moon, and Jupiter has four, so Mars, as the in-between planet, must have either 2 or 3 moons. Why Venus and Mercury have no moons is not addressed by this *logic*, but Ball's reasoning implies that one of two moon-number rules might be at play. One rule would be a predominantly doubling scheme, starting as 0, 0, 1, 2, 4. The other rule would run 0, 0, 1, 3, 4. Let us call these two sequences $S(N)$ and $B(N)$, respectively, where N is the planetary sequence number with $N = 1$ corresponding to Mercury, $N = 2$ corresponding to Venus, and so. Using the fitting condition that Earth must have one moon and Jupiter four, two moon-number generating rules can be established — these being the following:

$$\begin{aligned} S(N) = \text{Swift number} &= 0, \text{ for } N = 1, 2 \\ &= 2^*(N-2), \text{ for } N \geq 3 \\ B(N) = \text{Ball number} &= 0, \text{ for } N = 1, 2 \\ &= 1, \text{ for } N = 3 \\ &= 3, \text{ for } N = 4 \\ &= B(N-1) + B(N-2), \text{ for } N \geq 5 \end{aligned}$$

The Swift number effectively doubles at each step after the Earth ($N = 3$), while the Ball number, after Mars ($N = 4$), proceeds as a summation of previous terms. The Ball number sequence seems less elegant than the Swift number sequence,[9] having, as it does, more special starting conditions, although in 1726 when Swift was writing *Gulliver's Travels*, it was no less likely a *correct* sequence for planetary moon numbers than that afforded by the Swift number sequence (see Table 8.2).

With Asaph Hall identifying Phobos and Deimos (Figure 8.1), the fabled moons of Mars, in 1877, the Swift number sequence $S(N)$ becomes the preferred rule with respect to the available data, being exact all the way out to planet Saturn. At that time, one might have felt reasonably confident in predicting 16 and 32 as the final number of moons in orbit around Uranus and Neptune. The Ball number sequence would have been fine enough in 1726, but by 1877 it was out of date due to the discovery of Saturn's 8th Moon,

[9] This being said, after the first two terms, the Ball sequence can be cast in terms of the addition of two Fibonacci sequences, and this sequence is famed for its appearance in many natural and physical settings.

Table 8.2. Comparison of the Swift and Ball numbers for planetary moons (columns 3 and 4). Columns 5, 6, and 7 indicate the known moon count for the years 1726, 1877, and 1950. The last column indicates the number of known moons in orbit about the planets as of 2021, where the significance of spacecraft and space telescope exploration are clearly evident when compared to the numbers known in the pre-space era before 1950.

Planet	n	S(N)	B(N)	1726	1877	1950	2021
Mercury	1	0	0	0	0	0	0
Venus	2	0	0	0	0	0	0
Earth	3	1	1	1	1	1	1
Mars	4	2	3	0	2	2	2
Jupiter	5	4	4	4	4	11	79
Saturn	6	8	7	5	8	9	82
Uranus (1781)	7	16	11	—	4	5	27
Neptune (1845)	8	32	18	—	1	2	14
Pluto (1930)	9	64	29	—	—	0	5

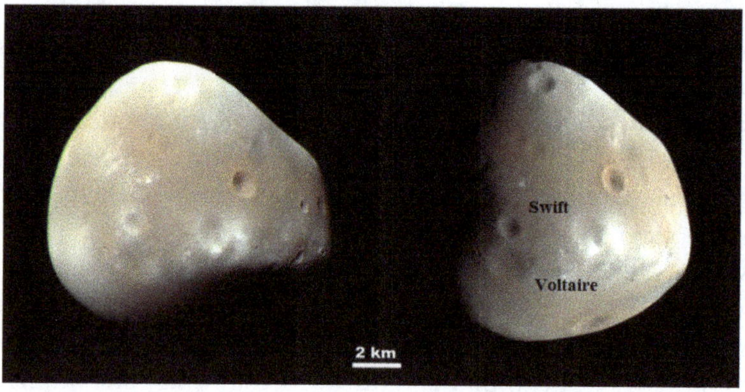

Figure 8.1. The Martian moon Deimos. The small, 1- and 2-kilometer-sized craters Swift and Voltaire are labeled in the righthand side image. Image courtesy of NASA.

Hyperion — discovered independently by William Cranch Bond and William Lassell in 1848. The fact that the Ball number for Mars is B(4) = 3 could still have been true post 1877, although it is now clear that there are no sizeable moons other than Phobos and

Deimos in orbit about Mars — which is not to say there have never been more moons in the past and that there might not be more moons in the future. Indeed, this latter point serves to remind us that the number of moons that orbit a particular planet can change over time and that accordingly a prediction such as B(4) = 3 for Mars cannot be ruled out as being incorrect until the actual moon count exceeds 3. Indeed, the exact origin of Phobos and Deimos is not entirely clear, but one scenario for their origin is that they are captured objects from the main-belt asteroid region. They may not have been captured at the same time, and there may have been more than two moons at earlier epochs — indeed, there is no specific reason that Mars cannot capture more asteroid moons in the future. Not only might Mars capture more moons in the future, but it is also clear that in the deep future, it is going to lose Phobos. The orbital semi-major axis of Phobos is decreasing by about 2 meters per century, and within the next 50 million years the moon will begin to break apart due to tidal forces, resulting in the formation of a disk of debris about Mars and material crashing into its surface. Deimos, in contrast, is slowly moving outward, but it will likely evolve, eventually, onto some stable orbit.

The Jovian planets are additionally thought to have captured a number of their moons at different times over the age of the solar system. Additionally, the retrograde orbit of Triton, the largest moon of Neptune, suggests that it is a captured Kuiper Belt object. Furthermore, Saturn has numerous moons with retrograde orbits (e.g., Phoebe), likewise Jupiter (e.g., Sinope), and these are most probably captured objects — either, cometary nuclei, asteroids, centaur, or Kuiper Belt objects.

It is not an unreasonable question to ask why it took astronomers so long to discover the moons of Mars. It was not, for example, a technological innovation that enabled their discovery since telescopes with the same (and even better) capabilities as the one used by Hall were available for use long before his 12 August 1877 discovery. The answer probably lies in expectation, with astronomers looking for one or more relatively large, well-separated, bright companions rather than two small and faint, close-in bodies. Once you

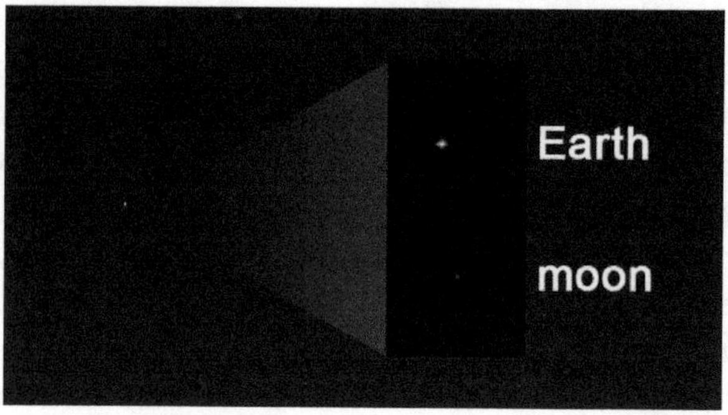

Figure 8.2. The Earth and Moon in the Martian sky, as imaged by the *Curiosity* Rover. A human observer standing on the surface of Mars would actually see the Earth and Moon as two distinct *star-like* objects. Image courtesy of NASA.

know they are there, then the moons of Mars are relatively straightforward to find. Indeed, pre-Raphaelite painter and amateur astronomer John Brett was to write in the 11 October 1877 issue of the journal *Nature* that he had easily found the new Martian moons with a 9-inch objective telescope.[10] Prior to the discovery of Phobos and Deimos, however, the primary moon systems that astronomers were familiar with were those of the Earth and Moon (Figure 8.2), along with the bright moons of Jupiter and Saturn. These three planets all sport very distinctive moon systems that are easily resolved by the eye or a small telescope, and they set the expectation that any moons around other planets should also be seen as prominent companions. Essentially, no one expected that planets might have small, relatively inconspicuous, close-in moons, and accordingly, they were not looking for such objects when viewing Mars through the telescope eyepiece.

There is absolutely no reason to expect that the number of moons in orbit about a given planet should vary in any systematic manner with respect to heliocentric distance — and indeed, the numbers do not. Accordingly, it must be concluded that neither the Swift number nor the Ball number sequences afford any reasonable (even logical) account of the number of moons that might accompany a

[10] Hall was using the United States Naval Observatory's 26-inch refracting telescope.

planet. There is simply no formula that can predict or even account for, the number of moons that might accompany a given planet (see, however, Chapter 9). Just like Swift, we can only guess what the number might be, and sometimes that guess might turn-out to be right. Most of the time, of course, non-informed guesses turn out to be wrong. While the moon count per planet may have escaped mathematical enumeration, it is still possible to ask if there is any specific orbital spacing rule that characterizes the separation of moons within a given satellite system — this possibility will be considered below (and again in Chapter 9).

That the technology available to astronomers, at a given epoch, has influenced the actual moon count for a given planet (Jupiter) is illustrated in Figure 8.3. Here the technological leaps highlighted are

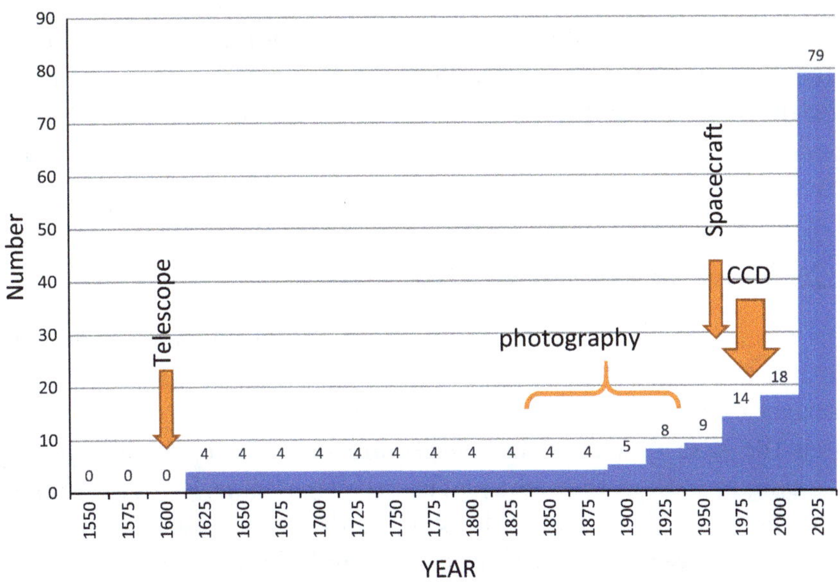

Figure 8.3. Jupiter moon counts in ¼ century time interval bins. Galileo reported the first four moons in 1610, and the total of 79 is complete up to 2021. The influence of introducing new technologies is evident under the headings of photography, spacecraft flybys, and charge couple devices (CCD). The discovery numbers are also related to astronomer interest in the topic and time granted for such survey projects on large-aperture telescopes.

those relating to the development of the telescope (from circa 1610), the application of photography (starting circa 1850), fly-by spacecraft exploration (beginning with the Pioneer and Voyager spacecraft in the 1970s), and the development of highly sensitive CCD cameras (starting circa 1990) [1]. In each step, the ability to find fainter objects is enhanced, and this enables the discovery of smaller bodies. Convolved with the effect of increased capabilities of detection is also an issue of astronomer interest. Indeed, looking is the ultimate selection effect, and if no one is interested in searching for new moons, then none will be found. Where the moon count for Jupiter will eventually stop is anyone's guess — even to this very day. The present object detection size limit at Jupiter is of order 800 meters across. At even smaller sizes, the number of moons could stretch into hundreds of objects. At some stage, of course, one has to also stop counting very small objects as being moons since once they are smaller than a millimeter or so their motion is strongly determined by interactions with the Sun's radiation field. The dust particles that constitute Jupiter's gossamer rings, for example, have a residency timescale of perhaps a few hundred to a few thousand years before being removed from the system.

8.2. How Many More Discoveries

It is a human conceit to focus (as much as we do) upon, and then glorify, the people who first discovered something. Certainly, new discoveries are interesting and important, but in astronomy, it is often the case that new discoveries are really a result of someone chancing to look in the right place at the right time. And, as often as not, many people might be looking in the same place at the same time — discovery is inevitable if one simply looks. It is worthwhile asking, however, if the universe contains an endless number of distinct discoveries. The answer, of course, is no — it is a finite universe (in extent and age) that we see around us, and it cannot contain an infinite number of distinct new phenomena. Where then does new discovery end? This was a question asked by astronomer Martin Harwit, and his thoughts are contained in the remarkable book,

Cosmic Discovery: the search, scope and heritage of astronomy (first published by The Harvester Press, Brighton, in 1981). The question of discovery, Harwit notes, is not just related to who looks and when, but is strongly influenced by what sort of instrumentation is available to view the universe with (as illustrated in Figure 8.3). There is a technology element to discovery, and there is, in the case of electromagnetic radiation, a wavelength dependency [1]. While astronomers have been able to view the heavens at optical wavelength, with telescopes since (about) 1608, they have only been able to view the X-ray universe since the early 1960s. Likewise, astronomers have only been able to study the universe with respect to gravitational waves since 2016. At some level, however, astronomers currently have access (not necessarily continuously and contiguously) to instruments that can make observations from the longest wavelength radio waves, through to the shortest wavelength gamma rays, as well as monitor neutrino, gravitational wave, and cosmic ray phenomena. Each time a new instrumental technique has become available (that is a new wavelength region opened-up to observation), Harwit notes that new discoveries generally follow — that is, some new kind of astrophysical object is either identified, or re-discovered, and seen in a new way. Now that there are essentially no new regions of the electromagnetic spectrum to view the universe from,[11] the question arises as to what, or more specifically how many, genuinely novel discoveries might there be to make in the future. This is a question of limits. Not limits to the imagination, nor limits with respect to present-day and even future observational abilities, but a pure limit as to what the universe and the laws of physics can produce and allow. In his book *Cosmic Discovery*, Harwit presents a list of 43 fundamental astrophysical phenomena (see also [2]) — this list includes such entities as planets, comets, asteroids, stars, galaxies, globular clusters, pulsars, black holes, gamma-ray bursters, radio galaxies, and interstellar magnetic fields. Using this list of fundamental cosmic phenomena, Harwit and Roger Hildebrand [3] have

[11] Which is not to say that the universe has been observed at all desired angular resolution and timescales.

attacked the question of future discovery by adopting a method used by biologists to estimate animal populations. The key idea is to estimate the size of a population from the number of triplicates, duplicates, and single discoveries found within a random selection. That is, the total number of possible discoveries is estimated from the number of independent re-discoveries made when using distinctly different observing techniques. The approach then is to consider sets of distinct observational techniques (e.g., optical telescopes, radio telescopes, spectroscopy, polarimetry, astrometry, and meteorite chemistry) and count which of the items in the fundamental phenomena list are distinctly observed by just one technique, by two distinct techniques, by three distinct techniques and so on. As with all such statistical methods, various assumptions have to be made, but the end result is an estimate of the total population of phenomena — which here is the total number of discreet and fundamental astrophysical phenomena in the observable universe. Harwit and Hildebrand run through the numbers and estimate that the total population of discreet astrophysical phenomena in the observable universe is set at about 130 ± 10. Remarkably, this indicates that only about 1/3rd of the distinct astronomical phenomena within the universe have been identified to date. This is good news for future astronomers and physicists since it indicates that they have plenty of work, surprises, and discoveries ahead of them. While some of Harwit's fundamental list phenomena (e.g., planets) have been observed by humans for thousands of years, it has been the application of technology, starting in the early 17th century, that has enabled astronomers to annotate about 1/3rd of the universe's novelty, and most of these novelties were discovered within the last 100 years. The future not only heralds the discovery of new fundamental phenomena (a further 2/3rds remain to be discovered), it implies that the universe is much more complex, mysterious, and convoluted than presently thought. Furthermore, the study by Harwit and Hildebrand is sobering in that it illustrates how statistical sampling techniques can take us from what it is that we do know, to what it is that we might one day know. And, indeed, their results indicate that the number of distinct astrophysical phenomena in the universe is finite.

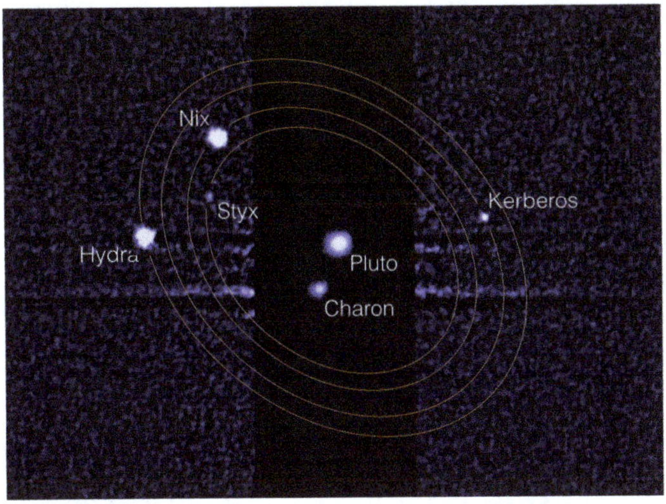

Figure 8.4. The Pluto–Charon (binary dwarf planet) system with its four moons (Hydra, Nix, Styx, and Kerberos). Image courtesy of NASA.

8.3. The Machinations of Pluto

It was the launch of the Hubble Space Telescope, in 1990, that eventually enable the moon count for Pluto to jump from zero to its present-day total of 5. The moon count was accumulated over a time interval of just 35 years. This being said, it could be argued that the moon count for Pluto should actually be 4, since Pluto and Charon (the latter first discovered in 1978) actually form a binary planet,[12] with the 4 moons found by the Hubble Space Telescope (between 2005 and 2012), being in orbit about the barycenter of the Pluto–Charon system (Figure 8.4). A close fly-by of Pluto by the *New*

[12]It is generally taken that the center of motion (barycenter) of a moon should reside within the interior of the primary planet. In the case of Pluto and Charon, the barycenter is located outside the physical body of Pluto, and accordingly the system should be classified as a binary planet. The additional problem with Pluto and Charon is that under the current International Astronomical Union (IAU) definition scheme, Pluto is considered to be a dwarf planet, where as Charon is not. Given a binary planet designation, however, then the Pluto–Charon system should be classified as being either a binary dwarf planet system or a binary Kuiper Belt Object. There are good reasons for considering the latter designation as being the better one.

Table 8.3. Table of orbital radii and periods for the Pluto–Charon system moons. The orbital radii (column 2) are given in units of the distance of Charon from the system barycenter (17,536 kilometers). The orbital periods (column 3) are given in days. The last column provides the ratio of the orbital periods to that of Charon.

Moon	Orbital radius	Period (days)	P_{moon}/P_{Charon}
Charon	1.00	6.387	1.00
Styx	2.42	20.162	3.16
Nix	2.77	24.855	3.89
Kerberos	3.29	32.168	5.04
Hydra	3.68	38.202	5.98

Horizons spacecraft in 2015 revealed no new moons in the system. Accordingly, the Swift and Ball numbers of 64 and 29 respectively are badly divergent with respect to their predictions — as, indeed, from earlier discussion, we would expect them to be.

It is not clear at the present time whether the Pluto–Charon system formed as a capture event between two Kuiper Belt Objects, or whether it resulted from the collisional break-up of a larger primordial object at some time in the early (formative) years of the solar system. Time will tell which of these scenarios plays out, but either way, Pluto is an intriguing oddity (planet-wise) that continues to test our understanding of solar system formation and structure.

The deduced orbital parameters for the Pluto–Charon binary are given in Table 8.3, and an intriguing numerical arrangement is revealed. The last column of the table is the relevant one here, and it indicates that the circum-binary moons orbit in near mean motion resonance (see Chapter 10) with ratios close to 1:3:4:5:6. That is, for every orbit of Pluto and Charon about the system's barycenter, Styx's completes just over 3 orbits, Nix just under 4 orbits, Kerberos 5 orbits and Hydra 6 orbits.

The remarkable arrangement of orbital periods is something that enables system stability, and it also allows for the construction of a straightforward mechanical model that describes the system's behavior. Mechanical models of the solar system have been made since the

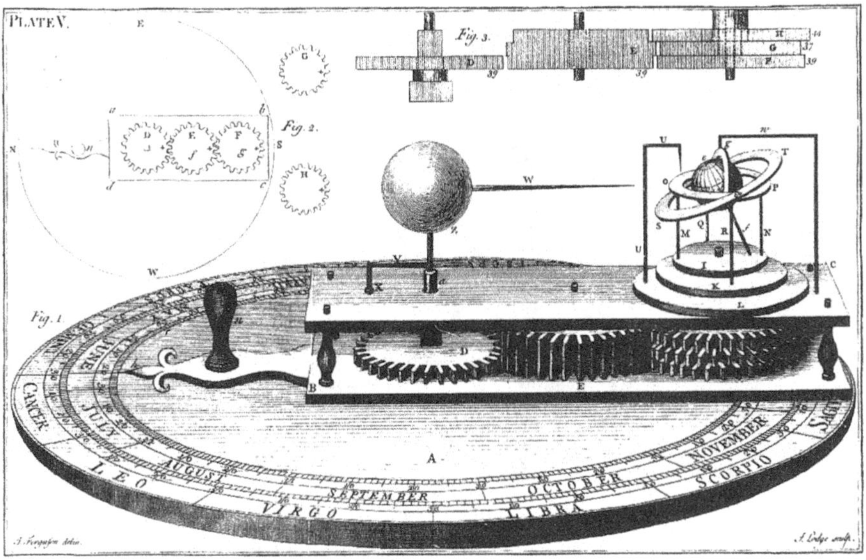

Figure 8.5. The Mechanical Paradox Machine by James Ferguson (c. 1750). This device is a tellurium, designed to explain the seasons, the length of day variation, and the times at which eclipse phenomena can be observed. The large central sphere represents the Sun, and the Earth (at far right) is surrounded by hoops that represent the ecliptic (horizontal ring) and the plane of the Moon's orbit (inclined ring) — the Moon itself is not included in this model. The gearing of this device allows for the parallel motion of the Earth's spin axis, as the support plate rotates about the Sun. It also mimics the retrograde rotation of the line of nodes for the Moon's orbit (this is the line joining the two points at which the Moon's orbit intersects the ecliptic).

early 18th century,[13] and some, the so-called orreries, tellurians, and cometaria are wonders of engineering design, gearing, and construction [4]. Such machines, for all their mechanical mystery (Figure 8.5) are the absolute anthesis of uncertainty and guesswork, and they embody strict predictability and strict system stability.

A (very) simple Pluto–Charon Moon Dial is illustrated in Figure 8.6, where gear ratios are used to pick out the appropriate orbital periods. The central small cog has 12 teeth and one complete rotation of this cog corresponds to the period of Charon (6.387 days).

[13] Recall the Antikythera Mechanism of ancient Greece, described in Section 4.4.

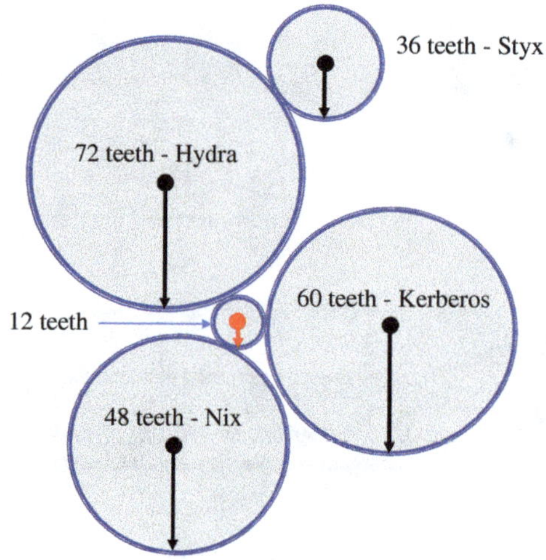

Figure 8.6. Basic gear design for a Pluto–Charon Moon Dial. One rotation of the 12-tooth driving cog corresponds to the orbital period of Charon about the system's barycenter.

The 72-tooth cog will complete 1 revolution for every 6 turns of the central cog, providing for the motion of Hydra; the 36-tooth cog tapping the 72-tooth gear will complete 1 turn for every 3 turns of the central cog, providing for the motion of Styx; the 60-tooth cog will complete 1 turn for every 5 turns of the central cog, providing for the motion of Kerberos; and the 48-tooth cog will complete 1 turn for every 4 turns of the central cog, providing for the motion of Nix. Other geartrain arrangements are certainly possible, but the point is, the motion of the Pluto–Charon system can seemingly be captured within the lock-tight machinations of wheelwork. Orbital dynamics, as it were, is described through the turning of imagined celestial clockwork.

There is something altogether satisfying about being able to construct mechanical models to describe physical phenomena — the gearing instills a sense of clarity and permanence. Such models play on the exactness expected of Newtonian imperialism. Indeed, William Thompson (Lord Kelvin) famously remarked that "I can never satisfy myself until I can make a mechanical model of a thing.

If I can make a mechanical model, I can understand it." Kelvin is here espousing the idea that models can both convey and stamp meaning onto reality — they are not just mimicking reality, they are transferring to the mind a tangible *image* of the ways and processes by which nature works. Somehow, the model literally contains the embodiment of understanding and provides a rigid, cause and effect, mechanistic and resolute viewpoint that can be transferred from one observer to the next. We understand the physical universe, according to Lord Kelvin, partly because we can make mechanical models of it, and these mechanical models are a tactile expression of the goodness of the underlying design. Unfortunately, Kelvin was wrong, and while mechanical model-making plays into the idea of Newtonian imperialism, with the world characterized by uniformity and exactness, they give a false impression of understanding what is actually going on, and they imply a designer. Mechanical models ape the appearance of physical phenomenon, but they do not explain why the system does what it does. The same, of course, can be said about mathematical equations. The equations capture the essence, but they are not the entity.

In a prize-winning essay concerning the stability of Saturn's rings, published in the *Edinburgh New Philosophical Journal* for 1855, Scottish mathematician James Elliot had this to say about orreries and mechanical models, "in them we look in vain for imitations of *Heaven's easy, artless, unencumbered plan*", further suggesting, "that they are seldom regarded in any other light than as mechanical curiosities, and are rarely used for *explaining* the subject of astronomy". These comments by Elliot follow in the same vein as those expressed by John Herschel, who in his influential and well-received *A Treatise on Astronomy* (published in 1833) described orreries as "very childish toys". This intentionally derogative appellation by Herschel essentially emerged from his earlier *Preliminary Discourse on the Study of Natural Philosophy* (published in 1830), where he argued that natural philosophy, in its true form, should be open and transparent, "divested, as far as possible, of artificial difficulties, and stripped of all such technicalities as tend to place it in the light of a craft and a mystery, inaccessible without a kind of apprenticeship". Herschel's comments are given some strength by the very name of Ferguson's device, The Mechanical Paradox (recall Figure 8.6), which invokes the idea of mysterious forces at play. They are further strengthened by the fact that orreries, and similar such models, only allow for circular motion,[14] and that only approximate periods of rotation and orbital separations are employed. For all this, Herschel is placing too much upon what these devices are really trying to do. They are not, and never were, intended to elucidate the physical principles of the phenomena — they are concerned with invoking a minds-eye view of system phenomena and characteristics. The Orrery was never intended to be a scientific or predictive instrument, and to call them "toys" essentially misses this point. It is impossible to capture the reality of planetary motion within wheelwork, but such wheelwork does have great pedagogical value. Indeed, James Clerk Maxwell realized the great value of mechanical models, and in his 1871 introductory lecture, as first, Director of the Cavendish Laboratory at Cambridge University extolled their use in what he called "experiments of illustration". Such demonstrations allowed, he argued, the student to meditate upon and visualize specific physical and/or complex motions removed from any obscuration and entanglement with other phenomena.

In the modern era, the idea of describing some new observational phenomenon or experimental result in terms of wheelwork and/or mechanical devices no longer seems to apply — although there is no fundamental reason for this. Instead, however, it is standard practice to describe such observations and results in terms of mathematical models, with these models being *explored* by computer simulations. In the pedagogical sense, the essential tension is between computational exactness and human understanding, and the former buys into the idea that natural phenomena can only be described mathematically. For all this, as discussed in Chapter 6, even a slavish (dogmatic?) insistence on using mathematical (and computational) models to describe dynamical systems is, in its own way, just as incapable of

[14]The less commonly constructed cometaria did employ elliptical gearing in order to mimic the first two Kepler's laws. See the author's book [5].

capturing all aspects of planetary motion as mechanical wheelwork. Even the exact equations of planetary interactions and dynamics cannot escape and/or predict over the chaos that is hidden in the unknown, and indeed, unknowable, initial boundary conditions.

8.4. What is a Planet Anyway?

There is an elephant in the room — it has been lurking, unattended in the background for the last seven chapters. In this case, it is not a Fermi elephant (recall Chapter 1) that we a dealing with, but, in terms of Douglas Adams's wonderful imagination,[15] it is an SEP, or, Someone Else's Problem. The problem, of course, is, what is a planet, and how should such objects be defined? Indeed, the modern take on this problem is, is it even possible to meaningfully define and distinguish a planetary body from other solar system objects? Historically, the meaning of the term planet has always been fluid, with the Moon and even the Sun falling under the label (Figure 8.7). Likewise, the first asteroids to be discovered were all initially called, and indeed, thought to be, planets [6]. The most recent attempt at forming a planetary status definition was that by the International Astronomical Union (IAU) in 2006, and this achieved little more than to raise the ire of all variously interested parties [7,8]. The current definition will certainly not hold for long, for a multitude of reasons, but the question is as follows: What will replace it, and indeed, is it actually meaningful to try and offer a definition? The IAU definition attempted to meld an ancient and modern perspective. Planets, it is argued, orbit the Sun, whereas moons orbit planets. This condition rules out such moons as Titan and Ganymede from being planets, even though they are both larger than Mercury. Planets are spherical due to their own self-gravity, but so too are some asteroids and many Kuiper Belt Objects. Planets have cleared the region of their orbits about the Sun of other objects — but it is not specified at what level, and over what actual physical region, this clearing takes place. This clearing condition effectively rules out any

[15] See, *Life, the Universe and Everything* (published in 1982).

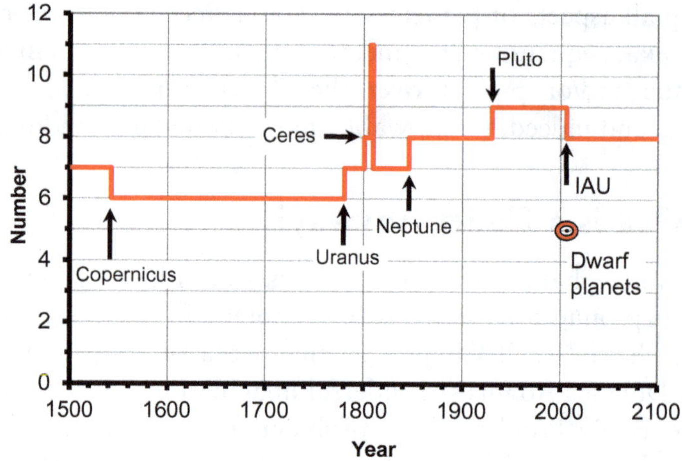

Figure 8.7. The fluid nature of planetary designation during the past five centuries. Prior to the publication of *De revolutionibus orbium coelestium* by Copernicus in 1543, 7 planets were recognized, these being the Sun, the Moon, and Mercury through to Saturn. With Copernicus, the Sun became a star and the Moon became a satellite, but the Earth became a planet. The object class of asteroids was established circa 1810 after the discovery of the first 4 such bodies (but see [6]). Pluto became a dwarf planet following the introduction of the IAU classification scheme in 2006 (Ceres was also reclassified as a dwarf planet at this time). Under the 2006 IAU scheme, the number of planets will now (and forever more) remain fixed at 8 objects — presupposing, that is, that Planet 9 does not exist.

asteroids, Pluto and other Kuiper Belt Objects from being included in the planetary pantheon. Indeed, if one takes into account the Trojan asteroids, then Jupiter is no longer a planet under the IAU scheme.

The three conditions for planetary status adopted by the IAU in 2006 were an attempt to keep the historical *status quo*, with the list of planets remaining small, and likely a number that will never change as we move into the far future, or at least not change by much.[16] The checklist developed by the IAU is largely based upon a parochial solar system view (i.e., objects orbiting the Sun), physical

---

[16] Planet 9, for example (if it actually exists), may yet be awaiting discovery in the outer regions of the Kuiper Belt.

size (the hydrostatic equilibrium condition), and orbital dynamics. Beyond the planetary status conditions, the current rules include the designation of dwarf planets, which are objects that orbit the Sun and are spherical due to their own gravity. There is one asteroid that satisfies this latter condition (the asteroid Ceres), and there are probably many Kuiper Belt Objects that will eventually fall into this category.[17] If orbiting the Sun is the only check-list characteristic that an object has, then it falls into the small solar system body category. This later group includes irregularly shaped comets, asteroids, and smaller Kuiper Belt Objects.

Given the IAU definition, the division where dwarf planets become small solar system bodies is not cleanly defined. The attainment of hydrostatic equilibrium (spherical due to their own gravity) is based upon composition, porosity, rotation rate, and collisional history. The 1000-kilometer-sized dwarf planet Haumae is predominantly composed of ice, but due to its rapid rotation has adopted the shape of an oblate spheroid; the 500-kilometer-sized asteroid Vesta fails to make the dwarf planet category, not through failing the hydrostatic equilibrium condition, but because of its past collision history — its southern polar region has, in the last several billion years, been destroyed through at least two massive impacts. Once, however, an object, irrespective of its composition, is larger than about 500 kilometers across then it will fall into either the dwarf planet or planet category (provided, of course, it is also in orbit about the Sun). In broad-brush terms, terrestrial, rocky planets can have masses up to about 10 times that of the Earth. Objects much more massive than this limit tend to capture extensive hydrogen and helium envelopes as they form, becoming Jupiter-like planets. Exoplanetary systems show the same diversity of planets as seen in the solar system, with an upper mass limit set at about 10 times that of Jupiter. Once an object is larger than this limit, it qualifies as a brown dwarf. This comes about since such massive bodies can briefly undergo deuterium fusion reactions within their interiors. Brown dwarfs, in some sense,

[17] The current dwarf planet number beyond Neptune is 4 objects, although there are many additional candidates.

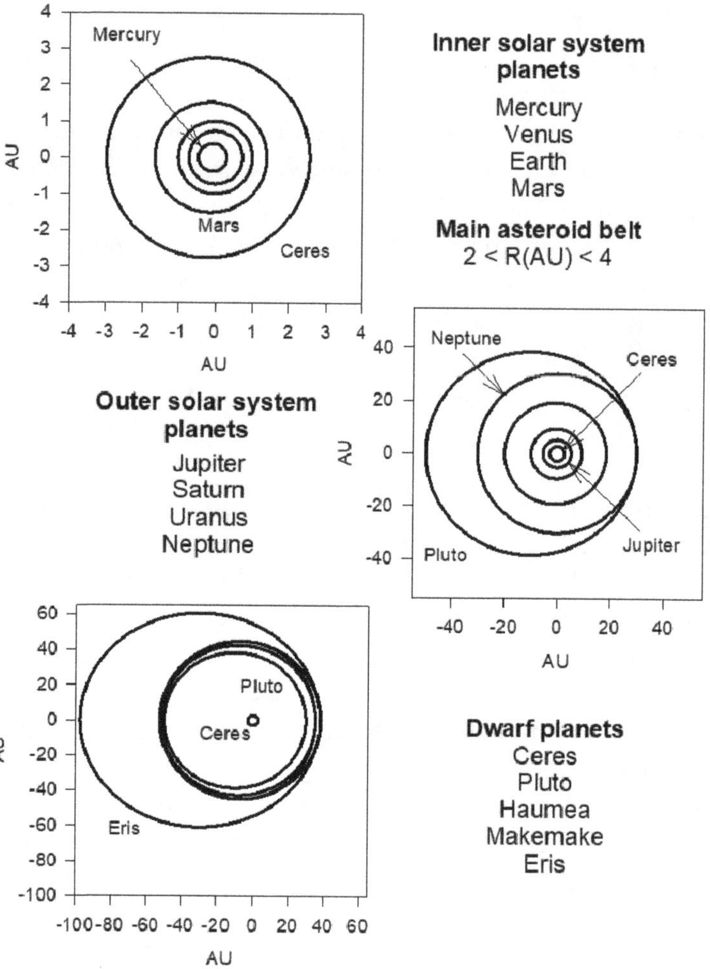

Figure 8.8. The scale of the planetary and dwarf planetary regions of the solar system.

are failed stars and can have masses up to about 100 times that of Jupiter. Above 100 Jupiter masses then an object can initiate hydrogen fusion reactions within its core and become a *bona fide* star.

In terms of its physical scale, the planetary and dwarf planet component of the solar system is illustrated in Figure 8.8. The recognized dwarf planets, specifically the orbit of Eris, take us to just 100 AU away from the Sun. Given that the outer boundary of the solar system (defined by the Oort cometary cloud[18]) is set at a distance about half way to the nearest stars, amounting to a distance of some 50,000 AU, the planets occupy just the inner 0.1% of the solar system's radius. It is in this small region, however, where all the formative action took place, the rest of the vast volume of the solar system was conquered through the machinations of gravitational scattering. Indeed, the Oort cloud contains the survivors of past close encounters with the newly forming Jovian planets within the inner solar system. These latter planets have sufficient gravitational clout that they can eject close-approaching asteroids and cometary nuclei right out of the solar system and into interstellar space. Those cometary nuclei that reside in the Oort cloud are the ones that didn't quite receive enough of a kick to eject them beyond the Sun's gravitational pull. The Oort cloud has also been sculpted, and indeed, gained its spherical profile from the random passage of nearby stars. These encounters tend to raise the perihelion point and circularize the orbits of Oort cloud members. Close encounters with other stars will also scatter Oort cloud cometary nuclei into the inner solar system, where they are seen as long-period comets, and eject others into interstellar space. Slowly, but surely, the Oort cloud is being depleted of its component members. The first interstellar objects to be detected in our solar system, an asteroid and a cometary nucleus that formed around two other stars in the galaxy, were found in 2017 (the asteroid Oumuamua[19]) and 2019 (comet Borisov). More such interstellar objects will, no

---

[18] Described by Dutch astronomer Jan Oort in the 1950s, the Oort cloud is composed of many trillions of cometary nuclei. These icy bodies formed within a few tens of AU from the Sun but were ejected from the inner solar system, by gravitational scattering, into the outer regions. Their orbits were subsequently circularized and raised by the passage of nearby stars and the gravitational field of the galaxy. The Oort cloud is very much a dynamical structure, not least with respect to the orbital distribution of the cometary nuclei within it but also its overall shape with respect to the ever-shifting background of nearby stars.

[19] The status of Oumuamua is still somewhat uncertain. It did not show the characteristics of a comet, but then it wasn't obviously an asteroid either. Perhaps the most unexpected characteristic of this object was its profile, it being spindle like with a length of at least several hundred meters, and a diameter of just a few tens of meters.

doubt, be found in the future, and their detection reminds us that the solar system is a vast, extravaganza of jumbled and interacting objects, that is intimately linked to the rest of the galaxy.

When developing their planetary definition scheme, the IAU committee assumed that it was astronomers that held the authority to make the recommendations.[20] The geologists, and astrobiologists, it would seem, had no specific role to play, even though it is this body of scientists that actually studies the physical properties of planetary makeup [9]. There is more, much more, to a planet than the computation of its orbit and the annotation of its dynamical history. Indeed, geologists can quite reasonably argue that a planet is any object that has undergone internal differentiation, and this means that many asteroids, Kuiper Belt Objects, and most large moons should be given a planetary designation. The fact that an object orbits the Sun is hardly a planet-defining characteristic (other than by historical precedent), and this leaves open the issue of exo-planets, as well as rouge planets — of which many hundreds are now known. It is easy to be critical, of course, and what the outcome of the 2006 IAU definition and voting process really reveals is that it is very difficult to agree upon and codify the physical and dynamic properties that make a specific body a planet rather than something else.

8.5. But is There Order in the Architecture?

Even a quick scan of Figure 8.8 suggests to the eye that there is order within the distribution of planetary orbits — the architecture appears organized and sound. Admittedly, the dwarf planet domain (from Pluto outwards) is beginning to look somewhat complicated, but in the inner and outer solar system domains, where we find the planets, the orbits seem to be nicely circular, and more or less evenly nested

[20] Mistakenly, it can be added, the IAU also felt that the public should play some role in the planet definition process. Given the complexity of the problem, it is difficult to see what non-experts could meaningfully bring to the table, and neither is it the purpose of scientific taxonomy to placate social norms and/or public expectations.

one inside of the other. How much of this architecture, however, can be explained in terms of necessity, and how much is due to pure luck and happenstance? At each step in the planet formation process, we have no doubt that no physical laws were ever broken, but again, and in spite of this law-abiding behavior, is it possible, we question, to predict from the very outset the final distribution of planetary orbits about any star — not just the Sun?

Once a planet has more than two moons, then a study of their radial ordering can be made, and indeed, the moon-rich Jovian planets do appear to show some remarkable orbital structure in their satellite systems. The architecture of satellite systems is just as intriguing as that displayed by the planets. Table 8.4 indicates the

Table 8.4. Planetary and moon spacing data. A lower size limit of 350 kilometers across has been set upon the moons for Jupiter and Saturn, but this has been dropped to 130 kilometers across in the case of the Uranian moons. Note, these moon size limits have no special and/or physical meaning, and accordingly, we are guilty of cherry-picking our numbers.

Central body	Companion	Orbital radius (km)	R(n)/R(n – 1)
Sun	Mercury	5.791×10^7	
	Venus	1.082×10^8	1.87
	Earth	1.496×10^8	1.38
	Mars	2.279×10^8	1.52
	(Ceres)	4.139×10^8	1.82
	Jupiter	7.783×10^8	1.88
	Saturn	1.426×10^9	1.83
	Uranus	2.871×10^9	2.01
	Neptune	4.497×10^9	1.57
	Pluto	5.915×10^9	1.32
			<ave.> = 1.69 ± 0.23
Jupiter	Io	4.212×10^5	

(*Continued*)

Table 8.4. (*Continued*)

Central body	Companion	Orbital radius (km)	R(n)/R(n – 1)
	Europa	6.709×10^5	1.59
	Ganymede	1.070×10^6	1.59
	Calisto	1.883×10^6	1.76
			<ave.> = 1.65 ± 0.08
Saturn	Mimas	1.855×10^5	
	Enceladus	2.380×10^5	1.28
	Tethys	2.947×10^5	1.24
	Dione	3.774×10^5	1.28
	Rhea	5.270×10^5	1.40
	Titan	1.221×10^6	2.21
	Iapetus	3.561×10^6	2.92
			<ave.> = 1.72 ± 0.63
Uranus	Portia	6.609×10^4	
	Puck	8.601×10^4	1.30
	Miranda	1.298×10^5	1.51
	Ariel	1.912×10^5	1.47
	Umbriel	2.658×10^5	1.39
	Titania	4.358×10^5	1.64
	Oberon	5.826×10^5	1.34
			<ave.> = 1.44 ± 0.11

spacing between the planets, and the spacing between the major moons (larger than 350 kilometers across[21]) of Jupiter, Saturn, and

[21] This limit is really a fudge, of course, since it carries no physical meaning with respect to moon formation. Furthermore, it is also the case that all satellite systems, at some level, are likely time variable, and certainly all the Jovian planets have many moons smaller than 350 kilometers in size.

Uranus. The moons of Neptune are not included in the table since it is taken that its moon system was strongly perturbed (or, indeed, largely disrupted) during the capture of its largest moon Triton from the Kuiper Belt.[22] Within the table entries, there is a notable similarity between the deduced average spacings $R(n)/R(n-1)$ for the planets and the various moon systems.

Table 8.4 indicates that the typical spacing between any two planets, and any two moons, is about $R(n) = 1.63R(n-1)$. This result is suggestive of the idea that the physical processes that controlled the growth of the larger satellites around the Jovian planets are similar to those which determine the formation of the planets about the Sun. Indeed, it seems evident that the spacings between the larger moons of the Jovian planets are described by their own Titius–Bode-like rules. This condition is evident, for example, in the Uranian moon system (Figure 8.9) where the orbital radius R can be expressed in terms of sequence number $N = 1, 2, \ldots, 7$, as an exponential (least square fit) equation:

$$R(10^6\,\text{km}) = \frac{1}{27}e^{2N/5} \tag{8.1}$$

For the first Uranian moon ($N = 1$) Portia, the best-fit equation provides an orbital radius that is 4% too big; for Ariel ($N = 4$), the formula provides an orbital radius that is 5% too small, and for Oberon ($N = 7$), the formula provides an orbital radius that is 4% too big. By its very construction (being least squares) equation (8.1) provides a good fit to the data, but what cannot be explained is why an exponential term should work so well, or why the specific numerical values of 27, 2, and 5 should appear. Indeed, the apparent good fit exhibited by equation (8.1) is a façade, and a multitude of other fitting options (e.g., by a polynomial equation, or as a power to base 10, and so on) could have provided an equally good, if not better, fit to the data. There is apparent order exhibited in Figure 8.9, but there is, in fact, no fundamental mathematical law on display. Indeed,

[22] That Triton is a captured moon is based upon its observed retrograde motion about Neptune.

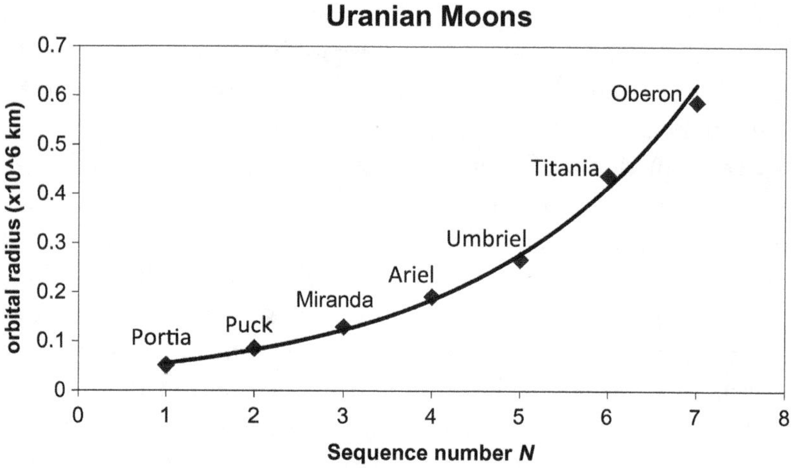

Figure 8.9. Orbital radius (in units of 1 million kilometers) versus sequence number for the larger moons of Uranus (diameters larger than 130 kilometers). The solid line is the least-squares fit given by equation (8.1).

Figure 8.9, a version of which is often seen in textbooks, and research papers is a sham. What is being displayed is a biased selection of the moons of Uranus. They are the largest of the 27 moons that are currently known to orbit Uranus. The diagram has been constructed on the totally arbitrary bases that the moons must be larger than 130 kilometers across. This size limit has no physical meaning behind it, and it simply ignores all the smaller moons that would completely destroy the apparent visual order in the diagram, and negate the apparent good fit to the system given by equation (8.1). Here, we learn an important lesson: the mathematics (that is equation (8.1)), as such, captures the apparent variation in the orbital radii of the selected moons but it does not explain why the variations came about. Furthermore, the mathematics seemingly inherent to equation (8.1) is only there because it was put there, not because it physically explains what is going on.

The entries in Table 8.4 and the variation seen in Figure 8.8 appear to suggest that the spacing between planetary orbits, as well as the spacings between (at least) larger moons, increase on average by about 63% at each step, and this prompts the following question:

Is there some universal orbital spacing rule at play? This was a question that James Challis[23] asked in a paper read to the Cambridge Philosophical Society on 8 December 1828, and he suggested that the answer was yes. The reason for this conclusion, however, was based upon false logic — or, being more generous, upon false assumptions. Challis noted that Bode's law, as he called it (see Chapter 9), had no underlying theoretical description, and specifically noted that "the theory of universal gravitation points to no such law". Accordingly, Challis argued that Bode's law must have been built into the solar system and that it represents the "original arrangement of the planets at the time they were first set in motion". From this premise, Challis argued that if such similar ordering rules could be found for the moons of the various planets, then this implied that some, "constantly operating cause" must be in play. In a triumphant tone, Challis writes, "I have met with success, the more surprising, that, though easily attainable, it had not been anticipated". With this statement made, Challis then goes on to argue why, in fact, the same distance rule does not apply to the various planetary moons. He concludes, however, that "The planets and satellites arrange themselves about their primaries at mean distances, which observe *with more or less accuracy* [my italics] the following progression:

$$a, a + b, a + rb, a + r^2b, \ldots$$

with the exception of the satellites of Saturn which assume a twofold progression of the same kind." To this, he adds, "it is probable that the value of r must necessarily be one of the terms of the series 1, 1½, 2, 2½, 3, ...", and that, "there is some degree of probability that the ratio of b to a may always be expressed by very simple numbers". The more Challis argues for some underlying law, the more

[23] Challis became Director of the Observatories at Cambridge University in 1836, and today he is perhaps best remembered for his missed opportunity to discover the planet Neptune, in spite of the fact that John Couch Adams (recall Chapter 6) had provided him with good positional data. This being said, the story is complicated, and Challis was not wholly to blame for the missed opportunity of discovery.

contrived his arguments become. Why are the rules, "more or less" accurate? Why does the rule break down for the moons of Saturn? Why is there a choice of r values, and what determines its specific choice? Why is the ratio of a/b expressible in terms of "simple numbers"? Effectively, Challis was doing nothing more than forcing a spacing rule on the moon orbits, and cherry-picking values for the three variables a, b, and r to get a good fit. There is no "constantly operating cause" at play, but there is the tenacity of a mathematician at work, using all the flexibility of mathematics to force an unstated, but probably pre-conceived result.

Although Challis could offer no physical reason as to why simple orbital progression laws might exist, in modern terms it could be argued that their existence is implied because the large moon formation process is basically the same as that of the planet formation processes, just on a smaller scale. It is now time to look much more closely at the Titius–Bode rule.

Informed decision-making comes from
a long tradition of guessing
and then blaming others for inadequate results.

Scott Adams

References

[1] Beech, M. On special epochs, the Copernican Principle, and future astronomy. *Journal of the Royal Astronomical Society of Canada*, **109**, 63–72, 2015.

[2] Harwit, M. Physics and astronomy — Will you join the dance? *Physics Today*, November, 173–187, 1981.

[3] Harwit, M. and Hildebrand, R. How many more discoveries in the universe? *Nature*, **320**, 724–726, 1986.

[4] King, H. C. *Geared to the Stars: The Evolution of Planetariums, Orreries, and Astronomical Clocks*. University of Toronto Press, Toronto, 1978.

[5] Beech, M. *The Wayward Comet: A Descriptive History of Cometary Orbits, Kepler's Problem and the Cometarium*. Universal Publishers, Boca Raton, 2016.
[6] Metzger, P., *et al.* The reclassification of asteroids from planets to non-planets. *Icarus*, **319**, 21–32, 2019.
[7] Draft IAU definition of "planet": https://www.iau.org/news/pressreleases/detail/iau0601/.
[8] Lecavelier des Etangsa, A. and Lissauer, J. J. The IAU definition of an exoplanet. https://arxiv.org/pdf/2203.09520.pdf.
[9] Runyon, K. D., *et al.* A geophysical planet definition. https://www.hou.usra.edu/meetings/lpsc2017/pdf/1448.pdf.

[5] Bacon, M. [illegible]
[illegible] Boca Raton, [illegible]
[6] [illegible]
[illegible]
[7] [illegible]
[illegible]
[8] [illegible]
[illegible]
[9] [illegible]
[illegible]

Chapter 9

The Titius–Bode "Law"

To separate the knowledge of Nature from the colours of rhetoric, the devices of fancy, or the delightful deceit of Fables.

— Thomas Spratt (1667)

9.1. Origins

The Titius–Bode[1] rule has without doubt dominated the historical narrative behind the deduced order of planetary orbits and spacings. It is a simple numerical rule, but it has a complex past, present and (most probably) future. It is a rule that has been dismissed as numerical nonsense by some, and hailed as a fundamental law by others. Given such extremes of thought, it is probably safe to assume that the truth lies somewhere in between — but exactly where it lies is still open to question.

The origins of the Titius–Bode rule, so-called, are obscure and the essential numerical sequence was known long before Johann Bode made it famous. One of the earliest textual accounts can be attributed to Scottish astronomer David Gregory, who in his *The Elements of Astronomy* (first published in 1715), noted that if the

[1] I will occasionally allow the expression of Bode's law to be used, since the Titius–Bode rule has been known under this guise throughout most of its history.

distance between the Earth and the Sun is taken as ten units, then the distance to Mercury and Venus are 4 and 7, that for Mars is 15, Jupiter 52, and Saturn 95. By inspection, this sequence begins to hint at the possibility of a geometric ordering, in the sense that the distance to Venus = 4 + 3, and the distance to Earth = 4 + 6. The next term in the sequence breaks down for Mars since the implied distance is 4 + 12 = 16 — not the 15 given by Gregory. Widening, however, the margin of acceptance between observations and prediction, Johann Daniel Titius (Figure 9.1) was the first author to turn the planetary spacing sequence into a geometrical progression. The key underlying numerical sequence is 0, 3, 6, 12, 24, 48, ..., with the distance calculation proceeding by adding 4 to each of these numbers, and then dividing the sum by 10. In this manner, the distance to each planet in astronomical units is approximately revealed.

Figure 9.1. Johann Daniel Titius (1729–1796). Image courtesy of Wikipedia commons.

Although given initially in words only, the progression presented by Titius can be written more conveniently in mathematical form

$$R(M) = 0.4 + 0.3 \times 2^M, \qquad \text{where } M = -\infty, 0, 1, 2, \ldots \qquad (9.1)$$

In this manner, $R(0) = 0.7$ gives the distance to Venus, $R(1) = 0.4 + 0.6 = 1.0$ gives the distance to Earth, and so on. To accommodate Mercury, a special value of $M = -\infty$ has to be considered, forcing, thereby, the powers of 2 term to zero. In this manner, the orbital radius for Mercury is: $R(-\infty) = 0.4 + 0 = 0.4$. A comparison between the predicted and observed orbital radii for the planets is shown in Figure 9.2.

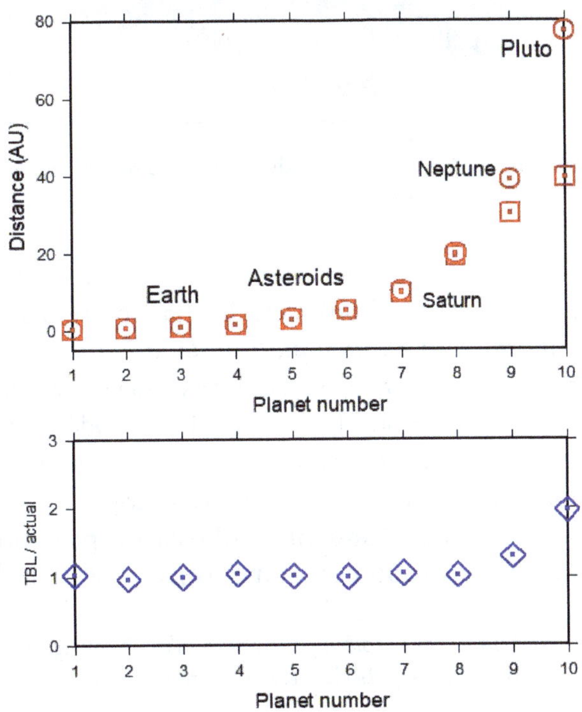

Figure 9.2. Upper curve: Comparison of the deduced (modern era values — squares) and TB rule predictions (circles) for planetary orbits from Mercury through to Pluto. Lower curve: The ratio between the observed and TB rule calculated values for planetary orbits. Note the X-axis in each graph is the planet number, starting with Mercury as planet 1 and Pluto as (dwarf) planet 10.

Figure 9.3. Portrait of Johann Elert Bode (1747–1826). Image courtesy of the University of Cambridge Repository.

Titius added a note concerning his planetary distance rule to a translation (published in 1766) that he had prepared of Charles Bonnet's astronomy text *Contemplation de la Nature*. History then informs us that Titius's planetary distance sequence was later added, as an un-referenced footnote, by Johann Elert Bode (Figure 9.3) to his astronomy text *Anleitung zur Kenntniss des gestirnten Himmels*[2] (second edition, published in 1772). It was largely through Bode's international prominence,[3] and his enthusiastic promotion of the sequence, that other astronomers came to learn of its utility, and it

[2]This translates to *Manual for Knowing the Starry Sky*. Although unreferenced in the 2nd edition, future editions of Bode's text did credit Titius for the origin of the rule.

[3]Bode was Director of the Berlin Observatory from 1786 to 1825. From there, he published is star atlas, *Uranographia* in 1801. Bode also founded, in 1774, and edited for 51 years, the *Astronomisches Jahrbuch*, a widely read and respected yearly volume relating to stellar and planetary positions.

was Bode who made great play of the *gap* located at $M = 3$ — this gap corresponding to the missing planet between Mars and Jupiter, with a predicted heliocentric distance of $R(3) = 0.4 + 0.3 \times 2^3 = 2.8$. Apart from the missing $R(3)$ planet, Bode's law (as it became generally known) provides a remarkably accurate algorithm for calculating the distance to the (then) known planets: Mercury through to Saturn (Figure 9.2). Indeed, using modern values for the deduced planetary distances, Bode's law provides planetary distances to within 5% of the actual observations. The least accurate agreement between the observed and the calculated distance is that for planet Mars ($M = 2$): the formula indicating a distance of 16 units, while the actual semi-major axis is equivalent to 15.2 units. Saturn, corresponding to $M = 5$, has an error term of 4.55%. Skipping over Earth, for which the formula value fits the observational value exactly (by definition of the units), the best fit between the observations and the formula is for Jupiter ($M = 4$), with an error of just 0.05% between the observed and the predicted distances. It is evident from Figure 9.2 that beyond Saturn the rule is becoming increasingly inaccurate — for Neptune, the rule predicts an orbital radius of 38.8 AU, which is 29% too big; for Pluto, the rule indicates an orbital distance of 77.2 AU, giving an estimate some 96% too big. For these planets, such problems, of course, were not evident to either Titius or Bode.

The history of the Titius–Bode law, in terms of the printed word, is illustrated in Figure 9.4. This diagram was generated from the Google Ngram webpage, with the search phrase being "Bode's law". The Ngram shows the usage and appearance of the specified search phrase, as a function of time, in the multiple millions of books that have been scanned by Google. There are many factors at play in interpreting the Ngram results, but in broad-brush terms, it appears that the law first began to appear as a regular entry in astronomy texts from about 1850 onwards. While the discovery of Neptune in 1845 is probably the cause of the peak located at circa 1860, the distinct peak circa 1930 appears to directly correlate with the discovery of Pluto.

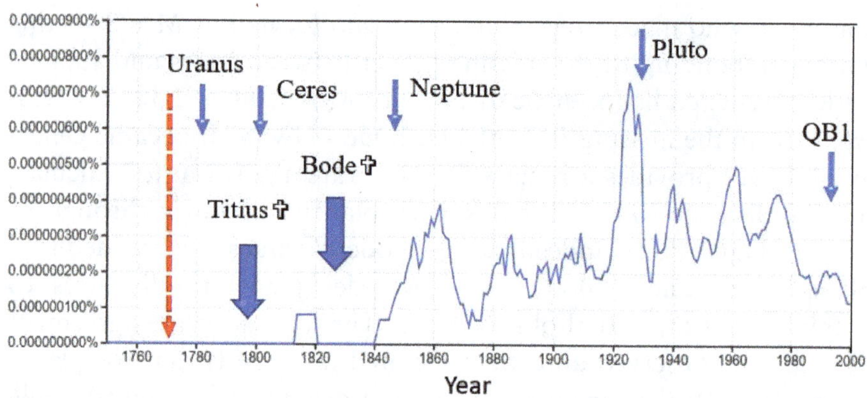

Figure 9.4. Google Ngram search under the heading "Bode's law". The search window covers the time interval between 1750 and 2000. The broad arrows indicate the years in which Titius and Bode died, whereas the thin arrows indicate the times of discovery for Uranus, 1 Ceres, Neptune, Pluto, and 1992 QB1. The long-dashed arrow indicates the publication date of Bode's 1772 text.

Bode's law is a survivor. It is included or at least mentioned in almost every astronomy book that is concerned with the solar system. For all this, however, it is an oddity, being a simple relationship for describing the orbital distances of the planets with respect to the Sun. I use the term oddity since it is an unreasonably successful sequence, and I use the term unreasonably successful because not only is it not a true mathematical law, neither is it named for the first person to describe it. These comments are perhaps trivial and pedantic, but the *law*, as it stands, poses many problems that have yet to be resolved.

9.2. Uranus Falls into Place

Just 9 years after Bode had introduced his footnote rule for finding planetary distances, on 13 March 1781, William Herschel, working with his work-horse 6.2-inch refracting telescope, swept-up the planet *Georgium Sidus*. This first telescopic discovery of a planet was a little delayed, however, since Herschel thought that he had found a comet. It was only later that the near-circular orbit derived for the object implied that a planet[4] has been detected. Bode was presumably both surprised and delighted by Herschel's discovery. The surprise is that no one had particularly expected to find a planet beyond Saturn.[5] The delight, however, would have been at the realization that the deduced distance to Uranus was close to that expressed by the law. Here was encouraging news — the law could accommodate the discovery of a new and unexpected planet, and this in turn added confidence to the expectation that a new planet must exist in the large, apparently empty region between Mars and Jupiter — corresponding to the $R(M = 3)$ location.

In the probable hope of garnering some form of royal patronage,[6] Herschel suggested that his newly discovered planet should be named *Georgium Sidus* in honor of King George III, but this cut against the tradition of applying classical names to the planets, and it also upset European sentiments. Indeed, it was Bode that suggested the planet be called Uranus. Likewise, it was Bode who determined the orbit of Uranus, showing that it was indeed moving in a characteristic (near zero eccentricity) planetary orbit.

9.3. The Gap at $M = 3$

Herschel's discovery of planet Uranus not only rocked the astronomical world, it gave credence to the Titius–Bode rule, fitting-in, as it did, at the $M = 6$ spot. This success spurred a resurgence of interest in finding the missing planet at the $M = 3$ position. Indeed, with the close of the 18th century it would seem that faith in the Titius–Bode rule was sufficiently strong that few doubted that a planet didn't exist at the $M = 3$ spot. One doubter, however, was the young

[4] Technically, Herschel was not the first astronomer to see planet Uranus, but it was his observations that resulted in its identification. Remarkably, John Flamsteed recorded the planet, as a star, on several occasions in the late 17th century, and likewise Pierre Le Monnier recorded it multiple times in the mid-18th century.

[5] This being said, no one had specifically argued that planets could not exist beyond the orbit of Saturn either.

[6] A hope that eventually paid off with Herschel being given the title of Kings Astronomer.

philosopher George Hegel who drew upon an historic president, set by Plato, to suggest that perhaps some caution was required. Writing in his doctoral thesis *Dissertatio philosophica de orbitis planetarum* (published in 1801), Hegel argued that while the Titius–Bode rule provided good estimates to the heliocentric radii of the known planets this in no manner required that there must be a planet at the $M = 3$ position. Indeed, the large gap between Mars and Jupiter could simply be a real gap. Hegel is, in fact, entirely right since there is no actual planet situated between Mars and Jupiter. In support of his notion that a real gap might exist between Mars and Jupiter, Hegel made specific reference to Plato's lambda sequence (described in the *Timaeus* — recall Chapter 3), where the Demiurge shaped the planetary orbits according to the series 1, 2, 3, 4, 9, 16, and 27, skipping over the number 8 in the even powers branch (recall Figure 3.2). To Hegel's point of view, the Titius–Bode rule might dictate where planets are to be found, but nature need not necessarily build a planet at every location so pin-pointed. Well, so much for philosophy. Indeed, before the ink of Hegel's thesis was barely dry a large body at the $M = 3$ location was discovered on January 1, 1801 — but it wasn't a planet and it wasn't alone.

Efforts to finally locate the supposed planet between Mars and Jupiter came to a head as the 18th century drew to a close. France Xavier von Zach spearheaded the charge and formed the Celestial Police, with various co-opted astronomers being set specific regions of the ecliptic to monitor for new, uncharted objects. Although working outside of von Zach's constabulary, an uncharted object, in the constellation of Taurus, was noted by Giuseppe Piazzi on 1 January, 1801, and initially, it was identified (just like Uranus beforehand) as a comet. This new body, however, was the first (and largest) of the asteroids, although originally it was awarded planetary status. By the close of 1801, the orbit of Ceres had been determined, and with a semi-major axis of $a = 2.767$ AU (modern value), it seemed that Ceres was indeed the long-sought-after planet at the $M = 3$ spot (corresponding to $a = 2.8$ AU) of the Titius–Bode rule. Estimates to the size of Ceres were soon made, however, and in 1802 William Herschel found a diameter of just 260 km across. Later, in

1811, Johann Schröter suggested a diameter of some 2600 km. The modern value comes in at 940 km across. Ceres, even with Schröter's over estimate, is a small world — indeed, much smaller than Earth's Moon. While the relatively small of size Ceres readily accounts for the difficulty of finding it (it is not visible to the unaided eye), it also makes one a little suspicious of calling it a planet. Under current IAU guidelines (recall Chapter 8) Ceres is classified as a dwarf planet, and it is a unique solar system object in that category, since all other dwarf planets have a different composition, and reside in the Kuiper Belt. Relative smallness is not the primary reason for denying planetary status to Ceres, and indeed, its status as the expected $M = 3$ Titius–Bode rule planet was soon brought into question. Within 6 years of the discovery of Ceres, three other large bodies were identified with orbits situated between Mars and Jupiter: these being the asteroids Pallas, Juno, and Vesta. Technically, Pallas provides the closest semi-major axis match to the required $a = 2.8$ AU for the $M = 3$ object, and it is only some 510 kilometers across. Rather than there being one small planet between Mars and Jupiter, the observations indicated that there were four. With the discovery of Pallas in 1802, Herschel suggested that a new class of objects — to be called asteroids — had been identified, and gradually, over ensuing years, astronomers accepted that these new objects were different to planets (principally in their diminutive size). While it seems that Herschel anticipated that many asteroids might eventually be discovered (and indeed, this was so) the fifth asteroid, Astraea, was not to be identified until December of 1845.

Soon after his discovery of Pallas, Heinrich Olbers suggested that it and Ceres were actually fragments of a larger planet that had either exploded (due to some unidentified internal force) or had been struck by a comet.[7] This same fragmentation idea could accommodate the discovery of Juno and Vesta, but with these objects in the mix, it was unclear how the various orbits had evolved — their orbital parameters (semi-major axis, eccentricity, inclination) were decidedly different from each other, and there was no common

[7] Recall, at this time, cometary nuclei were thought to be planetary-sized objects.

orbital interception point corresponding to the initial location of disruption. Unusually for astronomers, no alternative scenarios to the break-up hypothesis were developed, although it is now clear that it was the rapid formation and early orbital migration of Jupiter that quashed any chances of a planet forming in what is now the main belt asteroid region. Looking more toward theology than astronomy, Sharon Turner, in his work, *The Sacred History of the World* (first published in 1832), argued that the discovery of objects at the $M = 3$ location, "furnishes another impressive instance of the scientific plan and principles on which creation has been fabricated". Accepting that the four asteroids were indeed fragments of a planet destroyed by a comet, Turner argued that not only was the destructive power of comets part of the omnipotent Creator's plan for our planetary system, but that the Titius–Bode rule was, in fact, a predetermined condition, imposed by the *architect* before the solar system even formed.

Is it reasonable to conclude that the discovery of the first four asteroids can count as a successful prediction from the Titius–Bode rule? I would contend that no, it is not reasonable. If there had been a long, multi-year time interval between the discovery of Ceres and any other asteroids, then a case could probably be made for a successful prediction (quibbles about what a planet actually is being put aside). As it was, however, Pallas was discovered within a year of Ceres being identified, and other objects followed in relatively rapid fashion. There has, in fact, never been a planet situated at the $M = 3$ location, as predicted by the Titius–Bode rule, and accordingly it is simply incorrect to say that the rule provides for a successfully description of objects within the inner solar system, but read-on.

9.4. Finding Neptune

Neptune is often described as being the first planet to be discovered through the power of mathematical reasoning. And, indeed, its discovery was the result of human genius and hard-calculating graft (recall Chapter 6). The story behind the discovery of Neptune is epic, involving, as it does, two theoreticians, one flamboyant and overly self-confident, the other quiet and unassuming. The mental challenge was enormous, the calculations daunting and the whole idea was predicated on absolute faith in Newtonian physics. At stake was the motion of Uranus, and specifically the fact that its observed motion on the sky could not be accounted for in terms of the gravitational perturbations of the known planets. The deviations of motion that were observed with respect to the path of Uranus could, in principle, mean at least two things. It might indicate that at distances beyond about 20 AU from the Sun, Newton's law of gravity no longer worked as a straightforward inverse square law relationship, or it indicated that there was another planet beyond Uranus that was perturbing its path. Urbain Le Verrier (in Paris) and John Couch Adams (in Cambridge), two intrepid theoreticians, took the latter path, and rather than outright reject Newton's formulation of gravity, they decided to work backward from the observed perturbations to deduce the orbit for the required Uranus-perturbing body.

At the outset, both Le Verrier and Adams assumed the absolute immutability of Newton's inverse square law for gravitational attraction, and they reverse-engineered the observations of Uranus to determine the required orbit and position of the perturbing body — that is the orbit of the new planet. The calculations followed by Le Verrier and Adams required that several arbitrary characteristics of the perturbing body's orbit had to be made, and one of these was its orbital radius. So, what distance to choose? Both researchers adopt a value given by the $M = 7$ location of the Titius–Bode rule — accordingly, $a = 38.8$ AU. Fortunately, it transpires, the end prediction for the location of the perturbing object is not highly sensitive to the assumed value of its orbital semi-major axis. Accordingly, both Le Verrier and Adams were able to make relatively accurate predictions on where to look on the sky for the new, Uranus-perturbing, planet. Sure enough, on 23 September 1846 Johanne Galle, at the Potsdam Observatory in Berlin, was able to sweep-up the new planet, which in turn, and in keeping with classical tradition, was named Neptune.

The discovery of Neptune was a triumph of mind and mathematics over nature. Who could now doubt the power of human ability and mathematical reach? And, who could now doubt the veracity of Newtonian mechanics? For all this success, however, there was a fatality in the discovery of Neptune, and this was the demise of the Titius–Bode rule. Once it was discovered, then astronomers could deduce Neptune's orbital radius directly from observations. Finding a semi-major axis of $a = 30.07$ AU, it was soon clear that the Titius–Bode rule placement was in serious error. Indeed, the $M = 7$ distance was 29% to big, and the veracity of the Titius–Bode rule was clearly brought into question.

9.5. Accommodating Vulcan

During the time interval from circa 1860 to circa 1915 it was entirely possible to allow for the existence of a planet, planet Vulcan, interior to the orbit of Mercury. This planet, or ring of asteroids, was introduced by Urbain Le Verrier to explain the deduced anomalous perihelion advancement of Mercury. The question therefore arises as to the ability of the Titius–Bode rule to accommodate this proposed new planetary body. Evidently, this was a question considered by B. G. Jenkins, of Buccleuch Road, Dulwich, and writing to the journal *Nature* on 28 November, 1878, he reasoned that Vulcan could indeed, be accounted for. Jenkins returned to the original numerical sequence outlined by Titius and Bode: namely 0, 3, 6, 12, 24, 48, and so on, with each number in the sequence (except the second one) being twice that of its forerunner. To each of these numbers, 4 is added and the sum is divided by 10. What Jenkins argued was that the original sequence should start at –3, rather than 0, and then, the first planet from the Sun would be located at a distance of $(-3 + 4)/10 = 0.1$ AU. This Titius–Bode-derived orbital radius for Vulcan is within 29% of the 0.14 AU suggested by Le Verrier; a value no more at odds with the observations than Neptune's orbit (which also comes in with a 29% error term).

Having suggested an accommodation for the planet Vulcan, Jenkins's finished his letter to *Nature* with some speculations on the structure of the outer regions of the solar system. Starting with Ceres, in the main-belt asteroid region, Jenkins noted that moving inwards toward the Sun, there are three planets and then Vulcan or, more specifically to Jenkins's argument, the Vulcanoid asteroid zone. Moving outwards from Ceres, Jenkins, additionally noted, that there are another three planets, accordingly, by invoking a symmetry argument, he reasoned that beyond Neptune there must be, "a ring of asteroids at an average distance of 772 (that is, at a distance of 77.2 astronomical units]". Jenkins then continues, "Perhaps the optical instruments of the future may help to answer this question: is there a ring of asteroids beyond Neptune?". This invocation of a symmetry principle between the structure of the inner and outer planetary system pivoted on the location of the asteroid belt, is entirely spurious, but it has appeal to the *ordered* human mind (recall Section 6.1 concerning Kepler's polygonal ordering). Indeed, it is a mathematical ordering principle that is entirely artificial and is not dissimilar in form to Sharon Turner's (Section 9.3) finding evidence for the Omnipotent creator's signature in the *design* of the solar system.

Incredibly then, for so it would seem, Jenkins not only adapts the Titius–Bode rule to accommodate the existence of an intra-Mercurial planet (or Vulcanoid asteroid belt), but he also continues the rule beyond Neptune and anticipates, by more than one-hundred years, the discovery of the Kuiper Belt. Equally remarkable is the point that Jenkins's estimate of 0.1 AU for the location of the Vulcanoid asteroids is entirely consistent with present-day computer simulations. Indeed, numerical studies indicate that an orbital-stability zone exists in the region between 0.09 and 0.20 AU from the Sun [1, 2]. To date, however, no Vulcanoid asteroids have been detected, but this is largely be due to the extreme selection effect that places them very close to the Sun — a zone carefully avoided by both space and ground-based survey telescopes.

9.6. Reactions to the Titius–Bode Rule

Written according to the mathematical formulation given in equation (9.1), the Titius–Bode rule, when first introduced in 1772, was on a similar footing to that of Kepler's third law of planetary motion (introduced some 153 years earlier). It is a relationship justified entirely upon the basis that it approximately accounts for the observed planetary distances. Just as Kepler had no physical explanation for his third law: $P^2 = Ka^3$, so neither Titius nor Bode had a physical explanation for their planetary spacing rule. Historically, however, the difference between the two cases is that Isaac Newton was able (in his 1687 *Principia*) to accommodate and justify Kepler's laws under the umbrella of dynamics predicated upon the inverse square law of gravitational attraction. Long after Newton, when asked about the strange philosophical characteristics underlying quantum mechanical calculations, physicist Richard Feynman famously commented, that no one really understands quantum mechanics, but as formulated it gives the right results — the inference being, just "shut up and calculate". A correct result, as such, is a correct result, even if you don't know why the underlying theory works. The same could be said with respect to Newton's formulation of the gravitational force. Newton did not know how gravity worked, or what it was,[8] but calculations under his formulation could be made, and those calculations gave results that aggreged with the observations. Can the same be said of the Titius–Bode rule? Up until 1845, the answer to this question would have been a resounding yes. Not only did the rule successfully account for the spacings between all the then known planets, it was able to accommodate the discovery of Ceres, the first of the asteroids, at the $M = 3$ location, as well as the discovery of planet Uranus at the $M = 6$ location. The run of apparent successes, however, came to an abrupt end with the discovery of Neptune — at which time, and thereafter, the rule is found to be very much in error with respect to the

[8] Gravity is still a largely mysterious "force", although Albert Einstein, in 1915, introduced the concept of matter curving spacetime, showing that what we call gravitational attractions are really a manifestation of that curvature. Einstein's general relativity is a tremendously successful theory, and there is no hint of it not working correctly in those locations where it should apply. The profound problem, from a modern perspective, however, is how to meld general relativity and quantum mechanics into a theory of quantum gravity.

observed spacings. At other times, under different circumstances, the apparent failure of the law for $M \geq 7$ could be accommodated for by arguing that the failure was an indication of new formation-physics. After all, this is exactly what Einstein accomplished with general relativity — he didn't negate Newton's formulation for gravity, but accommodated it as a limiting condition in a more fundamental theory. The problem with accommodating the Titius–Bode rule into a grander scheme once $M \geq 7$, however, is that there is no theory to explain why it works for $M \leq 6$.

Not only is there no physical underpinning to the Titius–Bode rule, but what are we to make of the first term requiring $M = -\infty$? Indeed, it is an entirely ugly term. Starting in such a manner that it immediately weakens the law's appeal. There is nothing remotely appealing about a sequential jump of infinity between the first and second terms, only to be followed thereafter by successive jumps of one. The infinity jump is in no manner esthetically pleasing, and it makes one suspicious of the whole sequence. Not only does the sequence for M raise one's suspicions about the Titius–Bode rule — suspicious, that is, in the sense of it being a *real* law of nature, so too does the fact that the formula is so *clean*. By this latter *cleanness*, I mean that it includes simple numbers, 0.4, 3, and 2. Nature is rarely so accommodating — if it was, then the speed of light really would be exactly 3×10^8 meters/second (as nearly everyone assumes in most calculations) rather than the much more inconvenient 2.99792458×10^8 meters/second. This latter point is made largely in jest, since the constants of Nature are simply what they are, and they have absolutely no regard for human aesthetics and/or calculating convenience. Also, of course, the numerical values associated with the various constants of Nature are based on the actual units being used. But, for all this, the numbers in the Titius–Bode formula seem suspiciously chaste: why 0.4, and not 0.3999972, and why 2^M as opposed to $(1.7275)^M$ — the reason for the choice of these numbers will be discussed in the following. Mathematicians and theoretical physicists have historically talked about equations being beautiful, and, ignoring the ugly issue of the sequence number M, equation (9.1) is appealing enough (although, not obviously beautiful). But, of course, nature has no aesthetic appreciation, and time and time again, throughout the history of science, theories that started out being described as beautiful, become mired in approximations of one kind or another, tainted by messy constants, and brutalized by fudge-factors or guillotine terms. Appealing to the human sense of beauty in the description of a theory, or specifically in an equation, is a very definite mistake — it may stroke our collective human ego to use such a term, but it certainly does not guarantee physical correctness, and neither does it act as a useful guide toward distinguishing between competing theories (see Chapter 10). The appeal to beauty is sometimes described under the guise of Ockham's Razor, which is typically paraphrased as, "the simplest solution is most likely the correct one". The problem, of course, is that Ockham's razor is not a law of physics or, for that matter, mathematics — it is a tool of rhetoric. In reality, the simplest theory is the one that human beings (mathematicians and theoretical physicists included) think that they can actually get away with, with the tools that they have available, and thereby make some practical progress. Ockham's razor, simplicity, and/or beauty afford no guarantees with respect to separating out the good theories from the bad ones.

Since its introduction in the late 18th century, the Titius–Bode rule has both frustrated and fascinated astronomers. John Herschel in his highly influential *Outlines of Astronomy* (first published in 1849) described Bode's law in the following terms, "the progression is (rather rudely) that of the numbers 4, 4 + 3, 4 + 6, 4 + 12, etc." And by using the term "rudely" he hints at the fact that it is not an exact explanation of the observed spacings of the planets. Writing in 1882, Edmund Ledger in his *The Sun: Its Planets and Their Satellites* writes that with respect to planetary orbits, "we may ... allude in passing to the curious relation existing between them, first announced by Titius ... [and] afterwards brought into more general notice by Bode". While introducing Bode's law as being curious, Ledger goes on to write, "the truth is that Bode's arrangement hardly deserves to be termed, as it frequently is, a *law*, no theory having yet been suggested which gives a satisfactory reason for its existence .... Bode's so called law must at present be looked upon as little more than an example of a curious coincidence".[9] Arthur Searle in his *Outlines of Astronomy* (published in 1875) is relatively up-beat about the rule, and writes, "the comparative mean distance of all the principal planets as far as Uranus can be expressed pretty well by a rule commonly called Bode's law". Commenting upon the poor agreement with respect to Neptune, Searle argues that, "the law, however, may be hereafter explained by more general laws, so as to show why it does not hold so good in the case of Neptune". Just a few years later, in 1887, Elias Loomes, in his *Treatise on Astronomy*, notes that, "the discovery of Neptune has entirely refuted Bode's law of planetary distances .... Although this law represents pretty well the distances of the nearest planets, the error is quite large for Saturn and Uranus, and for Neptune the error amounts to more than 800 million of miles". Moving into the 20th century, the widely used *Astronomy* text by Russell, Dugan and Stewart (first published in 1926) contains the statement, "there is a curious approximation relation between the distances of the planets from the Sun, usually known as Bode's law ... it is not known whether Bode's law is a mere coincidence or whether it has some physical explanation which may be understood in the future. For Neptune, the law breaks down, and it really does so for Mercury, for which the law should give the distance 5.5 (adding half to 3, instead of 0 to 4)". The last part of the quotation taken from Russell, Dugan, and Stewart is based upon setting $M = -1$ in equation (9.1), which is perhaps no more perverse (and indeed, more logical) than taking $M = -\infty$ for the first term. In 1933, James Jean, in his *The Universe Around Us*, writes more negatively that, "it seems more than likely that it [the Titius–Bode rule] is a mere coincidence with no underlying rational explanation". Skilling and Richardson, in their *A Brief Text in Astronomy* (revised, and published in 1959), comment that "there is an interesting numerical device called Bode's law for fixing in mind the relative distances of the planets from the Sun. The name is misleading, however, for the relationship was not

[9] In contrast to the events that unfolded in Sherlock Holmes's *Adventure of Silver Blaze*, Ledger is essentially arguing that Bode's law is a false-positive law — akin, therefore, to a dog that didn't bark in the night.

discovered by Bode, and it is not a law". R. A. Lyttleton, in his *Mystery of the Solar System* (published in 1968), also criticizes Bode's law, writing, "Remarkable as the law may be, it is very difficult to assess its importance and status. My own view [and Lyttleton's views were certainly worthy of attention] is that too much significance may have been attached to Bode's law. In fact, it fails rather badly in the inner reaches of the solar system". This latter comment is a little odd, and one might assume that Lyttelton meant "outer reaches of the solar system", but, in fact, he has a point that was picked-up upon and explored more fully by Fred Hoyle in his *Astronomy and Cosmology: A Modern Course* (published in 1975). Hoyle writes, "Astronomers today differ markedly in the weight which they attach to it [the Titius–Bode rule]. Some have ever made it a cornerstone in attempting to develop a theory of the origin of planets, but others feel it to be a mere set of coincidences". In characteristic style, Hoyle begins to examine the veracity of the rule according to a list of 10 statements, to which *hit* and *miss* answers are evolved. It is worth looking at Hoyle's list in detail, and while they do not constitute any form of proof or indeed refutation of the Titius–Bode rule, the statements are highly penetrating. Hoyle writes the following:

1. The fit of Bode's law for Mercury is not meaningful because we start with 0.4 AU, deliberately chosen to agree with the value of Mercury.
2. The fit for Venus is also not meaningful because the 0.3 used in Bode's law can be regarded as another "free choice".
3. The fit for Earth is not particularly meaningful because we could have elected to use a multiplier other than 2.
4. Having fixed the numbers 0.4, 0.3, and the multiplier 2 for the first three cases, the agreement for Mars is a clear-cut success for the law.
5. The asteroids Ceres and Pallas seemed at first to provide another good success for the law, but by now many asteroids are known which disagree with it .... the position is thereby weakened.
6. Jupiter is a clear success.
7. Saturn is a clear success.

8. Uranus is a clear success.
9. Neptune is a serious failure.
10. Pluto is an even worse failure.

Hoyle's list reveals 3 nonmeaningful hits (Mercury, Venus, and Earth), 4 clear successes (Mars, Jupiter, Saturn, and Uranus), 1 semi-success (the asteroid belt), and 2 complete failures (Neptune and Pluto). In other words, Bode's law scores a poor 4 out 10 in terms of clear successes. This is less than impressive, and certainly calls the veracity of the "law" into question. Hoyle comments, "for myself, I have always felt Bode's law to be a general reflection of a process whereby the planets accumulated out of a swarm of much smaller bodies which at one time orbited the Sun .... Bode's law might be an expression of these spheres of influence". In some sense, Hoyle is arguing that the Titius–Bode rule is akin to the executive summary attached a complex set of interactions.

While Hoyle presented a lengthy discussion on Bode's law, it was by no means the last word on the topic. Writing in 1981, Peter Francis argued in his *The Planets — A Decade of Discovery*, that, "Bode's law is a bit problematic, and its significance is still not wholly understood, but it played an important part in early studies of the solar system". After presenting the formula for Bode's law [the same as our equation (9.1)], Francis comments, "this may seem to be merely a meaningless arithmetical coincidence, the result of artificial jiggling with numbers [i.e., Hoyle's points 1, 2, and 3]". Many astronomers have thought the same. The coincidence cannot be easily ignored, however, because ... it proved to have remarkable predictive powers [i.e., Hoyle's points 4, 6, 7, and 8]. Furthermore, however, after describing the discovery of Pluto, Francis notes, "Bode's law does not really dictate the design of the Solar System [i.e., Hoyle's points 9 and 10]". Also writing in 1981, in his *Orbiting the Sun*, doyen of planetary science Fred Whipple (Harvard University) noted, "the so-called Bode's law ... is not a physical law but only a convenient rule for recalling the distances of the planets from the Sun". To this he added, "no theoretical basis for the rule has been generally accepted".

Just a few years after Francis and Whipple were compiling their thoughts on the Titius–Bode rule, John Barrow and Frank Tippler in their classic text, *The Anthropic Cosmological Principle* (published in 1986), effectively dismissed the *law* as mere numerology — albeit a historically interesting piece of numerology. Interestingly, Barrow and Tippler draw their readers attention to another example of numerology, that developed by Johann Balmer for the spectral features observed in an excited hydrogen gas. Indeed, in 1855 Balmer suggested the wavelengths associated with what are now known as the Balmer lines of hydrogen follow the simple rule:

$$\lambda = \lambda_0 \frac{m^2}{m^2 - 4} \tag{9.2}$$

where λ_0 is a constant and $m = 3, 4, 5, 6, \ldots$ Balmer's formula, while derived empirically, proved some 60 years after its first publication, to be a critical part of Niels Bohr's first quantum mechanical model of the hydrogen atom. In the case of Balmer's formula, inspiration turned, once the correct physical theory had been developed, into an exact theoretical model. No such conversion, however, is likely to occur with respect to the Titius–Bode rule.

9.7. Refining the Rule

What are we to make of our brief review of the extensive literature relating to the Titius–Bode rule? To begin with, it is clear that the importance (or relevance) of the rule has clearly changed over time. Prior to the 1854 discovery of Neptune, astronomers appear to have placed considerable confidence in the rule — it accounted for the planets Mercury through to Saturn, and it partly inspired the search for a planet between Mars and Jupiter. The Mars–Jupiter gap (the $M = 3$ "planet") was thought to have been successfully filled with the discovery of Ceres in 1801, but this apparent success is now more reasonably interpreted as a failure of the rule, since there is no actual planet between Mars and Jupiter. The law successfully accommodated the discovery of Uranus by William Herschel in 1781. With

Johann Gale's confirmation of Urbain Le Verrier's prediction for the location of the Uranus perturbing planet, that is Neptune, in 1854, however, it was found that the rule had exceeded its reach. While Le Verrier and Adams used the Titius–Bode rule to close the otherwise unknown heliocentric distance parameter, it transpires that the error, amounting, as Elias Loomis noted in 1887, to some 0.38 AU (or some 57 million kilometers), was not actually critical to the evaluation of the sky-location prediction. It was soon realized, however, once the true orbital parameters for Neptune had been determined directly, that the Titius–Bode rule was in serious error. The discovery of Pluto (classified as a dwarf planet since 2006) by Clyde Tombaugh in 1930, the Titius–Bode rule was not just in error, it was downright wrong: the law predicted 77.2 AU, while the observations give 39.482 AU. With Pluto there can be no accommodation with the prediction given by the Titius–Bode rule when $M = 8$.

It is at this point that we are reminded of Russell's Reductionist Chicken. Introduced by philosopher Bertrand Russell in his *Problems of Philosophy* (first published in 1912), the story of the chicken mirrors that relating to the expectation (or faith) astronomers put (and continue to put) in the Titius–Bode rule. Russell's story relates to the problem of inductive proof, and describes how we can be easily fooled when using the constancy of past events as a guide to what will happen in the future. The point about the chicken is, as Russell writes, "The man who has fed the chicken every day throughout its life at last wrings its neck instead, showing that more refined views as to the uniformity of nature would have been useful to the chicken". In our case astronomers, buoyed by the apparent (though questionable) success of the Titius–Bode rule out to planet Uranus, simply assumed it would give the correct answer for any additional planets that might be found within the solar system. The induction is with respect to our expectation in the formula: it works for $M = -\infty$, it works for $M = 1$, it works for $M = 2$, it works for $M = 3$ and so on — it is not until we reach $M = 8$ that the reductionist neck of the Titius–Bode rule is finally rung.

While astronomy texts, right up to the present day, invariably present the original Titius–Bode sequence, other *refined* version of the law have been available since the early 1900s. One such reversion was presented by Mary Adela Blagg, and her first contribution on the topic appears in the April 1913 issue of the *Monthly Notices of the Royal Astronomical Society*. Blagg was a remarkable astronomer and mathematician, and one of the first women to be elected a Fellow of the Royal Astronomical Society (in 1916). Self-taught, she learned mathematics by borrowing and working through her brother's text books, but was inspired to begin studying astronomy after attending a series of University Extension lectures presented by J. A. Hardcastle (John Herschel's grandson). Primarily known for her pioneering and organizational work in selenography, Blagg also studied variable stars in collaboration with Herbert H. Turner (Savilian Professor of Astronomy at Oxford University). Using the title, *On a Suggested Substitute for Bode's Law*, Blagg effectively set out to perform a harmonic analysis on the observed minus the calculated values for planetary orbits. More than this, however, she also extended the analysis to look at the orbital arrangement of the moons associated with the Jovian planets. It is a lengthy paper, dense with tables comparing observed values for orbital semi-major axis, against averaged mean values and the predictions as set down by Bode's law in equation (9.1). Importantly Blagg finds that the best power law term to use is not 2^M, but $(1.7275)^N$ (I will switch to powers of N here since it is a different numerical sequence to that used for M), and remarkably (although recall Table 8.4) this power law progression applies not only to the planets but the moons of Jupiter, Saturn and Uranus as well. Deviations of the actual orbital semi-major axis' data for the planets and the geometrical progression are further corrected for by introducing a periodic function $0 \le f(\varphi) \le 1$, such that

$$R(N) = A(1.7275)^N[B + f(\varphi)] \tag{9.3}$$

where A and B are constants, and $\varphi = f(\alpha + N\beta)$, where α and β are constants, f is periodic function of 2π radians, and $N = -2, -1, 0, 1, 2, \ldots$ The harmonic analysis then conducted by Blagg indicated a form for the function f as

$$f = \frac{\cos\psi}{3 - \cos 2\psi} + \frac{1}{6 - 4\cos 2(\psi - 30^\circ)} \tag{9.4}$$

where $\psi = \varphi - 27°.5$. Blagg's formula is far from beautiful, but it provides a much-improved fit to the observed data — but then it was constructed to do so. Also, by using a power law in the form $(1.7275)^N$ and starting at $N = -2$, the ugly and unrealizable $M = -\infty$ needed to accommodate the orbital radius of Mercury, in equation (9.1), is avoided — which is not to say that starting at $N = -2$ isn't strange. For the planets, Blagg determined that the various constants were $A = 0.4162$, $B = 2.025$, $\alpha = 112°.4$ and $\beta = 56°.6$. Effectively, constant A sets the unit of distance, while constant B measures the deviation away from an exact geometrical progression. The larger the value of B so the less important becomes the $f(\varphi)$ term. In contrast, the smaller the value of B so the more important the $f(\varphi)$ term. Blagg's formula provides (to 2 decimal places) good correspondence for the planetary orbital radii out to Uranus, and the error between the observed and predicted distance for Neptune is just 0.2% — this error increases to 5.8% in the case of Pluto. While Blagg has created a wonderful Fermi elephant with equations (9.3) and (9.4), the end result is, once again, a rule that does not describe reality and it provides zero insight as to the underlying reasons for the observed planetary spacings.

Thirty-two years after the appearance of Blagg's paper, Donald Richardson (at the Armour Research Foundation in Chicago) performed a similar harmonic analysis of planetary and moon orbit spacings. Writing in the January 1945 issue of the journal *Popular Astronomy*, Richardson presented an expression of the form

$$R(N) = (1.728)^N \rho_N(\Theta_N) \tag{9.5}$$

where

$$\Theta_N = N\left(\frac{4\pi}{13}\right), \quad \text{and} \quad N = 1, 2, 3, \ldots. \tag{9.6}$$

Once again, ρ_N is a periodic function over 2π radians. Richardson, just like Blagg before him, found an appropriate set of constants and an expression for ρ_N that provided a good fit to the observed planetary and moon orbital spacing data. For all this, however, Richardson's formulation is no more accurate a fit in the outer solar system as that found by Blagg, being 0.76% in error at Neptune, and 5.7% in error at Pluto. Richardson has created nothing more than a Fermi elephant, and once again provides no physical insight concerning the reasons for planetary spacings.

To some extent, when it comes to the Titius–Bode rule, the conclusion of astronomers, after more than two-hundred years of deliberation is that one can take it or leave it. This is not particularly encouraging if one is looking to determine its true physical status. Indeed, the Titius–Bode rule falls amongst that mongrel pack of ideas which constitute the Demarcation Problem. That is, it is still far from clear if the law (*per se*) falls within the respectable domain of scientific enquiry, or if it is just a pseudo-scientific oddity (see later). In 1974, Michael Nieto (Los Alamos Laboratory) wrote on the Titius–Bode law in a special edition of the magazine *Pensée*. Published by the Student Academic Freedom forum, the special edition was concerned with the investigation of the outlandish ideas of Immanuel Velikovsky, and Nieto was asked to comment, both at a lecture and in writing, upon the idea that planetary orbits might have undergone recent reorganization. Velikovsky had suggested in his ridiculous (but best selling) 1950 text, *World's in Collision* that planet Venus had been ejected from Jupiter, as a comet-like object, some 17,000 years ago. Furthermore, Velikovsky argued, in passing close to the Earth, before settling into its present orbit (no actual collision was suggested) the wayward Venus caused a change in Earth's orbit and its spin-axes orientation, precipitating thereby numerous terrestrial catastrophes. Nieto comments that, "before we can know whether the subject of this talk means anything, we have to know whether the Titius–Bode Law of Planetary Distance" is indeed a *law*. If it is not a *law*, then clearly the law says nothing about anything, we can skip the rest of this talk, drop the discussion, and concentrate on something useful, like Watergate[10]". Nieto, as one might expect (by now), comes to no definitive conclusion as to the status of the *law*, but he brings out many useful points. In particular, Nieto picks up on the point of the law's accuracy. Using orbital data and mathematical fitting techniques, Nieto, echoing the earlier analysis by Mary Blagg and Donald Richardson, notes that the best-fit, geometric sequence for planetary spacing is[11]

$$R(N) = 0.34 \times (1.73)^N, \qquad \text{where } N = 0, 1, 2, 3, \ldots \qquad (9.7)$$

This particular formula oscillates, sometimes bigger, sometimes smaller, than the actual planetary distance sequence and it corrects for the ugly $N = -\infty$ term for Mercury in equation (9.1). The point, however, is why does the formula — that is equation (9.7) — which is tantalizingly close, not actually predict the exact values, sitting, as it were, "right on top of the planets"? Perhaps the oscillating correction term, as seen in equations (9.3) and (9.5) hint at the existence of new component to the law, with a periodic function being required to bring the predictions into the exact agreement so that

$$R(N) = 0.34 \times (1.73)^N \times f(N) \qquad (9.8)$$

The trick, however, or the main difficulty, is how to determine the form of the periodic function $f(N)$. Nieto notes that while adding in the periodic function $f(N)$ might well force a perfect mathematical fit, no progress as to the status of the law has really been made since there is no theory that predicts the form of the progression seen in equation (9.7). In the end, Nieto concludes that it all boils down to intuition and feeling, and unfortunately, "intuition presently goes both ways". For all this, Nieto places a positive spin on his Titius–Bode-like law — our equation (9.8) — arguing that the geometrical

[10]For younger, present-day readers, the Watergate Affair is now long behind us but was the political firestorm that resulted in the impeachment of U.S. President Richard Nixon in 1974, although technically he resigned prior to the enactment of the parliamentary vote.

[11]This is effectively our "uglier" — to be expected — version of equation (9.1).

progression part is reflective of the conditions under which the planets first formed in the young solar nebula, while the oscillating term, $f(N)$, is a result of the later gravitational re-organization of the planets into long-term stable orbits. Once again, such arguments may be correct, but there is still no underlying theory to explain the form of either the geometrical progression or the periodic function. And, finally, does the Titius–Bode law say anything useful about Velikovsky's suggestion concerning the supposedly recent planetary rearrangements, effectively the answer is no. If the *law* is a *law*, Nieto argues, then it is a fundamental law speaking to fundamental issues of planetary formation, and this would therefore weaken Velikovsky's claim for a recent rearrangement of the orbit of Venus.

Remarkable, strange, odd, and bizarre; all these words have been used to describe the Titius–Bode rule, and yet the fact remains that this simple numerical sequence replicates, to entirely unreasonable accuracy, the basic spacing between the planets from Mercury to Uranus. It is this unreasonableness that has intrigued astronomers (and amateur sleuths — see the following) over the centuries. The question still remains, however, is the rule just a numerical coincidence, a quirk of mathematics, or does it have some connection with physical reality. Could it be that within the natal embryo, out of which the planets were to form, some distinct process is at play that would make the Titius–Bode rule an inevitable outcome? Or is the rule a delusion — a sequence of numbers that has no meaningful connection with the physical processes that resulted in the observed structure of solar system at all?

9.8. The Result of Pure Chance

Randomness is not usually a word that one tends to associate with Newtonian clockwork, but it can be reasonably asked, given the entirely random conditions under which planetary masses accrue, and by which final orbits come about, if the Titius–Bode rule itself is nothing more than a random or chance sequence of numbers. Logically speaking, it makes little sense to ask if the distribution of planetary orbits is due to random chance, since throughout their entire history, during every second, from their first formation to the present day, they have evolved according to the deterministic laws of celestial mechanics. Of course, what we are dealing with is a sensitivity to boundary conditions, and perhaps a better question to ask of the Titius–Bode rule is, is it possible for a distribution of planetary orbits to evolve that cannot be explained by some sort of simple Titius–Bode-like sequence.

In an attempt to determine if a Titius–Bode-like rule is an inevitable outcome, irrespective of the planet formation, and orbital determination processes that operated in the early solar system, Wayne Hayes and Scott Tremaine (then both at the University of Toronto, in Canada) conducted a series of numerical experiments. As will be explored in Chapter 10, a necessary condition for the stability of a planetary system is that the planets never undergo repeated close approaches to each other. The closeness condition is often cast in terms of the Hill radius R_H (see Chapter 10 for a more detailed discussion on this) which depends upon the mass of a planet compared to that of the Sun. For two adjacent planets to remain on stable orbits, there separation must always remain greater than some low-number multiple of the sum of their respective Hill radii. To model this effect Hayes and Tremaine generated thousands of possible planetary sequences, and then separated out that subset of systems that would be unstable due to violating the Hill condition. Of the remaining (stable) systems, Hayes and Tremaine found that they could all be described by some Titius–Bode-like rule. What this says, is that there is really no physical significance to the Titius–Bode rule at all. What is important is the closeness of approach condition that prevails on the Hill radii associated with each planet in a given system. The other way of putting this is that Titius–Bode-like relationships are an emergent-like phenomenon of stable planetary systems. For all this, however, such rules do not tell us how a specific system formed, and nor that a planet must reside at every location implied by the rule — gaps can be real, and the system will still remain stable.

9.8.1. *What the exoplanets say*

The very first exoplanet, 51 Pagasi b was discovered by Michel Mayor and Didier Queloz, of the Geneva Observatory, Switzerland in 1995. Presently the number count for exoplanets is over 5000 bodies, with sizes and masses ranging from sub-Earth to larger than Jupiter (recall Section 8.4). A whole host of system architectures have been found, but of interest to us, here, amongst the systems with multiple planets are a number that exhibit Titius–Bode-like rules. The compact planetary system Kepler 11, centered on a Sun-like star discovered in 2011, provides one example where the spacing of the planetary orbits can be expressed in terms of the sequence number N, where $N = 1$ is the inner most planet (see Figure 9.5). In this case the rule is expressed in terms of powers of the fraction 4/3, with the orbits correspondingly growing by about 33% at each step. Interestingly, however, the rule is only a (reasonably) good fit to the observational data if a presently undetected planet (or gap) is taken to exist at location $N = 6$. It is presently not known if there is an

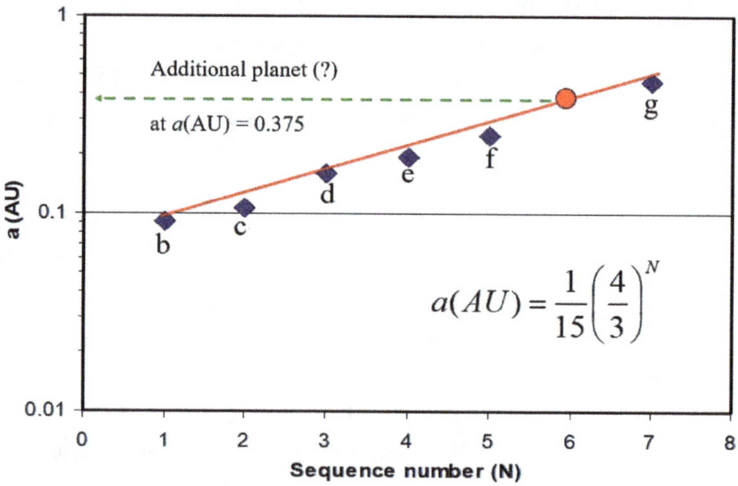

Figure 9.5. Orbital radius versus sequence number for the exoplanetary system Kepler 11. A Titius–Bode rule can be established with respect to the observed orbital radii provided an additional planet, or gap, is assumed to exist at sequence number $N = 6$.

$N = 6$ planet or even an $N = 8$ (and so on) planet reside within the system. Certainly, there is plenty of room for more planets beyond our putative $N = 7$ body (planet g), since compared to the scale of our solar system its orbital radius, at $a = 0.466$ AU, is just a little larger than the orbit of Mercury.

In 2013, two researchers at the Australian National Observatory, Timothy Bovaird, and Charles Lineweaver, took the "fitting" approach to its limit and conducted a comprehensive survey of all exoplanetary systems containing more than 4 planets. In total, 68 systems were identified and they set about finding a generalized Titius–Bode rule, in the form $a_n = aC^n$, $n = 0, 1, 2, 3, \ldots$ (where a and C are constants, and a_n is the orbital radius of planet n) to each. The numerical procedure they adopted was to examine the observed orbital properties relating to the planets in each system, then added test planets between known pairs to see if the addition improved the over all system fit to a Titius–Bode-like rule. In other words, they essentially forced the systems to obey a Titius–Bode rule (which reminds us of Eddington's statement that the mathematics is not there until we put it there). By this methodology, however, they predicted the existence of an additional 141 planets within the 68 candidate systems.[12]

Predictions are all well and good, but they need to be tested, and in 2014, this is exactly what Chelsea Huang and Gáspár Bokos, at Princeton University did. Using additional Kepler spacecraft data Huang and Bokos examined 56 of the systems for which Bovaird and Lineweaver predict that some additional 97 planets might reside. Only 5 new planets were detected, each in a position close to their predicted positions, and 1 new, non-predicted, planet was detected. The prediction success rate is a mere 5%. Indeed, Huang and Bokos looked at the various observational biases that exist in detecting a planet and predicted that the minimum number of planets that they would expect to find if the Bovaird and Lineweaver

[12] In similar vein to our result in Figure 9.5, Bovaird and Lineweaver predict a planet at $a_6 = 0.34$ AU in the Kepler 11 system. They also predict an additional planet with an orbital radius of $a_8 = 0.59$ AU.

predictions were correct, was 7. In other words, the predictions from the forced fitting of Titius–Bode rules to the known exoplanetary systems were at best questionable and, in reality, hopelessly wrong. This is not an indictment or criticism in any way of the work performed by Bovaird and Lineweaver, but rather it is a statement of the fact that exoplanetary systems do not support the idea that Titius–Bode-like rules are universal and meaningful.

9.9. Remarkability

One of the key precepts behind the Titius–Bode rule is that of remarkability. But remarkability, based purely on mental surprise, is no grounds for assuming profundity. Yes, it is remarkable that the Titius–Bode rule, based upon a simple sequence of numbers, does provide for a reasonably good description of the spacing of the planets (out to Saturn), but this is only so from a perspective of ignorance. In seemingly contradictory terms, however, the Titius–Bode rule is remarkable from both ends of the knowledge spectrum. It is remarkable if you have no idea how planets form, but equally, it is remarkable, in the modern era, that such a rule should come about given that it is generally now known exactly how planets form. Clearly, however, the Titius–Bode rule, by not providing for an exact description of planetary orbits, indicates that it cannot be a fundamental statement about reality. Indeed, all of the planetary ordering schemes that have ever been promoted are based upon the precept that since some simple mathematical rule provides a number close to the observed value for a particular planet, in some finite sequence of numbers, then the rule itself must have some physical meaning. This, of course, is nonsense — indeed, to use the apropos phrase of Jeremy Bentham — it is "nonsense upon stilts". We really have just one option when it comes to dealing with such a rule: it is wrong, it does not account for reality, and it should accordingly be discarded from all future astronomy texts. That the Titius–Bode rule has not been long discarded, however, being found in almost every general astronomy book published within the last 250 years, tells us much more about human nature than it does about planets within the solar system. The rule is nothing more than number play — a chimera generated from the ever-fecund world of mathematics (see Chapter 11). Simple numerical rules, like the Titius–Bode rule, rarely proffer physical insight, they can cause great confusion, and they promote wrong-headed expectation. And yet — and yet, damn it! — the Titius–Bode rule has provided astronomers with a wealth of inspiration. Perhaps, at best, one can think of the Titius–Bode rule as a mathematical version of a mnemonic, providing a means of recalling the approximate heliocentric distances to the planets. In the sciences, it would seem, that we have to live with, and even accept, inconsistency: wrong-headed concepts, incorrect reasoning, and wishful thinking can result in actual discovery and innovation. There is a moral here, and the moral is that just because some rule looks to be giving the right answer doesn't mean that it actually is (this will be a topic further explored in Chapter 11). We can confidently say, however, that even if the Titius–Bode rule had never been articulated, the main asteroid belt, Uranus, Neptune and Pluto would have certainly been discovered by astronomers in the same manner, sequence, and times that they were. The Titius–Bode rule has never been a vital driver of any actual discovery.

Having presented a brief history of explaining away the Titius–Bode rule, it seems appropriate to end this chapter by looking at some alternatives — would be usurpers of the rule's domain. Indeed, a great plethora of challengers to the humble Titius–Bode rule exist, and there is an additional abundance of rules that claim to reveal relationships between planetary spin rates, planetary magnetic fields, and planetary sizes and mass.

9.10. From the Pages of Journals

Every year, for so it seems, someone, somewhere, from almost every profession and walk of life, finds a new "law" that "explains" the orbital spacing of the planets — or some set of planetary characteristics (size, spin, or magnetic field). They are all chimeras, being nothing more than reflections of the human mind looking for simple order were none actually resides. The solar system and the multitude of objects within it is a complicated structure, and what order that is actually found evolved through countless random processes, selected according to Newtonian dynamics, over many millions of years. In order to understand the solar system, as well as planetary systems elsewhere, it would seem that a Darwinian-like outlook is required, where stability evolves (if at all), out of countless random interactions, selected for through Newtonian mechanical principles.

The journal *Icarus* was founded in 1962, and it has grown to become the pre-eminent journal of solar system studies. The authors guide to submitting an article to the journal is 14 pages long, but in the very first paragraph, just four sentences in, we find, "*Icarus* does not publish papers that provide 'improved' versions of Bode's law, or other numerical relations, without a sound physical basis". Indeed, here is the crux of the matter, and it is the term "sound physical basis" that is key. While the journal *Icarus* openly rejects articles concerning Titius–Bode-like numerology, the magazine *Popular Astronomy* (published between 1893 and 1951) seems to have actively attracted them. Indeed, all major astronomy journals, at one time or another, have published articles directly concerning the Titius–Bode rule, or concerning some similar such *miracle formula*. Finding numerical relationships is, in fact, easy, but finding actual physical meaning behind the numbers is very hard. For all this, however, let us briefly celebrate our human foibles (they do, after all, make life more interesting) and indulge in some exploration of 20th century imaginations from the cottage industry of Titius–Bode look-alike relations.

As the first example, we turn to an article by Mr. R. Sheward, of 21, Royal Mint Street, London, published in the *Astronomical Register* for October 1873. From the very outset, we encounter a problem (possibly editorial in origin) since the article is entitled Boch's law, which is presumably intended to read as Bode's law — indeed, it is the Titius–Bode rule that Sheward finds, "hardly satisfactory". With the desire to find a clearer relationship for planetary spacings, Sheward presents, "the result of [his] labours" in the form

Table 9.1. Planetary separation rule according to R. Sheward. By omitting the Earth, Sheward argues that the spacings of the planets follow a normalized sequence that runs 1, 1.9, 3.8, 6.0. See text for details.

Planet	Distance	Modern	Planet	Distance	Modern
Mercury	1.0	1.00	Jupiter	1.0	1.00
Venus	1.9	1.85	Saturn	1.9	1.83
Mars	3.8	3.90	Uranus	3.8	3.69
Asteroids	6.4 (mean of 50 orbits)	NA	Neptune	6.0	5.78

of a table — shown here as Table 9.1. Sheward begins by organizing the planets into two, minor and major, categories (in modern terms this corresponds to the terrestrial and Jovian planets).

Sheward says of his table (our Table 9.1) that, "this law itself speaks more eloquently than I can write". By this, he intends that if minor planets are normalized with respect to the orbital radius of Mercury, then the spacing sequence is 1, 1.9, 3.8, 6.4 (column 2 in Table 9.1). For the major planets, however, normalized to the orbital radius of Jupiter, we have spacings of 1, 1.9, 3.8, 6.0 (column 5 in Table 9.1). The close coincidence of these two spacing sequences is the symmetry law that Sheward is impressed by. It is not clear what values Sheward used for planetary distances, but columns 3 and 6 of Table 9.1 show how the scaling plays out with modern-day numbers. Perhaps needles to say, while we see a close, but not exact, agreement between Sheward's two sequences, we note the somewhat glaring lack of Earth in Sheward's table. Indeed, Sheward comments that the fact that the Earth breaks the symmetry of the two number sequences, and that it alone of all the minor planets has a satellite,[13] means it must be very special. Indeed, he goes on to use this uniqueness condition to single out Earth as the only, "habitable surface for the dwelling of man" in the solar system.

What are we to make of Sheward's ideas and conclusions? Certainly, being generous, we could say that a great deal of insight

[13] The Moons of Mars were not to be discovered for another 4 years — see Chapter 8.

has been extracted from a very limited data set. Being less generous, we must, of course, say his arguments are entirely specious — although we also acknowledge that Sheward was no-doubt being honest in his interpretations. Some insight is perhaps gained from the fact that in the year following his *Astronomical Register* article, Sheward published a book entitled *Prophetic Astronomy* (Charring Cross Publishing Co., 1874), in which he interprets biblical Genesis in terms of the creation of vast clouds of meteoritic matter. Stars and planets subsequently form through the collapse and accretion of these primordial clouds. Sheward also provides a prediction in his book that the solar system will come to an end in 1500 years (a very precise number) when it interacts with another star to form a binary system. To make sense of Sheward's ideas we have to realize that he had an agenda, and that it was his wish to find supposedly scientific evidence for the truth of biblical stories, and to establish the unique circumstances surrounding the domain of humankind. Sheward was not the first person to usurp mathematical reasoning and appropriate scientific arguments to justify the truth of the biblical creation account, and he will not be the last. His idea of normalizing planetary orbits according to two different domains (terrestrial versus Jovian), however, turned out to be something that many other would-be rule-makers readily adopted.

Using Jupiter as a pivot point is a characteristic of Butusov's Law [sic], introduced by Kirill Butusov in 1973. Described[14] by Ivan Kotliarov in the November 2008 issue of the *Monthly Notices of the Royal Astronomical Society*, this rule is based upon the observation that the product of the orbital radii of the planets symmetrical set about Jupiter is constant and equal to the square of the orbital radius of Jupiter. Accordingly, $a_6^2 = a_{6+i}a_{6-i} = 27.071$, where a_6 is the orbital radius of Jupiter, and $i = 0, 1, 2, 3, 4, 5$. This rule appears to naturally incorporate the dwarf planets Ceres and Eris (but can go no further outward unless planet Vulcan is invoked), and it provides orbital radii values that are within 11% of the observed values, although it is never more accurate than 2.5%. Kotliarov

[14] All of Butusov's publications on his rule are in Russian.

concludes that Butusov's formula "may seem to be just a number game" [and it is], but then offers journal readers the weak rejoinder that, if it is proved true, then we shall have to reconsider our understanding of planetary system formation. Once again, a pure numerical coincidence associated with a vague (never accurate) rule is being spuriously invoked as a reason to question standard cosmogonic theory. There is, indeed, nothing wrong with such questioning in principle, we do not know everything about how planets and planetary system form — far from it — but it is certainly clear that Butusov's formula is not a good reason to abandon current thinking and methodologies.

9.11. A Miscellany of Orbital Oddities

The examples considered in this section are by no means exhaustive of the total number of papers published on Titius–Bode-like laws during the 20th century. Rather, I have picked out a few of the more interesting examples that at least try to offer the reader a little bit more than just the results of some number play — although that is invariably what they do present. Some of the examples are surprising, others are mildly pleasing (even amusing), and yet others are downright bizarre, but assuredly, they are all wrong.[15] Wrong, that is, with respect to providing any useful insight as to the origin and physical make-up of the planets, and/or the origin of the observed structure within the solar system. We are running, here, of course, the fine line between reasonable and unreasonable speculation — science is full of speculation, and it is a vital part of making scientific

[15] By saying that these ideas are wrong is not an arrogant, backward-looking, blinkered, modernist interpretation of history. It is a pragmatic statement of fact. None of the ideas to be presented have gained any traction with respect to other researchers, and none of them have produced any advancement in our collective understanding of planetary systems. Furthermore, while I will be quite critical of the arguments employed by a number of named authors, no disrespect or deep ridicule is intended — rather I am simply attempting to probe, pull, and reassess what it is that has actually been written down and implied.

progress. But, like many things in life, while it is not possible to say where the boundary between reasonable and unreasonable lies, we can usually tell when the divide has been crossed.

Let us begin with the seemingly remarkable result presented by Clement R. Merrill of Airdrie, Alberta, Canada. Writing in the May 1924 issue of *Popular Astronomy*, Merrill investigates the sizes of planetary systems. "In a general way", Merrill writes, "the planets of greatest mass and highest axial speed possess the most extensive satellite systems". Here is our first problem, Merrill's statement for proceeding, while sounding reasonable, is, in fact, just an assumption and not based upon any physical argument. Furthermore, Merrill goes on, "it should be possible to determine the size of a planetary system from certain elements of the planet's physical condition, such as mass, density, axial velocity and the inclinations of the planet's axis". In this manner, Merrill introduces a system constant defined as follows:

$$(\text{Mass} \times \text{Density} \times \text{Axial velocity} \times \text{Inclination})_{\text{planet}} = \text{Constant} \times \Sigma(r_{\text{moons}}) \tag{9.9}$$

No rhyme or reason is given as to why this formula should work, or why the size of a planetary system should be determined by the axial (spin) velocity of the planet or its axial inclination. And, indeed, they do not depend upon these terms. Furthermore, ignoring the strange scaling that Merrill applies to various quantities, it is not at all clear why it is the sum of the orbital radii of the moons that are needed, rather than, say, the orbital radius of the outermost moon — which presumably better defines the size of the system. None-the-less, Merrill pushes-on and produces a table claiming to show how accurately the formula fits the available data for the multiple-moon supporting planets (indeed, with zero error in some cases). No explanation is presented for the reasons why some planets have no moons, and like-wise the orbital summation term for a given planet's satellites assumes that all of the associated moons have been discovered. In the case of Jupiter, Merrill uses a term summed over eight moons (the outermost one being Sinope). The discovery of just one additional moon in any of the systems that Merrill claimed were described with 100% accuracy, would ruin his argument — but no account is given for this possibility. Indeed, in the case of Jupiter, for example, the moon count changed in 1938 (with the discovery of Lysithea), and those for Uranus and Neptune in 1948. In summary, Merrill's analysis simply makes no physical sense. It is predicated on false, indeed, non-physical assumptions, and it makes unjustified claims to its accuracy and applicability. In the end, it is just delusion predicated on fudge-factors, and number play.

While Merrill's ideas are predicated upon false assumptions and incoherent physical arguments, the article by Albert E. Caswell (Physics Department at the University of Oregon), in the 5 April, 1929 issue of *Science* is predicated on a reasonable hypothesis. Indeed, Caswell notes that he has discovered a simple relation between planetary distances, that bears similarities with the quantized electron orbitals of the Bohr atom. Specifically, "*the mean distance of the planets from the Sun are proportional to the squares of simple integral numbers*" [Caswell's italics]. In 1929, quantum mechanics was very much in its ascendency, and the atomic model developed by Niels Bohr in 1913, gave physical credence to the empirical law found by Johann Balmer for the emission lines of hydrogen (recall equation (9.2) above). Caswell's relation for planetary radii, from Mercury through to Neptune (including the main belt asteroids when $n = 8$) reads the following:

$$R\ (\mathrm{AU}) = 0.0425n^2, \quad \text{where } n = 3, 4, 5, 6, 8, 11, 15, 21, 27 \quad (9.10)$$

Caswell gives no specific reason for why the constant should be 0.0425, and there is no clear structure in the sequence required for n, and we effectively encounter Wittgenstein's finite rule paradox again. For all this, with a healthy dose of rounding, the formula accounts for the orbital radii of the planets to an accuracy of about 3%. Impressed by the numerology, Caswell comments, "on the whole the agreement is good". In many ways, Caswell carries our interest right to the end — that is, until he suggests that the results "can scarcely be accidental", which of course, with the complete lack of any physical underpinning, is exactly what they are. Caswell, however, returns to the issue in an article published in the January 1932 issue of *Popular Astronomy*. Here, he expands upon his analysis of the orbital radii of Jovian and Saturnian moons, again finding formulae similar to (9.10), but with different n-sequences and constants. In this second article, Caswell also evaluates orbital energy, and orbital angular momentum for each planet, but can find no clear relationship between them, and no reason is found for the "squares of simple integers" rule. Essentially Caswell's second paper simply fizzles out, with no physical justification for formula (9.10) being presented. While Caswell briefly discusses the role of resonances in establishing stable and unstable zones within a multi-planet system (see Chapter 10), the ideas that he presents give no clue as to where the underlying physical mechanism for the quantization of planetary orbits might be found. One suspects that Caswell was really following a fad, or fashion, for developing quantization rules, rather than drawing upon any specific inspiration from quantum mechanics. Caswell's 1929 *Science* article did, however, catch the attention of J. B. Penniston, in Seattle, Washington. Writing in the 16 May, 1930 issue of *Science*, Penniston indicates that he had had the same idea as Caswell in the mid-1920s, but had since modified it to a form

$$R\ (\text{AU}) = n(n + 1)/20 \qquad (9.11)$$

where the n values are taken from the third diagonal in Pascal's triangle representing binomial coefficients. The sequence at play is 1, 3, 6, 10, 15, 21, 28, 36, 45, 52, ..., and Penniston argues that the planets Mercury out to Neptune (including Ceres) correspond to the sequence n = 3, 6, 10, 15, 28, 55, 91, 190, 300. Here, of course, is an immediate problem: the n values are simply cheery-picked to give the right (approximate) answers, and others are just ignored. Furthermore, why should the third diagonal of Pascal's triangle have anything to do with determining planetary orbits? Penniston provides no physical justification for his formula or selection process, and he is doing nothing more than playing with numbers.

Mr. J. Miller, of 61A Holborn Viaduct, London, begins his 5 February, 1938 *Nature* letter with the comment that Bode's law, in spite of having issues with respect to accounting for the orbits of Mercury[16] and Neptune, should not be "regarded as merely a coincidental truth". To this end, he suggests subtracting out the orbit of Mercury from the calculations, in other words, use the orbit of Mercury as the starting point rather than the Sun. With this maneuver in place, Miller then notes that the distances to the other planets can be written as a geometric sequence with

$$R_N \text{ (AU)} = 0.33 \times 2^N, \qquad \text{where } N = 0, 1, 2, 3, 4, \ldots \qquad (9.12)$$

Accordingly, R_0 (AU) = 0.33 is equal to the distance between the orbit of Venus and Mercury, while R_1 (AU) = 0.66 corresponds to the distance between Earth and Mercury, and so. The approximation given in (9.12) becomes worse and worse as N increases — for the distance between Saturn and Mercury, the formula gives R_5 (AU) = 10.56, rather than the required 9.21 AU; an error of some 15%. For Neptune, the percentage error between the required and predicted distance R_7 rises to 42%. For all this, Miller stoically ignores the problem.[17] "This general relationship seems to apply also to the orbits of satellites", Miller intones, and applies his formula to the satellites of Saturn, again, subtracting out the orbit radius of the first satellite Mimas. In this manner, he bravely makes a prediction that a satellite should exist between Iapetus and Phoebe. Furthermore, the Editor (1st Baronet, Sir Richard Gregory) of *Nature*, clearly being in a very lenient mood, published another communication from Miller in the 8 October 1938 issue. In this communication, Miller reiterates his initial scheme, now described as being "purely empirical and only very approximate", and extends his satellite

---

[16] By this, Miller is referring to the odd requirement that in order to account for the orbit of Mercury, the M-exponent in the Titius–Bode formular, our equation (9.1), must be set to $-\infty$. The problem with Neptune is, of course, the large difference between the law-predicted and actual heliocentric distances.

[17] Miller is not constructing a Fermi elephant, but he is certainly ignoring the elephant in the room of his argument.

predictions to Jupiter (placing a moon between Callisto and Himalia), and Uranus (placing a moon between Titania and Oberon). Miller's articles are remarkable for the whole range of errors and problems that they include. First, his rule is nowhere near as accurate as the Titius–Bode rule. Second, there is no clear physical reason to base the unit of measure upon the radius of the first planet, or moon, as the case may be. Not only this, Miller refers to his rule as being empirical. This, in fact, is not the case. It is a purely assumed mathematical rule that has no bearing on the actual observations. The other issue, of course, is that Miller assumes that the innermost satellite of each of the Jovian planets has actually been found — which, of course, they had not in 1938. With respect to the predictions of the various "missing satellites" in the gap indicated for Jupiter, we now know of 3 moons between Callisto and Himalia, none of which are at the orbital radius predicted by Miller. Likewise, 3 moons fall in the gap Miller identified for Saturn, and no moon is found, as predicted, between Titania and Oberon in the Uranian system.

As noted earlier, it is easy to be critical, but Miller's two articles (the only two astronomy articles he appears to have written) are remarkable. They are remarkable in that while clearly based upon number play, having no physical underpinning, and providing a dreadful fit to orbital radii of the planets and their satellite systems, they were actually published. This remarkability, however, perhaps says something about the scientific community in the late 1930s. Indeed, this was a time when researchers were looking for, and finding through quantum mechanics, geometric rules, and/or quantization rules as the organizing principle of physical structure. These same organizing principles, it is conjectured, extend to the structure of the solar system itself. Indeed, this reflects a return to Medieval thinking where the microcosm was seen as representative of the macrocosm.

Picking up the dangling threads left by Albert E. Caswell, and following the math-play of Miller, George W. Todd, Professor of Experimental Physics at King's College, Newcastle-upon-Tyne, set out to examine the consequences of positing quantized planetary orbits. Writing in the 5 March 1938 issue of *Nature*, Todd bravely questions the strict applicability of the inverse square law of gravity. Furthermore, noting that, "Bode's law can no longer be regarded as fortuitous", Todd sets out to find a force law, "which will give all three laws of Kepler, Newton and Bode for [a] stable orbit". Accordingly, Todd introduces a force $F(\mu) = \mu^2\Phi(\mu)$, with $\mu = 1/r$, where r is the heliocentric distance. In the pure Newtonian case $\Phi(\mu)$ = constant. Considering the stability conditions for a circular orbit, however, Todd introduces the condition that stable orbits can only exist if $\mu\Phi'(\mu)/\Phi(\mu) = 0$, where $\Phi'(\mu)$ is the differential $d\Phi(\mu)/d\mu$. With all this straightforward classical mechanics in place, Todd argues that the standard Newtonian $\Phi(\mu)$ = constant, is not necessarily the only formulation to provide for stability. This is a true enough mathematical statement. What if, Todd questions, we take $\Phi(\mu) = \cos^2(1/\mu)$? In this case, the orbital stability condition holds true whenever $1/\mu = r = n\pi$, where $n = 0, 1, 2, 3, \ldots$ and this would allow for a stable quantized set of orbits to come about. For all his mathematical play, however, Todd can take his analysis no further, and he ends defeated, noting only that, "if a shape for $\Phi(\mu)$ can be found" which satisfied the stability condition when $1/\mu = r = 3 \times 2^{(n-2)} + 4$, $n = 1, 2, 3, 4, \ldots$, "then a law of force giving Kepler, Newton and Bode will have been discovered". The end is rather inglorious, mathematically sound, but not very helpful. Todd has outlined what mathematical conditions must apply to any modified gravity law, but he has not shown that any such modification can be found. We, the readers, are once again left in the lurch. Todd's analysis is interesting, however, in that he is prepared to question the universal applicability of the inverse square law of gravity. Certainly by 1938, Einstein's theory of general relativity was well established, and shown to be experimentally sound. And, general relativity yields Newton's inverse square law of gravitational attraction in the limit that $2GM/Rc \approx 0.0$. Accordingly, the result that Todd is searching for cuts not only across standard theory, but runs counter to well established experimental results. In itself, such conflict, or questioning of established ideas, is not a bad thing, or a specific problem — it is what science is all about. Indeed, some present-day forms of modified gravity posit that in the very low acceleration domain that gravity effectively behaves as an inverse distance (rather than inverse square) law. Such modifications seek to explain the observed properties of galaxy rotation curves, and negate the necessity for any specific form of dark matter. Todd, however, is effectively calling for a complete and utter re-writing of physical theory (Newtonian and Einsteinian) in order to satisfy an arbitrary (indeed, bogus) planetary spacing rule.

While Miller forced planetary orbits to obey a strict geometric sequence [equation (9.12)], giving spacing distances relative to the orbit of Mercury, ophthalmologist E. Voilas in Cairo, Egypt, found greater significance in allowing the rule to apply to just a part of the planetary system. Writing in the December 1940 issue of *Popular Astronomy*, Voilas opens his discussion by chiding his very readers, writing, "astronomers, surely, must have found very little interest in the problem [to replace Bode's law] for otherwise they could not have failed to discover the new law; the very law that we are submitting below for their consideration". Once again taking the orbit of Mercury as given, and the starting point for his calculations, Voilas argues that Mercury, Venus, and Earth are described by the arithmetic sequence R (AU) = $0.3n$, $n = 0, 1, 2$. The orbital radii of the Earth through to Uranus, including the asteroid belt, are described according to the rule: R (AU) = 0.6×2^n, $n = 0, 1, 2, 3, 4$. And finally, the orbital radii of Uranus, Neptune and Pluto are described as R (AU) = $9.6n$, $n = 0, 1, 2$. Significantly, according to Voilas, the spacings are described by overlapping arithmetic sequences (Mercury to Earth, and Uranus to Pluto) and a geometric sequence (Earth to Uranus). The formula fit, compared to the observations, is 2% too big at Saturn, 3% too small at Neptune, and 2% at Pluto — which is admittedly not bad, but then, nor is it, for a supposed fundamental new law, exact. Voilas continues his analysis by considering the orbital radii of the moons of Jupiter, Saturn and Uranus, and finds, by allowing for copious "missing" moons, that overlapping arithmetic and geometric sequences can be found to [approximately] described each system. Voilas, like Miller, and Penniston before, is engaged in number play, but finds significance in the fact that a close accounting for orbital separations can be obtained through the judicial use of arithmetic and geometric sequences. Once again, however, no account is given by Voilas for why the ordering sequence should start at the orbit of Mercury, and/or why the planetary system should be organized according three domains, each with their own ordering progressions.

Directly following Voilas's article in the December 1940 issue of *Popular Astronomy*, is another article, this time on "an astronomical relationship", by Fred C. Bond of West Allis, in Wisconsin. While Voilas opened his article by mildly criticizing astronomers, Bond opens his by warning his readers against the "dangerous rut of orthodoxy". Indeed, he continues, "the most significant discoveries are those that do not seem to fit into the accepted scheme of things, because they alone point the way toward new and different truths". Using the discovery of cosmic rays, and the success of general relativity as examples, Bond presents the fruits of his labor that, "should be warmly welcomed of the consciousness of the true scientist". The "highly significant relationship" found by Bond relates to the ratio of the sidereal periods of the planets divided by the number of revolutions made by the Earth as it orbits the Sun. Given $U = 0.3662422$, the number of sidereal days in a year divided by 1000, Bond presents the reader with the following equation:

$$y = \frac{\text{Log}\,P}{U} + 7 \tag{9.13}$$

where y is what Bond calls the space number. The variable term is the logarithm to the base 10 of the orbital P in years. The space number, according to Bond, "appears to have a fundamental cosmic importance", since, as he notes, with judicial rounding up and rounding down, y is an integer value (except that is, for Pluto). The integer sequence for y runs as 7, 8, 9, 10, 11, 12, 13, and (then strangely) 13.5, corresponding to the Sun, Mars, the asteroids, Jupiter, and on through to Pluto. Bond also includes in his list the Moon, for which $y = 4$, and the Earth for which $y = 0$. Interestingly, Bond also places the stars at $y = 19$, by which he means the precession of the equinoxes, which has a period of 26,300 years. Furthermore, adopting a somewhat Ptolemaic outlook, Bond excludes from his list Mercury and Venus, because they appear, to a terrestrial observer, to orbit the Sun and because they have no moons of their own. Bond builds upon the result that the y, space number, and values in his list are all close to being integer, "from a mathematical standpoint it is scarcely conceivable that this consistent and regular series arrangement [the integer sequence of y values] could be the result of chance alone". Indeed, Bond insists that the result, "indicates a deep-seated and fundamental cause, whose nature is still obscure". With these last words, of course, Bond loses his argument, and all we are left with is a statement of his own sense of incredulity. Yes, it is remarkable, interesting, entertaining, and even surprising that the y values are all close to being integers, but there is absolutely nothing in the material presented to suggest that some fundamental result concerning planetary orbits and/or the stellar realm has been found. Once again, we have a confusion born of a mathematical coincidence and an idea presented with no physical reasoning or underpinning behind it. Bond's result is leading us nowhere.

9.12. A Miscellany of Oddities

Not only is there a vast cottage industry concerning alternative forms of the Titius–Bode rule, there is a long history of finding the relationship between the physical properties of the planets. The formula developed by Clement R. Merrill [equation (9.9)] is one example of this latter exercise, but others have been more specific in scope. Donald Richardson, for example, who we encountered earlier in Section 9.7, concerning the Blagg–Richardson formulation of the Titius–Bode rule, presented a complex formula for planetary mass as a function of its ordinal number. Writing in the December 1944 issue of *Popular Astronomy*, Richardson explains, that the mass of each planet in the solar system is given by the following equation:

$$M(n) = 6.80590\rho_n, \quad \text{with } n = 1, 2, 3, 4, \ldots \tag{9.14}$$

where $M(n)$ is expressed in units of $46.3685 \times 10^{-10}\ M_\odot = 9.227 \times 10^{21}$ kg, and where the function ρ_n, "equals the length of the radius vector of a simple circle-ellipse function corresponding to a polar angle of $\pi n/5$". One is immediately wary of such a statement for the form of any empirical function entering into a description of the solar system — wary simply because it carries no physical meaning. The equation for ρ_n is a complicated trigonometric expression, which is also forced to change at the ordinal number $n = 5$, the orbit of the asteroids. Indeed, Richardson argues that the mass unit being used in his formula corresponds to the mass of the asteroid belt. While Richardson casts his ideas in terms of planet growth by accretion within a nebula disk, his analysis is nothing more than the creation of yet another Fermi elephant that has been forced to agree with the observations. His analysis ultimately fails because it provides no physical insight as to why the various planets have the masses that they do, and no explanation for why the asteroid belt should act as a fundamental unit of mass is provided.

Picking up the idea discussed by Clement Merrill in 1924 (but without reference to Merrill), Carlos Jaschek (Observatorio Astronomico, Argentina), writing in the December 1960 issue of the *Observatory* magazine, once again considered the size of satellite systems. Merrill, recall, had found significance in the fact that the larger the mass of a planet so the greater the number of accompanying satellites, and the greater the size of the satellite system. Jaschek, finding equal significance in this accordance, derived a power law relationship of the following form:

$$R = 3.3 \times 10^{-13}\, M^{0.82} \tag{9.15}$$

where R is the radius of the outermost satellite's orbit in centimeters and M is the mass of the planet in grams. Table 9.2 provides a comparison of observed and predicted numbers.

Just like Merrill earlier, Jaschek has assumed in his analysis that all of the satellites within the various planetary systems have been found, but this is patently not true of the Jovian planets, for which numerous small, distant moons have been discovered since 1960. Looking at the fractional error terms (column 6 in Table 9.2) one can immediately reject Jaschek's formula as having any utility

Table 9.2. Comparison of Jaschek's formula [equation (9.15)] against observations. Columns 1 and 2 indicate the planet and the planet's mass. Column 3 is the predicted radius of the outermost satellite, while columns 4 and 5 give the actual radius of the outermost known satellite in 1960 (when Jaschek was writing) and the present day (2022). The last column indicates the ratio of the observed to predicted radius for the outermost satellite in 2022.

Planet	Mass (grams)	*R* (centimeters) predicted	Outer 1960	Outer 2022	Factor
Earth	6.046×10^{27}	1.992×10^{10}	3.847×10^{10}	3.847×10^{10}	1.93
Mars	6.417×10^{26}	3.166×10^{9}	Deimos 2.346×10^{9}	Deimos 2.346×10^{9}	0.74
Jupiter	1.900×10^{30}	2.223×10^{12}	Sinope 2.368×10^{12}	Kore 2.421×10^{12}	1.05
Saturn	5.685×10^{29}	8.268×10^{11}	Phoebe 1.293×10^{12}	LVIII 2.611×10^{12}	3.16
Uranus	8.682×10^{28}	1.771×10^{11}	Oberon 5.835×10^{10}	Ferdinand 2.043×10^{11}	1.15
Neptune	1.024×10^{29}	2.027×10^{11}	Nereid 5.504×10^{11}	Nesso 5.005×10^{12}	24.69

and/or meaning. The same conclusion must be drawn with respect to a follow-on paper to Jaschek's written in the December 1961 issue of the *Observatory* magazine by M. Friedjung, of Ruislip, in England. Friedjung claims that a better correlation between satellite system size and planetary mass can be found if one expresses the orbital radius of the outermost satellite in units of the orbital radius of the inner most satellite. Just as Jaschek's formula fails because of the discovery of more outermost Jovian moons since 1960, so Friedjung's formula doubly fails through the discovery of more innermost Jovian moons since 1961.

The search for laws linking such physical characteristics as mass and satellite system size is essentially doomed to failure — in spite of not unreasonable expectations — because of the observational difficulties in establishing a true satellite inventory. Indeed, casting the problem in terms of 3-body dynamics, including the planet, outermost moon, and Sun, Jaschek finds (and it is still true) that no system is even remotely close to occupying its maximum Hill-sphere (see Chapter 10) radius. For all this, Titius–Bode-like laws may yet exist between other system variables. One such law is that described by Christopher T. Russell (University of California) in the 9 March 1978 issue of *Nature*. This, so called, Magnetic Bode's law concerns an apparent correlation between the value of the magnetic dipole moment M_{surf} and the spin angular momentum L of a planet. Expectation would suggest that these two quantities might well show some correlation, since the dynamo model for magnetic field generation is dependent upon the rotational speed of a planet's core. For all this, the dynamo model is still not fully understood, and it operates in a complex manner according to the physical conditions that apply in a planet's core. Setting the magnetic dipole moment to be $M_{surf} = B_{surf}R^3$, where B_{surf} is the surface magnetic field strength, and R is the radius of the planet, and taking the spin angular momentum to be that of a rotating sphere, so $L = kMR^2/T$, where, k is a constant, M is the mass of the planet, and T is the spin period, so the observational data suggests that

$$M_{surf}(\text{Gm}^3) \approx 4 \times 10^{-9} L^{0.83} \ (\text{kg m}^2 \text{ s}^{-1}) \qquad (9.16)$$

The relationship, however, is far from convincing. Indeed, the predicted magnetic dipole moments for the Sun and Saturn are too small by a factor of 10, that for Mars too small by a factor of 100, and too small by a factor of 1000 for Venus. It is the observational data for Mercury, Earth, Jupiter, Uranus, and Neptune, that drives the relationship form of equation (9.16). Clearly, as Russell points out, there is more to determining the surface magnetic field strength of a planet than its rotation rate, size, and mass. Wisely, Russell comments that "no physical significance" is attached to the relationship found. It is, at best, an approximate scaling law, and indeed, a scaling law that points to its own limitations. The limitation is that it does not take into account the physical state of a planet's interior. Furthermore, one can reasonably quibble about why relationship

(9.16) is even called a Bode-like law,[18] since the formula is not related to a sequence number, or ordering term in heliocentric distance. For all this, in the modern era, the idea of the Magnetic Bode's law has been generalized to the Radiometric Bode's law. In the latter case, it is the power of the radio emission from a planet (with a magnetic field) that is taken to scale as the power of the incident parent-star's wind. Once again, the relationship is a power or scaling law, not a Bode-like law. This particular scaling law, however, indicates that Jupiter-like exoplanets might be directly detectable with radio telescopes working in the frequency range of 10 to 1000 MHz.

As a final example in this section, let us consider an article by cardiologist Dr. Gürler Iliçin, of Sevgi Hospital in Ankara, Turkey. Writing in the *Quarterly Journal of the Royal Astronomical Society* for December 1994, we are informed by Dr. Iliçin that, "for some time I have been searching for numerical relationships which link planetary data and these turn out to be more common than would be expected just by chance". The relationship that is presented concerns the ordinal number n of the planets, Mercury ($n = 1$) through to Neptune ($n = 8$) (ignoring the main belt asteroid region), the orbital period P and mass M of the planets, and the spin period T of the planets. Specifically, defining the following terms:

$$G_n = \sqrt{M_n / T_n} \tag{9.17}$$

$$K_G(n) = \frac{G_n G_{n+2}}{G_{n+4} G_{n+6}} \tag{9.18}$$

$$K_P(n) = \frac{P_n P_{n+2}}{P_{n+4} P_{n+6}} \tag{9.19}$$

and upon inserting numerical values, it transpires that: $K_G(1) \approx K_G(2)$, and $K_P(1) \approx K_P(2)$. Iliçin notes that various expressions for K_G

[18] It would appear that the term was coined by Charles F. Kennel in 1973 with reference to the book *Physics of Planets*, by Vasilii Moroz, published in 1967. Moroz specifically assumed in his discussion on the decimeter radio emission from Jupiter that the magnetic moment of a planet was proportional to its mechanical moment. See NASA Technical Translation F-515 (1968), p. 350.

and K_P are within 3% of each other, and remarks that while the values for the orbital periods show a steady progression as n increases, the same cannot be said for the mass or spin periods. With a final comment that "I present it [i.e., that $K_G(1) \approx K_G(2)$, and $K_P(1) \approx K_P(2)$] merely as a curiosity", the reader is literally left wandering. The result is nothing more than number play — a hobby admitted to by the article's author — and, indeed, even if any physical significance could be attached to the three defined terms (of which there is none) then what would such a result tell us about the solar system and planetary formation? Well, of course, it would tell us absolutely nothing.

9.13. The Golden Ratio

There is a long and dubious popular history surrounding the golden ratio (also called the golden mean, the golden section, and the golden proportion) — it is one of those numbers, just like π and e, that seems to crop up almost everywhere in nature, in art, in architecture, in mathematics, and in music, and accordingly, people have eulogised about its apparently miraculous and synergistic properties. For all this, however, it must also be said, that in spite of the mystical baggage that accompanies it, many interesting applications, and mathematical theorems can be associated with the golden section [3].

Mathematically, there is nothing profound about the definition of the golden ratio Φ — it simply concerns the ratio of two numbers a and b such that $\Phi \equiv a/b = (a + b)/a$. And, this definition unravels to yield the quadratic

$$\Phi^2 - \Phi + 1 = 0 \tag{9.20}$$

giving $\Phi = \frac{1}{2}(1 + \sqrt{5}) = 1.6180339887...$ as the positive root. The golden ratio appears to be a constant of geometry, and just like π and e it is a transcendental number. Johannes Kepler wrote that "geometry has two great treasures", the theorem of Pythagoras and the golden ratio. Of these two he described the Pythagorean theorem as a measure of gold, but the golden ratio he described as a precious jewel. It is worth noting, that the second solution to equation (9.20) gives $\Phi = \frac{1}{2}(1 - \sqrt{5}) = -0.6180339887...$, a number which apparently nature has no call for.

An association between the golden ratio and planetary orbits was made by Mr. H. L. Spahr, of Atlanta, Georgia, in the July 1933 issue of *Popular Astronomy*. Spahr began with some pure speculation and suggested that the material out of which the planets eventually formed had been "ejected or pulled" from the Sun along the path of a logarithmic spiral. This spiral arc then fragmented and formed planets at radii according to the geometrical series 1, d, d^2, d^3, d^4, d^5, ... , with the ascending ratio term d "being the famous golden section" (i.e., $d = \Phi$). Having thus been formed at these specific broken-spiral locations, Spahr then argues that the planets evolved to their presently observed orbital radii. Furthermore, arguing that if we could only determine which planet had migrated the least, then we would be able to determine how much each of the other planets had moved since they formed. Unfortunately, Spahr tells us, there is no way of knowing this, and he simply assumes in his further discussion that the Earth's orbit has changed the least, and that this will be taken as the unit distance (i.e., he sets $d^2 \equiv 1$ AU). The solar system is accordingly arranged as indicated in Table 9.3. Boldly, seeing an apparent break in the planet sequence between Jupiter and Saturn, Spahr postulates a gap containing a missing planet (or asteroid ring) at some 6.854 AU from the Sun. Clearly, Spahr's sequence is not a particularly good fit to the observed orbital data, being some 6% to big at Mars, some 19% to small at Jupiter, some 16.5% to big at Saturn, and too 3.3% to small at Neptune.

It is interesting to note that while in 1933 it may well have been thought a brave step (even foolhardy) to predict a new planet (or asteroid zone) between the orbits of Jupiter and Saturn, some present-day scenarios, so-called Jumping Jupiter models, posit exactly this [4]. This "jumping" planet, however, is ejected out of the solar system as the outer solar system planets migrated to their currently observed locations. Not only did Spahr speculate on the possibility of their being additional planets in the solar system, he also sought to take his

Table 9.3. Comparison of Spahr's golden-section sequence for planet orbital radii against the observed values. Columns 1 and 2 indicate the planet and its position in Spahr's golden-section sequence. Columns 3 and 4 compare the predicted and the actual orbital radii. Spahr has scaled the sequence so that Earth's orbit radius is equal to 1 AU.

Planet	Sequence	Series	Actual
Mercury	$1/d^2$	0.382	0.387
Venus	$1/d$	0.618	0.723
Earth	1	1.00 (assumed)	1.00
Mars	d	1.618	1.524
Asteroids	d^2	2.618	—
Jupiter	d^3	4.235	5.203
(missing)	d^4	6.854	—
Saturn	d^5	11.09	9.521
Uranus	d^6	17.94	19.20
Neptune	d^7	29.03	30.02

formula to smaller and smaller dimensions, bringing the "microcosm under the same law as the macrocosm". In this manner, Spahr speculated (in a very roundabout manner) about the speed of electrons in their orbitals, suggesting that outer electrons should, "travel faster than the inner" ones. Here, Spahr is speculating, in a similar manner to Caswell in 1929 (as discussed above) with respect to the Bohr atomic model, and thinking of electrons in a purely particle sense. In many ways, Spahr's reasoning is based upon the maxim, "in for a penny, in for a pound", assuming what in the modern era would be called a self-similar scaling principle, operating on all scales, from the atomic to the cosmological, although Spahr did not attempt to take his sequence beyond the scale of the solar system.

Following in the speculative footsteps of Spahr, chemist Jan Boeyens (University of Pretoria, South Africa), in an article appearing in the December, 2009 issue of *Physics Essays*, discussed a commensurability model that sets out to explain the distribution of planetary orbits, as well as galactic, and atomic structure. Indeed, Boeyens claimed, without any supporting justification, that, "the most telling argument in favor of an orderly evolution of the solar system is still the Bode-Titius [sic] law". Be this statement as it may, Boeyens goes on to develop an argument which suggests that the scale of planetary orbits is governed by, "a simple numerical pattern based on the golden ratio". Table 9.4 is a summary of Boeyens' results, which seemingly offers no clear explanation as to why a Titius–Bode-like rule should come about. Boeyens proceeds by scaling planetary orbits to that of Jupiter's, noting that the inner planets, Mercury to Mars, have orbits that scale as some fraction of $\tau = 1/\Phi$. The Jovian planets, Jupiter to Neptune, in contrast, have orbital radii expressed as pure integers. This result, in spite of what Boeyens claims, does not remotely justify the use of the golden section — rather it appears that there is a close, but somewhat vague coincidence between two columns of manipulated numbers. Boeyens is spinning a story, admittedly an intriguing story, but it is a story that has no real substance, and it is a story that ultimately fails to carry any conviction. Indeed, it is an old story that runs parallel to the narrative presented by "D. Rookes, c/o Ardilla Books, 29 Gloucester Crescent, London NW1", nearly 40 years earlier. Writing in the 29 August 1970 issue of the journal *Nature*, Rookes solicits comments on his "interesting discovery". Specifically, Rookes normalizes the orbital radii to that of Jupiter, and then identifies various integer fractions (see Table 9.4) that provide, "an almost perfect prediction for planet solar distances". For Mercury, for example, Rookes finds the fraction is 1/16, giving an orbital radius of 5.203/16 = 0.325 AU — which is 16% to small. Finally, indicating that the significance of his findings is admittedly obscure, Rookes closes with the comment that "the full explanation is contained in the publications *Genesis One*, *Two* and *Three*" — which is equivalent to saying "and God calculated so".

According to Boeyens' analysis, the spacing of the various planets (as per Table 9.4) is organized according to the following rule:

Table 9.4. Orbital radii normalized according to that of planet Jupiter. Columns 1 and 2 provide the object name and its recognized orbital radius. Column 3 indicates the orbital radius normalized to that of Jupiter. Column 4 indicates the values adopted by Boeyens, while column 5 indicates the values given by Rookes. **Key:** (*) Boeyens asserts that this is the "mean orbit of the asteroid belt". This is an overly generous value, and the weighted mean orbit is closer to 2.7 AU — but this would destroy the supposed connection with τ.

Object	Actual (AU)	Normalized to Jupiter	Boeyens sequence	Rookes fraction of Jupiter's orbit
Mercury	0.387	0.074	$\tau/8 = 0.077$	1/16
Venus	0.723	0.139	$\tau/5 = 0.124$	1/7
Earth	1.000	0.192	$\tau/3 = 0.206$	1/5
Mars	1.524	0.293	$\tau/2 = 0.309$	1/3
Asteroids*	3.216	0.618	$0.618 \approx \tau = 1/\Phi$	—
Jupiter	5.203	1.00	1 (assumed)	1
Saturn	9.521	1.830	2	2
Uranus	19.20	3.690	4	4
Neptune	30.02	5.770	6	6

$$\text{Terrestrial planets: } r\ (\text{AU}) = k\tau/n, \quad \text{where } n = 8, 5, 3, 2, \tau,$$
$$\text{and} \quad k = 5.203\ \text{AU}$$
$$\text{Jovian planets: } r\ (\text{AU}) = kn, \quad \text{where } n = 1, 2, 4, 6$$

No clear justification is presented for the rule change after the asteroid belt or the reason for scaling to Jupiter's orbit, and indeed, the accuracy of the model is not particularly impressive. The model's prediction is 4% too big at Mercury (better than that found by Rookes), 5.5% too big at Mars, 9.3% too big at Saturn, 8.4% too big at Uranus, and 4% too big at Neptune. The original Titius–Bode rule provides a better over-all fit to the observed orbital radii. Boeyens, like Rookes some 39 years before him, and Spahr 37 years earlier still, is doing nothing more that playing with numbers, and making far too much play of apparent numerical coincidences.

Curious, profound, unexpected, inexplicable, and beyond coincidence — all these terms have been used by one author or another to describe their new planetary ordering rule. Sadly, these labels are exactly what the results of their work are not. Invariably, a closer look at what has actually been written reveals that no new physics, insight, or profound rule has been found — the works are all, with virtually no exceptions, number play and coincidence. Sometimes the number play is naive, at other times it is hidden behind the obscurity of complex arguments where the authors are doing nothing more than constructing a Fermi elephant tethered in place by the mystical chain of Gleipnir.

9.14. Ever Hopeful, Ever Lost

People can have the strangest of hobbies, and it would appear that for some their all-consuming passion is to find a numerical sequence that describes the spacing of planetary orbits within the solar system. It is not the worst hobby that can be imagined, but it is one doomed to failure. Doomed, that is, if one expects to make some fundamental discovery. Indeed, for some, the hobby can turn into an *idee fix*, and this is less healthy, and this is the reason why the journal *Icarus* has its particular editorial policy — it saves wasting time on unnecessary correspondence, and arguing with people who have anchored viewpoints.

Does the Titius–Bode rule or, indeed, any of its variants discussed above, rise to the level of a physical theorem? The straightforward answer to this is, no. Do these same ideas, if true, challenge our accepted views of standard theory? The answer to this is, yes. In a direct and meaningful way, if any of the arguments discussed above were actually true then astronomers and physicists have greatly underappreciated (and, indeed, do not understand in a fundamental and profound way) the formation of planets and planetary systems, or for than matter cosmology and atomic theory. The fact is, the Titius–Bode rule, and its many numerological variants are fundamentally flawed when it comes to representing some form of physical reality. This is not to deny, however, that, in the sense of the Strong

Law of Small Numbers (see Chapter 7), they might latch-on to, by pure chance, a descriptive numerical sequence. Indeed, the Personal Incredulity Principle (PIP), as introduced by Richard Dawkins, is often invoked by proponents who have discovered a new variant of the Titius–Bode rule. That is, a set of numerical coincidences is seen as being necessarily remarkable and profound. The PIP being invoked, rather than simply accepting the much more probable argument that nothing more than a chance similarity between two sets of numbers has been identified.

Written and spoken languages, just like mathematics, are rule-based systems, and with language and words, one can construct incorrect statements that do not break any linguistic rules. One can write, for example, that this sentence is 27 words long. The sentence (as just written) breaks no linguistic rules, but it is entirely incorrect — it contains 11 words, one number, two commas, and one full stop. The same kind of contradictions can be seen in art and mechanical diagrams. Figure 9.6 shows the famous painting by Rene Magritte entitled *This is not a pipe*. Yes, it shows a pipe, but it is, as Magritte points out, not a real pipe — it is a 2-dimensional painting of a pipe and a very different object to the 3-dimensional pipe that one can smoke. On a similar theme, Figure 9.7 shows a 2-dimensional, mesh rendition of a Klein bottle. This 2-dimensional

Figure 9.6. *Ceci n'est pas une pipe*, painting by Rene Magritte (1929). Image courtesy of the Los Angeles County Museum of Art.

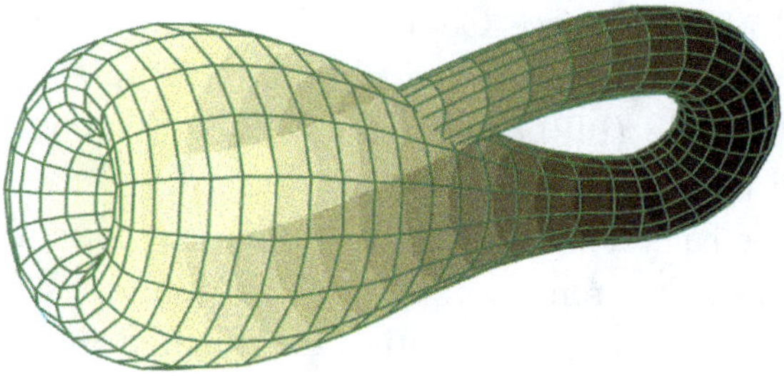

Figure 9.7. A two-dimensional rendition of the Klein bottle. Image courtesy of Wikimedia Commons: https:// commons.wikimedia.org/wiki/File:Klein_bottle.svg

surface represents a non-orientable surface with no boundary,[19] and technically it can only be constructed (that is it is embedded) within 4-dimensional space.

On paper (and in the human mind) one can not only envisage objects (such as the Klein bottle) that are impossible to realize in an actual physical sense, but also design impossible perpetual motion machines, and construct globally ambiguous diagrams. Cut off from physical reality, and the strict obeyance of the laws of thermodynamics, perpetual motion machines can run (in the imagination) forever in a self-sustaining manner (Figure 9.8). The diagrams of such perpetual motion machines contain no apparent contradictions, they are logical and precise, but no such machine will ever work in reality. Indeed, it is one of the strangest properties of the human mind that it can imagine, and even comprehend, impossibilities, that is, it can *picture* objects and machines that cannot possibly exist or work in the real world. Not everything that the human mind can envisage has a counterpart in reality, and the same statement follows with respect to mathematical constructions.

Along with perpetual motion machines, the human mind can also construct and seemingly make some sense of impossible objects. Such objects possess a global ambiguity, that is they cannot be

[19]The more familiar Mobius strip, which is also a 2-dimensional, none-orientable surface is perhaps better known than the Klein bottle. In the case of the Mobius strip, however, it does have a boundary and it can be embedded within 3-dimensional space.

Figure 9.8. Perpetual motion engine designed by Robert Fludd (1618). In this design, water pouring from a water reservoir turns a water wheel (at far left). This wheel in turn drives a mill wheel (through the central vertical shaft). The mill wheel also turns an Archimedean screw (middle right diagonal) which transports water back to the reservoir. In this manner, the power source (the falling water) is continually replenished, and the reservoir never runs dry.

physically constructed, but intriguingly they show localized order. An example of this is the Penrose Triangle (Figure 9.9) [5], where each vertex of the triangle is ordered and geometrically sound, but when viewed as a whole, the apparent localized order disappears in a contradiction, with the eye and brain, struggling to make a stable

Figure 9.9. The Penrose Triangle. An impossible 3-dimensional object with local order at each vertex.

interpretation. This example is different to the Klein Bottle, shown in Figure 9.7, since the triangle's construction ambiguity cannot be resolved in 4-dimensional (or any other dimensional) space. Dutch artist M. C. Esher is famed for his highly imaginative use of such structures in his etchings [5, 6]. The existence of optical illusions, such as the Penrose Triangle, implies that the human mind can accommodate objects that reside outside of reality, but it is far from clear that such phenomena should be expect to have a real existence in the supposed Platonic realm.

No one (I assume) purposely sets out to produce nonsense or pseudo-science, but it is evident that nonsense and pseudo-scientific ideas abound in the world [7]. This is indeed, an odd state of affairs, since it implies that there is some governing-body with the authority to decree when something is nonsense and/or pseudo-scientific. Technically, there is no such body, other than the collected response of invested practitioners over time. Most examples of pseudo-science are never actually proved wrong, at least at a level acceptable to the promoting parties. Rather, they simply fade away when no one takes any more notice of them. This is not to say that pseudo-scientific, and even plain wrong ideas (examples being phrenology, and astrology) can't have a long run of popularity, and even apparent moments of legitimacy. While no one accepts the claims of phrenology today, many (a majority?) of people still read their daily horoscopes, or believe in ghosts, and any number of the junk-ideas that can be read on social media outlets. We can only conclude that humans are gullible, and that the human brain is largely ill equipped to deal with the true nature of reality. This is why most people turn away from the sciences — they are hard, and they are not built upon simple, common-sense rules and expectations; the history of science from Aristotle to the present day informs us of this fact.

Given that science and pseudo-science occupy separate domains of credibility, and only that the former actually has some contact with reality, can, we ask, a boundary be drawn between them? On the one side of the division, it is clear that science is being conducted, while on the other we have mumbo-jumbo. Furthermore, if such a division can be made, on what side of the division does the Titius–Bode rule sit? Problematically, while it seems clear that a division between science and pseudo-science must exist, it is not, apparently, a hard division, but rather it is a fuzzy zone. While all of the ideas presented in Sections 9.11, 9.12, and 9.13 clearly fall on the pseudo-science side of the boundary, an argument could be made that the Titius–Bode rule itself falls on the mercurial boundary itself. For all this, however, the Titius–Bode rule is typically presented and argued for in a pseudo-scientific manner. Indeed, the Titius–Bode rule, and all the other relationships described above, provide no means by which their dictates can be tested, verified, and/or falsified. Not only this, it is invariably the case that the numerical results from Titius–Bode-like rules, once presented, are never a precise match to the actual observations — in any other area of *bona fide* scientific study, such poorly fitting results would be taken as a clear sign of an incorrect or failed approach. Furthermore, it is also typically the case that the final argument of the author(s), upon presenting their near coincidences, assumes that the apparent closeness must be a sign of something fundamental going on. If only scientific understanding was so straightforward, the universe so simple, and nature so easily read.

9.15. So, is there a Theory?

Michael Nieto ended his wonderful book *The Titius–Bode Law of Planetary Distances* with an optimistic, but open-ended statement; arguing, "much more interesting work lies before a *finis* can be written on this venerable (yet often damned) Law of planetary astronomy". He is partly right. It is unlikely, however, given the vast cottage industry that has developed around the Titius–Bode rule, that it will ever be considered as solved. Furthermore, it is also the case that there is no indication that the Titius–Bode rule will legitimately aspire to become the Titius–Bode Law. This is not to say that no theoretical framework will ever be developed, it may yet, but what is certain is that any such framework will be highly complex, highly contorted, and extremely subtle. If this is the future in store for the Titius–Bode rule, then so be, and bring on the future.

Writing in the early 1970s, Nieto explored three avenues by which the Titius–Bode rule might be theoretically explained. These are presented as electromagnetic, gravitational, and nebular theories. The first theoretical framework was developed during the first decade of the 20th century and was predicated on the idea that the Sun had a very strong magnetic field. In this manner, positive ions emitted from the young Sun's outer regions, with specific and discreet energies, would spiral outwards to their appropriate cyclotron radius set according to their specific mass and charge. This mechanism produces regions of higher density enhancement about the Sun in which the planets eventually formed. It's a neat idea, but totally inadequate as a model for planet formation. First, the Sun's magnetic field has never been strong enough to produce a limiting cyclotron radius, and second, the model is not capable of explaining the chemical makeup of the planets, and it cannot account for the division between terrestrial and Jovian planetary types. The gravitational theories developed to explain the Titius–Bode fail for a vast number of reasons, some of which we have explored in this chapter (see also Chapter 10). The main gravitational theories relevant to explaining the Titius–Bode rule were developed in the mid-19th century and differed according to where the material from which the planets formed came from. One set of theories assumed that the planetary material was captured by the chance encounter of the Sun with interstellar streams. The streams themselves are composed of meteoritic material. The other set of gravitational theories assumed that the planet-making material was ripped from the Sun's outer layers due to the chance close passage of another star. Neither of these theories carries any weight in the modern era. Additional gravitational sorting and sifting theories have been proposed (see Chapter 10), but none of these have proven convincing beyond an *ad hoc* level. The third group of theories considered by Nieto relates to the solar nebula hypothesis, and while these, again, have not resulted in an explanation for the Titius–Bode rule, they at least have resonance with present-day theories concerning planet and host-star formation ideas. Indeed, it is now clear that planetary systems form as an integral and natural consequence of star formation. The interstellar material that collapses, under gravity, to produce a new star, as a result of rotation and the conservation of angular momentum, settles into a small (of order 100 AU across) flattened accretion disk about the central protostar. Material flows through the disk to produce the central star but eventually evolves into the birth domain for planets (Figure 9.10).

If there is ever going to be a theory accounting for the Titius–Bode rule then it will have to be predicated upon the magnetohydrodynamic equations that describe an evolving, extended, rotating, heated, turbulent, viscous, magnetized disk of gas, ions, ice, and dust. This is not the domain for the mathematically faint of heart and nor is it a domain accessible to present-day computation — the numerical methods necessary to solve the equations in a consistent manner, without gross approximations being made, simply do not exist at the present time — and may never exist. Some progress can be made, however, if one allows for physical simplifications, and if one artificially imposes the decoupling of specific physical processes from the analysis. In this manner, for example, the magnetic field and ion transportation component might be ignored, with just the fluid flow equations being considered. This simplifies the mathematical problem at hand, but it still fails to produce a readily tractable set of equations.

Figure 9.10. Gas and dust disk around the star HD163296. The series of bright concentric rings are separated by (darker) low-density regions where planet formation is in progress. Image courtesy of ALMA.

To account for the inevitable appearance of Titius–Bode-like laws, the mathematical problem at hand is that the equations describing the structure a finite flattened disk of gas and dust necessarily requires the appearance of density enhancements within the disk at intervals spaced according to a law of the form $R_n = R_0K^n$. Even when gross physical and mathematical simplifications are introduced within the analysis, however, illustrating such an inevitable outcome is far from straightforward, but some progress (perhaps) has been made in this respect by François Graner and Berengere Dubrulle [8, 9]. Indeed, in their analysis, Graner and Dubrulle strip the problem down to its bare bones and consider the possible configurations of an infinitely thin, cold-gas disk rotating about a central mass.

The results established by Graner and Dubrulle are tantalizing, but they hardly suggest a label change from Titius–Bode *rule* to Titius–Bode *law* Indeed, their analysis shows that, in the highly simplified situation, density enhancements occurring at locations corresponding to a geometrical progression are possible. While this result might at first seem encouraging, with Titius–Bode-like rules being shown as possible solutions to the equations, the analysis does not prove that they are the only solutions possible. This may seem like nit-picking, but that is the whole point of mathematical proofs — to be nit-picky. Furthermore, there is absolutely no guarantee that the analysis applied to the simplest possible set of equations will apply to the more realistic set of equations appropriate to a real and actual circumstellar disk. So, is there a theory behind the inevitable appearance of Titius–Bode-like laws within all planet-forming systems? At this stage, the answer is clear, and this is that there is no physical theory of circumstellar disk evolution, with respect to the size of planetary orbits, that requires the necessary appearance of Titius–Bode-like spacing rules. Furthermore, it is now clear that where planets might start forming in a specific natal disk is not necessarily where they will find a final and stable resting place. The manner in which planets can both chaotically, and systematically re-arrange their orbits will be explored in the following chapter.

References

[1]. Beech, M. and Peltier, L. The vulcanoid asteroids: Past, present and future. *American Journal of Astronomy and Astrophysics*, 5, 28–41, 2017.

[2] Levenson, T. *The Hunt for Vulcan*. Random House, New York, 2015.

[3] Acherson, D. *The Wonder Book of Geometry — A Mathematical Story*. Oxford University Press, Oxford, 2020.

[4] Morbidelli, A., *et al.* Evidence from the asteroid belt for a violent past evolution of Jupiter's orbit. *The Astronomical Journal*, **140**, 1391–1401, 2010.

[5] Penrose, L. and Penrose, R. Impossible objects: A special type of visual illusion. *British Journal of Psychology*, **49**, 31–33, 1958.

[6] Beech, M. *The Physics of Invisibility — A Story of Light and Deception*. Springer, New York, 2012.

[7] Gordin, M. *On the Fringe — Where Science Meets Pseudoscience.* Oxford University Press, Oxford, 2021.
[8] Graner, F. and Dubrulle, B. Titius-Bode laws in the solar system I: Scale invariance explains everything. *Astronomy and Astrophysics*, 282, 262–268, 1994.
[9] Dubrulle, B. and Graner, F. Titius-Bode laws in the solar system II: Build your own law from disk models. *Astronomy and Astrophysics*, 282, 269–276, 1994.

Chapter 10

Eccentric Bounds, Kirkwood Gaps, and Least Interactions

Kepler, as described in Chapter 6, sought an explanation for the spacing of planetary orbits in terms of the spheres that could be inscribed and circumscribed about the Platonic solids. These same polyhedra, he also argued, offering a natural explanation as to why there are six planets (Mercury through to Saturn) — that is, the five solids allowed for the construction of six very specific zones. The discovery of Uranus in 1781, of course, negated Kepler's polyhedral correspondence. Furthermore, through the 17th and 18th centuries, there grew an increasing suspicion among astronomers that the large gap between the orbits of Mars and Jupiter must contain a planet of some sort. Indeed, Kepler suggested as much in 1596. Such thinking, begs the question, however, as to what options are available for the placement of such a planet, and indeed, other planets within the solar system. This question will be considered in the following, first from a purely geometric, non-collision, perspective, and second from one including gravitational interactions.

10.1. Eccentric Bounds

The orbit of a planet, as announced by Kepler in 1609, is in the shape of an ellipse with the Sun located at one of the focal points (recall K1 in Chapter 6). Just like the Platonic polyhedra described in Chapter 3, the mathematical properties of the ellipse have been studied for many centuries, but all that we need to consider here are the essential terms. Accordingly, an ellipse is described by its semi-major axis a and eccentricity e. The eccentricity is a pure number derived according to the ratio of the position of the focal point OF from the center of the ellipse and the semi-major axis (Figure 10.1). That is, $e = OF/a$, and for an ellipse $0 \leq e < 1$. If the focal points are placed at the very center of the major axis, $OF = 0$, so a circle is obtained, and the circle is accordingly a special ellipse — an ellipse of zero eccentricity. As the distance of the focal points increases ($OF \rightarrow a$) from the center, however, the eccentricity approaches unity.

The ellipse, and its special case of the circle, are members of the large family of long-studied curves called conic sections. The other family members are the parabola (described according to $e = 1$) and the hyperbola (described according to $e > 1$). We need not be concerned with the parabola and hyperbola here, but they do have direct relevance to the study of cometary orbits.

The heliocentric distance $r = F_2P$ (see Figure 10.1) is described in terms of the semi-major axis a, the eccentricity e, and the true anomaly ν

$$r = a(1 - e^2)/[1 + e\cos(\nu)] \qquad (10.1)$$

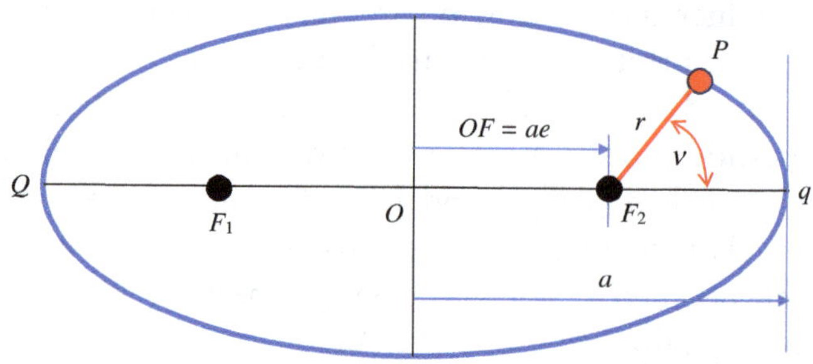

Figure 10.1. The ellipse is described in terms of the semi-major axis a and the offset of the focal points $F = F_1 = F_2$ from the center O, with the eccentricity being defined according to the ratio $e = OF/a$. For a planetary orbit, K1 indicates that the Sun is located at one of the focal points (the other is empty), r is the heliocentric distance, and the angle ν is the true anomaly. Additionally, the locations q and Q correspond to the perihelion distance and the aphelion distance, respectively.

From equation (10.1), a number of important orbital locations can be described. The closest point to the Sun, the perihelion distance q of a planet, for example, occurs when the true anomaly is zero, and accordingly

$$q = a(1 - e) \tag{10.2}$$

Likewise, the greatest distance of a planet away from the Sun, its aphelion distance Q, occurs when the true anomaly is equal to 180 degrees. Accordingly,

$$Q = a(1 + e) \tag{10.3}$$

With the two expressions (10.2) and (10.3) in hand, a geometrical stability condition can be set up between any two planetary orbits. Essentially, the condition to be established will be that which avoids direct encounters. Ignoring planet–planet gravitational interactions (just for the moment), the stability condition is simply that the orbits do not overlap. On this basis the planets will always avoid being physically in the same location at the same time, that is, they cannot collide. Given two planets one (the inner most) having semi-major axis a_1 and eccentricity e_1, the other (the outermost) having semi-major axis a_2 and eccentricity e_2, the collision-stability condition will be satisfied provided the aphelion point of the inner planet is smaller than the perihelion distance of the outermost planet (Figure 10.2), that is, $q_2 = Q_1$. Substituting now from equations (10.2) and (10.3) and using Kepler's third law (K3 in Chapter 6), the non-collision stability conditions require that

$$\frac{P_2}{P_1} = \left(\frac{1+e_2}{1-e_1}\right)^{2/3} \tag{10.4}$$

Having established the mathematical requirement for the non-collision stability condition in equation (10.4), an analysis of the planetary orbits within the solar system can be made. Indeed, the question

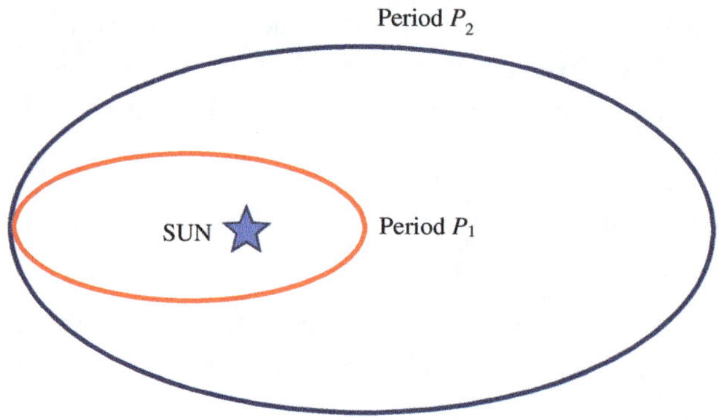

Figure 10.2. The geometry of the non-collision stability condition appropriate to two planets. The innermost planet (small ellipse) has orbital period P_1 and the outermost planet (large ellipse) has a period P_2. The non-collision stability condition is that the aphelion distance of the innermost planet is smaller than the perihelion distance of the outermost planet.

is as follows: How well do the actual planetary orbits avoid satisfying equation (10.4)?

Taking each planetary pair, Mercury–Venus, Venus–Earth, and so on, and assuming an egalitarian rule[1] such that $e_1 = e_2$ we now proceed to evaluate e_{max}, the maximum eccentricity that each orbit can have (for the given orbital periods P_1 and P_2) before collisions can geometrically occur. The numbers exhibited by the planets in our solar system are displayed in Table 10.1.

It has been said, more than once, over the years that there is nothing more boring to look at than a table of numbers. There is much truth behind this statement, but Table 10.1 does provide us with some useful insight, and it is worth more than a brief scan. Each column of the table hints at order, and remember it is the order that we are looking for. Scanning along column 2 of the table, for example, the eye catches a similarity between the characteristic values for the ratio P_2/P_1. With the exception of the Mars–Jupiter

[1]There is, of course, no absolute requirement of equality between the two eccentricities, but remember we are both imposing and looking for order in this analysis.

Table 10.1. The non-collision stability limit. Column 1 indicates the planetary pairs; column 2 provides the ratio of the orbital periods of the planetary pairs; column 3 shows the evaluation of equation (10.4) under the condition that $e_1 = e_2$ — this gives e_{max}; column 4 shows the actual orbital eccentricity of the individual planets.

	P_2/P_1	e_{max}	e (actual)
Mercury	—	—	0.2056
Venus	2.55	0.30	0.0068
Earth	1.63	0.16	0.0167
Mars	1.88	0.21	0.0935
Jupiter	6.31	0.55	0.0481
Saturn	2.48	0.29	0.0533
Uranus	2.85	0.34	0.0507
Neptune	1.96	0.22	0.0040

pairing, this ratio is characteristically about 2. Indeed, not including the Mars–Jupiter pairing, $P_2/P_1 = 2.23 \pm 0.43$. So, here we learn two things, first that for typical planetary pairings the orbital period approximately doubles from one planet to its next nearest neighbor. Second, we learn that this approximate period-doubling relationship fails badly between Mars and Jupiter, where the period ratio is some 3 times higher than the average of the other planetary pairings. This suggests that there is something happening between the orbits of Mars and Jupiter that requires further investigation. Column 3 of Table 10.1 converts the orbital period ratios to their equivalent e_{max} values [via equation (10.4)]. Again, excluding the value for the Mars–Jupiter pairing, we find $e_{max} = 0.25 \pm 0.06$. Characteristically, therefore, each planetary pair can occupy orbits (each of the same eccentricity) up to a maximum eccentricity of $e =$ 0.25 before geometric collisions can occur. The exception, again, is the Mars–Jupiter pairing for which the eccentricity of each orbit could be as high as 0.55 before collisions could occur. Column 4 of Table 10.1 shows the actual eccentricity of each planetary orbit, and it is readily seen that with the exception of Mercury, planetary orbits are almost circular (that is $e \approx 0$), and that in no case does any e_{actual} exceed e_{max}. In other words, the stability condition, based upon geometric collisions alone, is well satisfied within our solar system. None of the planets is positioned such that a collision is going to occur at the time of closest approach to its next nearest neighbor.

In terms of looking for order, Table 10.1 provides both comfort — the Earth is not remotely likely to collide with either of its nearest neighbors, Venus and Mars — and intrigue. The intriguing part, however, indicates the existence of a gap. The gap in this case is related to the Mars–Jupiter pairing. The question is as follows: Why does this particular pairing stand out so much in Table 10.1 as being different from the other planetary pairings? Having found a gap, in this case, a break (or significant difference) in the sequence of tabulated numbers, the question that comes to mind is as follows: Does the sequence break indicate the potential existence of another planet, or even planets, between Mars and Jupiter? Indeed, taking $e_{max} = 0.25$, two additional planets could be squeezed in between Mars and Jupiter, and direct collisions would still be avoided. These two hypothetical planets will have orbital periods of $P_{M1J} = 4.04$ years and $P_{M2J} = 8.69$ years respectively. But, of course, why stop at the Mars–Jupiter pairing? What about possible planets interior to the orbit of Mercury and planets beyond the orbit of Neptune? A rule is a rule, after all, so let's apply it.

We have reasoned on the basis of a scanning Table 10.1, with an eye set toward finding and imposing order, that there appears to be a non-collision stability rule at play in the solar system. This rule can be stated as follows:

> **Max-equal-*e*-Rule:** The spacing between adjacent planets in the solar system is such that their orbital periods are close to the ratio $P_{out}/P_{in} = 2.15$, corresponding to the condition that the orbital eccentricity of planetary pairs is such that $e_1 = e_2 = e_{max} = 0.25$ — see equation (10.4).

This rule is by no means exact (it is a rule of thumb), and it does not fully apply to the actual planets within the solar system, but it does allow for the generation of a number of hypothetical planets (planets that might be squeezed into existing orbital gaps), and it

Table 10.2. Hypothetical solar system with planets spaced according to the maximum equal-eccentricity rule for adjacent orbits. The comments given in column 4 are expanded upon in the text.

Planet	a (AU)	P (years)	Comments
M1	0.140	0.052	3:1 resonance with Mercury Vulcanoid asteroids
M2	0.232	0.112	2:1 resonance with Mercury Hypothetical planet Vulcan
Mercury	0.387	0.241	
Venus	0.723	0.615	
Earth	1.000	1.000	
Mars	1.524	1.881	
M1J	2.557	4.042	3:1 resonance with Jupiter Ceres at: $a = 2.77$ AU, $P = 4.61$ years
M2J	4.227	8.690	4:3 resonance with Jupiter M2J with $e = 0.25$ just overlaps the orbit of Jupiter ⇒ direct collision possible
Jupiter	5.203	11.86	Kirkwood gaps
Saturn	9.521	29.39	Sculpting of main belt asteroid orbits
Uranus	19.20	84.08	
Neptune	30.02	164.4	Pluto located at 2:3 resonance, with $a = 39.48$ AU, $P = 247.9$ years
N1	49.99	353.5	2:1 resonance with Neptune (Kuiper Belt Cliff)
N2	83.28	759.9	Scattered disk Close to the $N = 8$ location of the TB rule at 77.2 AU

suggests that the solar system could be structured according to the sequence shown in Table 10.2. In this table, two planets have been inserted interior of Mercury (these are labeled M1 and M2), two planets have been inserted between Mars and Jupiter (labeled M1J and M2J), and two planets beyond the orbit of Neptune (labeled N1 and N2).

In Table 10.2, a whole series of additional planets have been invoked, and the million-dollar question is as follows: Do any of them exist? Furthermore, it follows, that if none of them exist, the question is "why". The full discussion of Table 10.2 will take some time, and several features of the solar system will need to be described. I will begin with a discussion of hypothetical planets M1J and M2J and then move on to consider M1 and M2. Indeed, recall, the region containing M1J and M2J corresponds to the large radial gap, between the orbits of Mars and Jupiter, which has historically attracted the attention of astronomers. From a modern astronomical perspective, it is also the transitional region between terrestrial and Jovian planets; a gap where one might expect interesting things to be observed. But, for all this, it is assuredly a region of the solar system that is devoid of any planets — so, "why is this gap real" is the question that arises.

10.2. Kirkwood Gaps

In 1866, American astronomer Daniel Kirkwood described an oddity at play in the solar system. He was writing at a time 65 years after the discovery of the first asteroid, Ceres, and at a time when an additional 86 asteroids had been discovered. His announcement was made at a meeting of the American Association for the Advancement of Science in Buffalo and concerned the appearance of gaps in the semi-major axis distribution of asteroid orbits (Figure 10.3). In this case, the gaps indicated locations where no asteroids were found, and what Kirkwood realized is that the gaps fell at locations where the orbital period of an asteroid would be some simple fraction of the orbital period of Jupiter. In this manner, if Jupiter has an orbital period P_J then gaps can be found those locations where $nP_{ast} = mP_J$, where n and m and small integer numbers and P_{ast} is the orbital period of the asteroid. A distinctive dearth of asteroids, for example, is found at the heliocentric distance of 2.5 AU, and this location corresponds to $n/m = P_J/P_{ast} = 3/1$, indicating that the asteroid makes three orbits about the Sun for every one of Jupiter. Other distinctive gaps are found at $n/m = 5/2$, $7/3$, and $2/1$. These gap locations

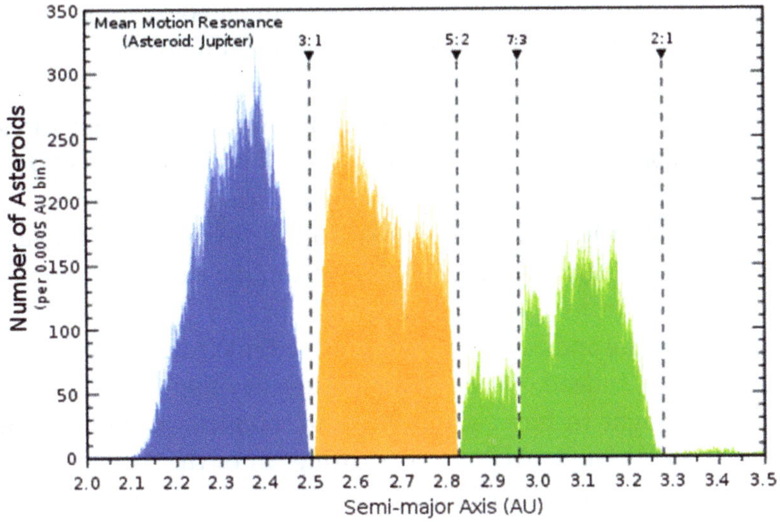

Figure 10.3. Main belt asteroid distribution — number versus heliocentric distance. Kirkwood gaps, where the number of asteroids drops to near zero, are seen in those regions centered on the 3:1, 5:2, 7:3, and 2:1 resonance locations with respect to Jupiter.

correspond to resonances, where periodic gravitational *tugs* by Jupiter occur at the same location within an asteroid's orbit — rather than at random locations within the orbit. Importantly for the future orbit of an asteroid, the former *tugs* add up over time, while the latter ones even out. Detailed calculations reveal that as a consequence of the periodic gravitational *tugs* in a resonant orbit, the eccentricity of an asteroid's orbit gradually increases. By increasing the orbital eccentricity, equations (10.2) and (10.3) indicate that the perihelion and aphelion points of the asteroid will move inwards, toward the Sun, and outwards, toward Jupiter, respectively. This effectively removes the asteroid from occupying a near circular ($e \approx 0$) orbit with a semi-major axis equal to the resonance location. The slow, but irrepressible action of resonances is highly effective at clearing gaps in the distribution of main belt asteroids.

The gravitational forcing term associated with an unstable resonant configuration is not especially large and is largely determined by the mass ratio of Jupiter and the Sun. An asteroid located at the 3:1 resonance position will have an orbital radius of about 2.5 AU (see Figure 10.3), and accordingly, at its closest approach to Jupiter, the slight gravitational *tug* from the planet will be of order ratio of the $F_{Jupiter} \approx (M_{Jjupiter}/M_{Sun})F_{Sun} \approx 0.001F_{Sun}$. These small periodic *tugs*, in the same location of the asteroid's orbit, toward Jupiter, in magnitude just 1/1000th that of the Sun, add up over time, and slowly but surely act to clear the asteroid away from the resonance location. Detailed calculations indicate a clearing time of just a few tens to hundreds of thousands of years are involved in clearing the 3:1 resonance location. Indeed, many of the meteorites that have landed on Earth are thought to have acquired their eccentric orbits (and perihelia smaller than 1 AU) after drifting into the 3:1 resonance region — the region often being described as the meteorite escape hatch of the asteroid belt.

Resonances are also responsible for sculpting gaps within planetary ring systems, such as those seen within Saturn's rings, and within the Kuiper Belt region. In the latter case, Neptune is the key planet, and, for example, the so-called Kuiper Cliff at about 48 AU from the Sun is situated at the 2:1 resonance location (see Figure 10.4). This concordance also explains why there is no planet corresponding to N1 at about 50 AU (see Table 10.2 and Figure 10.4). Planet N2 (see Table 10.2 and Figure 10.4) could possibly exist (as a dwarf planet) at 83 AU, but it would reside in the scattered disc region. Indeed, the scattered disk, beyond the Kuiper Cliff, remains a region of unclear origin. In principle, there is no reason why an object, moving on a low eccentricity orbit, should not be found within the region beyond 50 AU — but, as seen in Figure 10.4, the observations indicate a dearth of low eccentricity objects. Something has effectively stirred up the eccentricity of the objects that orbit in the scattered disk. The question, of course, is what. It is generally taken that the objects in the scattered disk formed much closer toward the Sun, but obtained their high eccentricity orbits as a consequence of the outward migration of Neptune (see the following section) — indeed, they were literally scattered there. There are, however, other factors that may have been in play, and it has been suggested that the orbits exhibited by

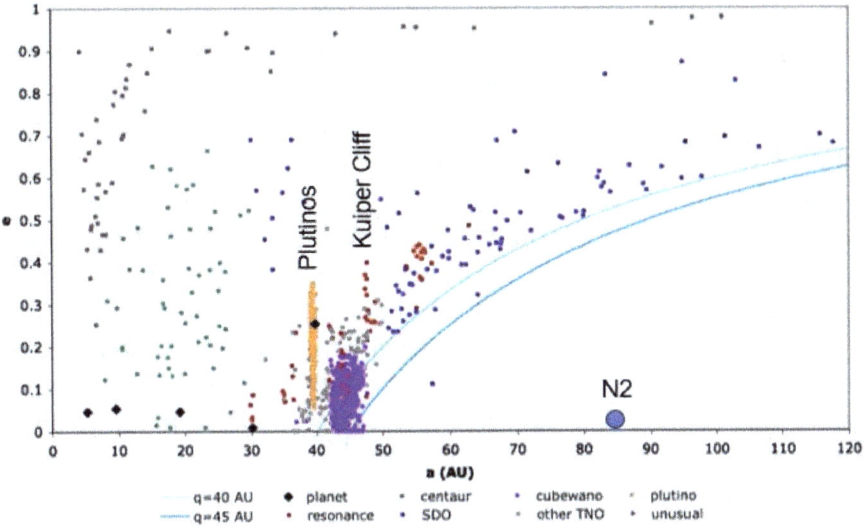

Figure 10.4. Eccentricity versus orbital semi-major axis for various solar system objects. The planets Jupiter through to Neptune are shown by black diamonds, and the classical Kuiper Belt Objects (a.k.a. Cubewano's) are shown as dark blue dots. The Kuiper Cliff, at about 48 AU, is located at the 2:1 resonance with Neptune. The vertical column of orange dots at 39 AU indicates the Plutino population with a 3:2 resonance with Neptune. The two curves to the right show the loci of objects with perihelion distances corresponding to 40 and 45 AU. The dot labeled N2 is placed according to Table 10.2.

scattered disk objects may have been sculpted by a small Earth or Mars-mass planet located beyond the orbit of Pluto, or by a close encounter between the Sun and another star, or even a close fly-by of the Sun by a rogue planet. On this basis, it may just be possible that our hypothetical N2 planet (or the TB rule $N = 8$ body — see Table 10.2) may yet exist,[2] or may have once existed in the distant past. Our hypothetical planet N2 is not the same as the recently postulated Planet 9, which has been invoked to explain the unexpected argument of perihelion clustering seen amongst the so-called Extreme Trans-Neptunian Objects (ETNOs). Only a few dozen ETNOs are

[2] Reasoning on the basis of Pascal's wager, I should add, that you read this existence prediction here first.

known, and they orbit the Sun with perihelia well beyond the orbit of Neptune on highly eccentric orbits. Intriguingly, however, it appears from the observational data that they have their perihelia situated in the same region of space. This observation, if actually true for all ETNOs, is suggestive of some form of gravitational sculpting. It may also, of course, be an example of the law of small numbers at play (recall Chapter 7). In this respect, Michael Brown and Konstantin Batygin, both at Caltech, have suggested in a series of publications that the orbits of ETNOs are shaped according to the presence of a 5 to 10 Earth-mass planet (a so-called super-Earth[3]) located in the outer regions of the solar system, with a putative orbit having a semi-major axis between 400 and 800 AU, and a moderately high eccentricity between 0.2 and 0.5 AU [1]. These values place the perihelion location somewhere between 200 and 600 AU from the Sun. At aphelion, it will be located between 500 and 1000 AU from the Sun — the orbital period will additionally be somewhere between 10,000 and 15,000 years. It is not entirely clear, given that it actually exists, where Planet 9 might have formed. Possibilities include capture from another star in the solar system's natal gas cloud, the capture of a rogue planet,[4] or within the solar system, forming close in toward the Sun and then being ejected outward. Extensive searches have not, as of yet, identified any candidate objects for Planet 9 — but the search for it, and more ETNOs continues apace. Alternative scenarios to Planet 9 have been developed to account for the apparent argument of perihelion clustering exhibited by the ETNOs (should this clustering actually be real, and indeed, many researchers have suggested that it is the result of observational selection effects [2]). The situation is gloriously unclear, calling for much more observational study, and the greater probing and development of theoretical models. Perhaps needless to say, Planet 9 has no contact with the

[3]This mass was chosen on the basis that super-Earths are abundant in exoplanetary systems.

[4]Such Free-Floating planets are observed, and while not in orbit about a parent star, such objects must have formed in the dust-gas disk around a protostar, and have then been ejected into interstellar space.

Titius–Bode rule, it not being accounted for as a former inner solar system object, nor as a captured object.

Unlike the situation in the 2:1 and 3:1 resonances, it turns out that the 3:2 resonance configuration can be either stable or unstable depending on how the objects are arranged. Indeed, with respect to Jupiter, the Hilda asteroid family is located at the 3:2 resonance location (at about 4 AU from the Sun). This latter family of asteroids completes three orbits about the Sun for every two orbits of Jupiter, but the asteroids remain in a stable configuration since the location of successive close approaches with Jupiter is spread uniformly around the orbit of the asteroid, and because the asteroid and Jupiter are in conjunction when the asteroid is at aphelion. Such configurations can only come about because of special capture conditions, and in general a 3:2 resonance will not remain stable. The other solar system situation in which a stable 3:2 resonance has come about is that between Neptune and Pluto. In this case, the resonance stops close encounters between Neptune and Pluto from occurring when Pluto is at perihelion. Indeed, there are a whole host of Plutinos in a 3:2 resonance with Neptune (see Figure 10.4) — Pluto just happens to be the most obvious of these objects. That stable 3:2 resonant pairings do occur in the solar system tells us, in no uncertain terms, that planet migration must have occurred at some stage in the early solar system (see the following). In terms of our hypothetical planets (recall Table 10.2) a reason is now evident as to why there is no planet located at M1J. Indeed, the specified location of M1J corresponds to the 3:1 resonance location with Jupiter, and accordingly, it falls within a gap-clearing zone. The other hypothetical planet allowed under the Max-equal-*e*-Rule, planet M2J falls at a location corresponding to a 4:3 resonance with Jupiter, and this, again, is an efficient clearing zone. Not only this, but M2J also falls in that zone where strong gravitational perturbations from Jupiter will be experienced.

Attempts have been made to explain the Titius–Bode rule according to a self-organizing principle built around the location of stable and unstable resonance locations. Markus Aschwanden (Lockheed Martin, at Palo Alto) has suggested [3], for example, that the arrangement of planets in our solar system follows a scaling rule that runs as follows:

$$\frac{R_{i+1}}{R_i} = \left(\frac{P_{i+1}}{P_i}\right)^{2/3} = \left(\frac{H_{i+1}}{H_i}\right)^{2/3} \tag{10.5}$$

where R_i and P_i correspond to the orbital radius and orbital period of planet *i*, and where, according to Aschwanden, the most dominant harmonic resonances are those given by (H_{i+n}/H_i), with $n = 1$, 2, or 3, and specifically (3:2), (5:3), (2:1), (5:2), and (3:1). The first two terms in equation (10.5) are no surprise, and nothing more than Kepler's third law — the third term, however, is predicated on the identification of numerical coincidences. In the case of the orbits of Venus and Mercury, for example, we have $(R_2/R_1) = 1.8682$, $(P_2/P_1) = 2.5535$ (using the numbers presented in Table 10.2). The choices we have for the dominant resonances given by Aschwanden are 1.5, 1.6667, 2.0, 2.5, and 3.0. Notice, none of the resonance ratios are exact equivalent to (R_2/R_1), but that for (5:2) is at least close. The problem here, however, is that the 5:2 resonance is generally unstable (see Figure 10.3), so, on this basis, one would not necessarily expect the orbit of Mercury to be stable. What this tells us is not that there is a strong or necessary self-organizing principle at work, but rather that the planets are simply organized according to randomly selected small number ratios in their orbital period. This is not to say that resonances are not unimportant in sculpting solar system structures — they are (see the following). Furthermore, however, the rule outlined by Achwanden carries no specific predictive power, there will always be some integer pair of numbers that are capable of approximately explaining the ratio of periods of adjacent planets, but, for all this, there is no absolutely imperative that a given pair of planets within a system must occupy some specific, small order resonance location.

Achwanden's claim of a fundamental self-organizing principle is not convincing, and indeed, it simply re-iterates a result presented some 70 years beforehand, by Archie Roy and Michael Ovenden, in

1954 [4]. Specifically, Roy and Ovenden studied the relationship between the period ratios of adjacent pairs of planets P_i/P_j within the field of integer fractions A_i/A_j. Defining the difference equation

$$\epsilon_{ij} = \left| \frac{P_i}{P_j} - \frac{A_i}{A_j} \right| \tag{10.6}$$

Roy and Ovenden investigated the distribution of ε_{ij}, with respect to the 46 pairs of integers A_i and A_j that can be selected from the integers between 1 and 7. From their study of the distribution of ε_{ij} values, perhaps not surprisingly, they concluded that the distribution of planetary orbits was "unlikely due to chance" — although they did have a decidedly strange definition of the term chance. Their interpretation of "by chance" essentially meant any system that did not form according to a Titius–Bode-like rule. What Roy and Ovenden have really shown in their analysis is that the period ratio of adjacent planetary pairs can be readily explained by low-order integer ratios, which is not the same as saying that planets are preferentially arranged (or formed) according to the selection of low-order resonances.

While Roy and Ovenden introduced the term chance to distinguish between those systems that are described by a Titius–Bode-like rule from those that are not, it is the case that chance can dictate the appearance of planetary systems that satisfy distinctive resonance chains. The now classic example of this is the exoplanet system associated with the star TRAPPIST-1a. In this case, seven planets are known to orbit an M-spectral type star, with a mass just 10% that of the Sun, and a size about that of Jupiter. The remarkable feature of the planetary orbits is that they are arranged in a resonance chain that runs approximately as 8:5, 5:3, 3:2, 3:2, 4:3, and 3:2. These are all relatively low-order integer ratios, and the system formation and its subsequent long-term stability requires some form of explanation. The most likely explanation for the organization seen in the TRAPPIST-1 system is that of planetary migration, and coupled-planet-disk interactions. While it is not credible that the planets formed where we actually see them now, it is the case that system evolution can effectively select-out and somehow maintain[5] stable resonances between planetary pairs. For all this, there is a simple linear rule for the orbital radii in the TRAPPIST-1 system, with the orbital semi-major axis being given according to the sequence number $N = 1, 2, 3, \ldots, 7$, as follows:

$$\log[a\ (\mathrm{AU})] = -2.0666 + 0.124N \tag{10.7}$$

while the linear rule provides a reasonable fit to the seven known planets, there is absolutely no guarantee that it will successfully predict the location of any 8th planet (should such a planet exist) — we are once again in the thrall of Wittgenstein's rule following paradox (recall Chapter 6).

10.3. Molchanov Frequency Chains

That adjacent objects within a planetary system can occupy orbits that have a stable small-order resonance structure is certainly observed, but it is not a necessary condition for orbital stability. Indeed, with a little perseverance, additional relationships between planetary orbital periods can be found. An example of such a number play at work is that presented by Albert M. Molchanov in 1968 [5]. Writing in the journal *Icarus* Molchanov argued that, "the structure of the solar system is determined by a table of resonance relations". Rather than using the orbital period P as the unit of measure for each planet, Molchanov considered the orbital frequency $\omega = 2\pi/P$, and from these terms, he established a series of linear relationships (see Table 10.3). To read Table 10.3, the first relationship is that between Mercury, Venus, Earth, and Mars, for which we have

$$\omega_{\mathrm{Mercury}} - \omega_{\mathrm{Venus}} - 2\omega_{\mathrm{Earth}} - \omega_{\mathrm{Mars}} = 0 \tag{10.8}$$

[5] Detailed numerical studies indicate that planetary-pair resonances do not, in general, hold once the disk has dissipated. Systems like TRAPPIST-1 are the exception rather than the rule.

Table 10.3. Molchanov frequency chains, providing a whole series of linear relationships between the orbital frequencies of the planets. See text for details.

	Mercury	Venus	Earth	Mars	Jupiter	Saturn	Uranus	Neptune	Pluto
Mercury	+1	–1	–2	–1					
Venus		+1		–3		–1			
Earth			+1	–2	+1	–1	+1		
Mars				+1	–6		–2		
Jupiter					+2	–5			
Saturn					+1		–7		
Uranus							+1	–2	
Neptune							+1		–3
Pluto						+1		–5	+1

The second relationship is between Venus, Mars, and Saturn, with

$$\omega_{\text{Venus}} - 3\omega_{\text{Mars}} - \omega_{\text{Saturn}} = 0 \tag{10.9}$$

Having constructed his table of relationships, Molchanov argues that "the presence of even one resonance is surprising, and double and triple resonances are extremely improbable. However, nature does not limit itself to these very convincing reasons but seemingly is drawn to the opposite extremes". Indeed, Molchanov claimed that, "the result obtained cannot be explained by chance". The problem, of course, with such a claim is that such relationships can easily be explained by chance — it is just a matter of one's outlook.

The relationships discovered by Molchanov (Table 10.3) are remarkable and provide model frequencies within 1% of the actual values. The problem, however, as pointed out in numerous critical responses to Molchanov's paper is that the relationships are not unique and that they fail to provide any physical rationale for their existence. Why, one can readily ask, should the orbital frequency of Venus bear any relationship with those of Mars and Saturn (our equation (10.9)). Indeed, one can find an alternative relationship that runs as follows:

$$2\omega_{\text{Venus}} - 3\omega_{\text{Earth}} - 3\omega_{\text{Jupiter}} = 0 \tag{10.10}$$

The other alternative relationship is that which exists between Neptune and Pluto, for which the 3:2 resonance indicates: $2\omega_{\text{Neptune}} - 3\omega_{\text{Pluto}} = 0$. This entry is, in fact, glaringly absent from Molchanov's table. Indeed, it was such reasoning that led Stanley F. Dermott and others to study the Molchanov relationships more fully. Writing in the January–March 1969 issue of the *Monthly Notices of the Royal Astronomical Society*, Dermott reasoned that the existence of alternate equations brought Molchanov's ideas into question. Furthermore, Dermott was able to find similar sets of linear equations for planetary systems having randomly generated periods. In short, Molchanov's claim of having found an improbable result was overstated and, indeed, just plain wrong. Albert Einstein is quoted as saying that, "nature hides her secrets because of her essential loftiness, but not by means of ruse". In the case of Molchanov's result, Einstein was wrong — nature, or more specifically numbers, are indeed devious. For all of the planetary harmony and interaction found by Molchanov, what he had actually found was a numerical chimera.

10.4. Introducing Gravity

The geometric stability condition discussed in Section 10.1 ignored the effects of any gravitational interaction between the planets. This can be readily accommodated, however, by what is called the Hill sphere radius. Introduced by American astronomer George Hill the radius of interest encloses that region in which gravitational interactions cannot be ignored. The Hill radius can be estimated by equating the orbital angular speed Ω_P of a point-mass (small moonlet) moving around a planet, to that of the planet's angular speed Ω_S about the Sun.[6] This will be the radius at which the Sun's gravitational influence equals that of the planet. In this manner, given a planet of mass M_P, with heliocentric orbital radius R_P (we shall assume $e = 0$, circular orbits), the Hill radius R_H will be given by the relationship $\Omega_P = \Omega_S$, giving

[6] Technically speaking, it is a 3-body gravitational problem.

$$\frac{2\pi}{P_{\text{moon}}} = \frac{2\pi}{P_{\text{planet}}} \tag{10.11}$$

which upon substituting for the orbital periods from Kepler's 3rd law [recall equation (6.4)] gives

$$\sqrt{\frac{GM_{\text{P}}}{R_{\text{H}}^3}} = \sqrt{\frac{GM_{\text{Sun}}}{R_{\text{P}}^3}} \tag{10.12}$$

giving the Hill sphere radius as

$$R_{\text{H}} = R_{\text{P}} \sqrt[3]{\frac{M_{\text{P}}}{M_{\text{Sun}}}} \tag{10.13}$$

A full 3-body gravitational interaction analysis introduces a factor of 1/3 into the cube root term in equation (10.13), but this factor need not concern us here. The region within which a planet's gravitational influence is important, according to equation (10.13), depends upon the mass of the planet and its heliocentric distance. Table 10.4 illustrates the Hill radii for the various planets within the solar system.

Detailed numerical simulations indicate that the effective clearing zone about any given planet amounts to at least several Hill radii, with NR_{H}, where $N \sim 4$ to 5. Under this expanded domain condition,

Table 10.4. Hill radii R_{H} (in astronomical units) for the planets within the solar system, and the objects that each of the planets will tend to influence.

Planet	R_{H} (AU)	Sculpting influence
Mercury	0.0022	Vulcanoid asteroids
Venus	0.0098	Vulcanoid asteroids
Earth	0.0144	Meteoroid streams
Mars	0.0105	Inner edge of Main belt asteroids
Jupiter	0.5129	Main belt asteroids/cometary orbits
Saturn	0.6077	Centaurs
Uranus	0.6764	Centaurs
Neptune	1.1204	Kuiper Belt

a second reason is found for the nonexistence of hypothetical planet M2J (as argued for under the Max-equal-*e*-Rule). That is to say, any planet (such as out M2J) that happened to be located within $N = 2$ Jupiter Hill radii would undergo strong gravitational interactions with Jupiter, and soon be ejected from the solar system, or at the very least be rapidly re-orbited.

10.4.1. *Hill's gravitational relaxation*

Hill investigated the origins of the Titius–Bode rule in a letter published in the 28 February 1970 issue of the journal *Nature*. Specifically, he set out to test the idea that the planets relaxed into their observed orbits following an early period of strong gravitational interactions. In this manner, he numerically simulated the orbital evolution of a series of planetary systems with initial starting distances and orbital eccentricities chosen at random. As each system evolved, due to mutual gravitational interactions, he examined the variation in the orbital periods of each adjacent planetary pair as a function of time. What he found was that over large fractions of simulated evolutionary time, planetary pairs tended to have period ratios situated reasonably uniformly between 9:4 and 8:3, with a secondary, smaller tendency, to fall near a period ratio of 3:2. From his numerical simulations, Hill concluded that in spite of a large variation in the initial planetary mass distribution and starting orbital conditions, systems tended to evolve, via mutual gravitational interactions, into quasi-stable states in which planetary pairs had orbital periods that were a small integer fraction of each other, with no specific requirement that resonances locations had to be selected for.

10.4.2. *What do the exoplanets tell us?*

Just as the planets within the solar system are found at distances well separated in terms of their Hill radii, so too, it turns out, are exoplanetary systems. These results are evidenced in a recent study and review conducted by Lauren Weiss (University of Montreal) and co-workers of some 900 planets contained in 355 multi-planet

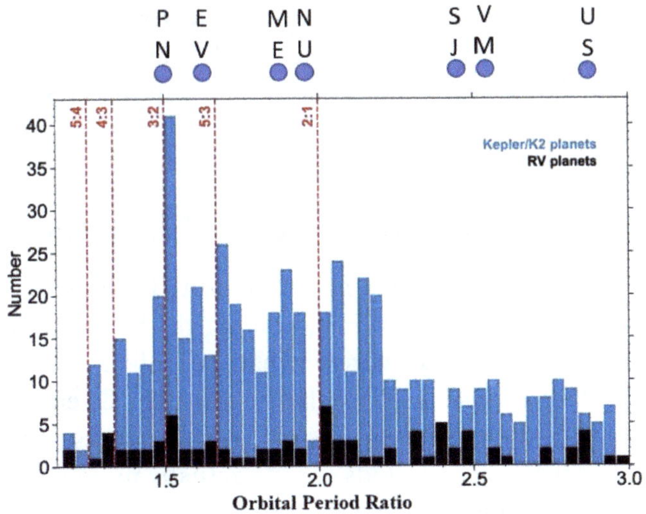

Figure 10.5. Distribution of orbital period ratios for adjacent pairs of planets in multi-exoplanetary systems. The bins are 1/24 wide, and the positions of several mean motion resonances are shown by vertical (red) dashed lines. Note the peak corresponding to the 3:2 captured resonance systems, and the minimum at the 2:1 resonance location. *Kepler* satellite data shown in blue, and radial velocity data shown in black. The orbital period ratios for solar system planets are shown along the top of the diagram (see Table 9.1). Image adapted from Deleuil *et al.* (2020) [7].

systems [6]. Looking at planetary masses, and orbital spacings, Weiss *et al.*, deduced the following:

- Planets tend to be separated by at least 20 mutual Hill radii.
- Planets are rarely closer than 10 mutual Hill radii.
- There is a correlation between planet size and planet spacing.
- In 65% of planetary pairs, the outermost planet is larger than the innermost planet.

These trends clearly indicate that gravitational interactions and system dynamics must play an important role in determining final planetary arrangements, but with planetary spacings being found that avoid close Hill radii encounters, with NR_H being such that $N > 10$–20. In other words, planets within stable systems like to avoid each other, and the more massive a planet, the further away are its companions. Likewise, it would appear from the multi-planet *Kepler* satellite data (Figure 10.5) that planetary pairs only rarely evolve into stable resonant-locked orbits — but such orbits are not excluded. The most common pairing being found is the 3:2 stable, but special capture, resonance (recall Section 10.2).

10.5. The Principle of Planetary Claustrophobia

Writing in the journal *Vistas in Astronomy*, Ovenden continued, in 1975, the analysis begun by Hill, coming to the same basic conclusions, but used them to derive what he called The Principle of Planetary Claustrophobia (PPC). In this manner, Ovenden noted that within an *N*-body gravitating system, such as the solar system, the bodies only spend short periods of time actually interacting in a strong and dramatic manner. Indeed, most of the time the objects are far apart and interact only mildly. Building upon this, Ovenden argued that the PPC dictates that planets should spend the majority of their time far apart, interacting as little as possible. In this manner, the PPC will favor orbits that are close to stable regions, but not necessarily orbits that are in resonance. While resonances will lead to sculpting and may result in stable orbits being found, the PPC seeks to minimize the gravitational effects of any one planet on all its companions.

As with the analysis presented by Hill, that by Ovenden is amenable to numerical experimentation, and the resultant PPC can be justified in physical terms. The justification of the PPC follows, in fact, from what has been termed the Principle of Least Interaction Action (PLIA). In this latter sense, the "Action" is a term represented by an integral taken along the path of system evolution, as it moves from some initial starting position to some end location at some specific finish time. The action A is defined in a very general manner by the system Lagrangian, L, where the Lagrangian is defined in terms of the kinetic, K, and potential, V, of the system, with $L = K - V$. For a free particle, for example, just the kinetic term applies, and $L = ½mv^2$, where m is the particle mass, and v its velocity. In general, then, the action A is given as follows:

$$A = \int_{Tstart}^{Tend} L\,dt \tag{10.14}$$

where T_{start} is the start time and T_{end} is the finish time. What the PLIA calls for is that a system evolves so that $\delta A = 0$, where the δ symbol indicates a small change (the more technical way of saying this is that the PLIA is a variational principle). Equation (10.14) is all very well and good, of course, but it completely lacks any predictive power. There is no clear way of showing that a system obeying a PLIA must evolve into one in which the orbital period of planetary pairs is described by low-order integer ratios, or indeed, exhibit a Titius–Bode-like rule spacing of final orbits.

What we can take away from the work of Hill, and Ovenden is that gravitational interactions are important in shaping planetary systems (a conclusion that no one ever doubted), but the manner in which the interactions and evolution take place is, depending upon unknowable circumstances, both violent and rapid as wells as subtle and slow. Gravitational interactions shape planetary orbits by wielding both a sledge hammer and a delicate planishing tool.

10.6. Vulcan and the Vulcanoids

Returning once more to Table 10.2, we have yet to deal with the two hypothetical inter-Mercurial objects M1 and M2. These hypothetical bodies, recall, are allowed in terms of the Max-equal-*e*-Rule, and the question is as follows: Have any such objects ever existed? It is certainly possible that objects could have formed interior to the orbit of Mercury, and remained on stable orbits over the current age of the solar system. This latter result is demonstrated in the detailed numerical calculations presented by Evans and Tabachnik in 1999 [8]. Indeed, they found that the zone between 0.1 and 0.19 AU from the Sun allows for highly stable orbits to exist — that is an object placed (or located) in this zone could in principle remain there for billions of years. The likelihood of such bodies existing, however, is open to question. Certainly, out hypothetical objects M1 and M2 fall close to the 3:1 and 2:1 resonance locations with respect to Mercury, and these are unstable resonances. It is entirely possible, however, that asteroid-like objects might have formed, and indeed may still exist in the region between 0.06 and 0.2 AU. The inner edge at 0.06 AU is set by a sublimation condition, with any stony object smaller than about 50 kilometers across occupying such an orbit being completely evaporated over the age of the solar system (4.56 billion years). Observational surveys indicate that no objects larger than about 20 kilometers across are likely to exist, at the present epoch, interior to the orbit of Mercury, but it is entirely possible that a small population of asteroid-like objects, the so-called Vulcanoid asteroids, does exist in the region interior to the orbit of Mercury. For all this, it can be safely stated that no planet (or even dwarf planet) sized objects exist in those regions indicated for our hypothetical M1 and M2 bodies.

10.7. Making the Solar System — A Modern Take

The field of planet formation has undergone dramatic growth during the last half-century. The growth is a result of both observational revelations, and the development of applicable physical theory — especially so in the area of numerical experimentation. Travel only a few generations of astronomers back in time, and planets were thought to be ultra-rare, and literally add-on features to just a few fortunate stars. It was commonly supposed that planets formed within a looping strand of material ripped from the outer layers of one star by the close passage of another; or that planets, having formed somewhere (and somehow) in the interstellar medium, were fortuitously captured by a star during a close encounter. Planets, then, had no natural place in the star formation process. In the modern era, the exact opposite applies. Planets are abundant; indeed, they are more numerous than the stars, and they are the natural byproducts of star formation. Planets, it is now clear, form within the circumstellar disks of gas and dust that surround newly forming stars (Figure 10.6 and recall Figure 9.10).

This is not the place to fully review the details of star formation and the processes that result in the collapse of gas clouds within dense cores of molecular clouds [9]. What interests us here is the fact that planets grow within the flattened disk of material (gas and dust) that forms around a protostar, and specifically, we are interested in

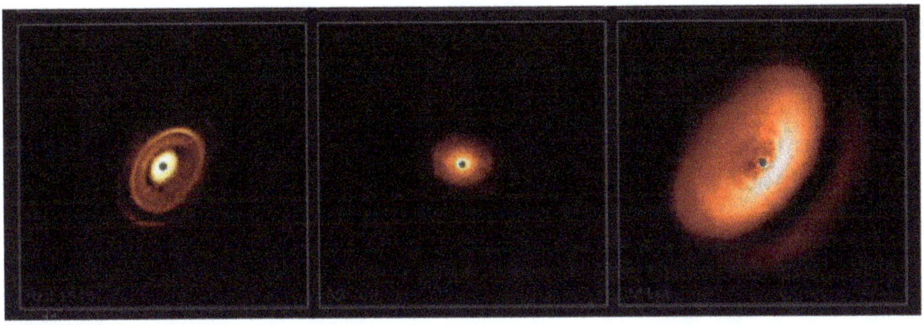

Figure 10.6. Three examples of protoplanetary disks, composed of gas and dust, surrounding newly forming stars. The concentric rings seen within the disk shown to the far left represent evidence for modification caused by ongoing planet formation. Image courtesy of ESO.

those processes taking place within the disk during the last few millions of years of its existence. At this stage, the protostar has basically acquired its final mass and is well on its way to becoming a *bona fide* star.[7] The planets grow within the disk through accretion, following the maxim that the rich get richer, it is the larger planetesimals and protoplanets that get ever larger over time. It is a chaotic smash and grab process, but ultimately a few dominant bodies begin to emerge. What sort of planets emerge from a protoplanetary disk will be dictated by the mass and extent of the disk, as well as the dominant make-up of the smaller planetesimals from which they are being formed. The disk mass literally determines how many in number, and how massive the planets might ultimately be (it is the raw building material). In parallel, what type of planet that can form in a disk is governed by the temperature of the disk. In this latter respect, the temperature in the disk will decrease with distance from the central protostar. This temperature gradient within the disk dictates the overall chemistry and determines what sorts of solid building materials can exist. Close toward the innermost edge of the disk, the temperature is high since it is located close to the parent

[7]That is, a star is taken to be an object generating energy within its interior through nuclear fusion reactions.

protostar, and in this region, only materials such as rock and iron can exist in a stable solid form. Further out, as the temperature drops, ice can begin to exist in a solid stable form, and further out still carbonaceous molecules of numerous kinds form. The key transition region is that surrounding the so-called ice line. In the solar nebula, it is estimated that the snow line (really a circle) was located at about 3 AU away from the proto-Sun. Inside the ice line, the predominant building material is rock and iron. Beyond the ice line, the predominant building material is composed of various ice. There will be some radial mixing of materials, but this basic composition divide, pivoting at the ice line, will dictate whether a Jovian planet or a terrestrial planet will ultimately form.

Many details are yet to be understood with respect to the planet formation process, but the essential idea for the terrestrial planets is that they form from the accretion of objects (planetesimals, and proto-planets) made predominantly of iron and stone, while Jovian planets grow through the initial formation of a massive icy core that begins to accrete gas directly and rapidly from the disk. The threshold for producing a Jovian planet is the formation of a core having a mass above about ten times that of the Earth. Once this threshold is passed then the core can begin to accrete and hold on to a massive gas envelope, with the final mass of the planet being determined by the amount of gas available for accretion in the disk, and the disk lifetime. Estimated growth times (based upon numerical simulations) for the various planets in our solar system are typical of the order of a few 100,000 years for Jupiter and a few 100 million years for Earth. The details are obscured by the fact that we do not know (nor will ever know) what the mass of the solar nebular was. In contrast, we do know how long the solar nebula lasted — about 10 million years. This latter number is determined through the laboratory analysis of materials contained within primitive meteorites.

While the scenario of accretion growth, within a disk of gas and dusk, along with the ice line division separating the formation zones of terrestrial and Jovian planets is, no doubt, broadly correct, there are still many details of planet formation that have yet to be understood. Furthermore, it is now clear that in many cases the location where we see a planet, in its orbit about its parent star, is not the same location in which it formed. Indeed, the very first exoplanet to be discovered, 51 Pegasi b, in 1995 indicated this fact perfectly — it being a half-Jupiter mass planet located in an orbit having an orbital period of just 4.23 days. This so-called hot Jupiter could not have formed so close to the parent star 51 Pegasi, which is a Sun-like star, but must have formed beyond the ice-line (at least about 3 AU from its parent), and then migrated inwards. The question, of course, is what drives the migration, and additionally, what stops it. Over 5000 exoplanets have now been discovered since 51 Pegasi b, and there appears to be a whole spectrum of planetary configurations, with hot Jupiters orbiting close-in, and cold Jupiters orbiting far our — both objects being problematic: the one in an orbit too hot for it to have formed there, the other in an orbit too far out for it to acquire a mass appropriate to a Jovian planet during the lifetime of the nebula disk. What is evident is that once planets have begun to form, they interact with the disk itself, and they gravitationally interact with each other. These two processes seek to re-arrange orbits, moving the planets away from the locations in which they originally began to form. Migration will not only result in gravitational scattering events, re-distributing orbits, or even ejecting objects (planetesimals, and planets) from the system, it will also effect the conditions under which other planets will form within the disk, robbing them of material, for example, that they might otherwise have accreted.

To understand what might have taken place in the solar nebula during the time of planet formation and migration, one has to make recourse to numerical modeling. This is now a standard practice, but it is a practice that provides numerous pitfalls and problems. Numerical simulations, of course, are only as good as the equations relating to the applied physical theory, the starting conditions applied, and the adopted simplifications (invariably) needed in order to make the computations tractable. Fortunately, in an N-body simulation of gravitational interactions, the dynamical theory is well understood, and the equations to be solved can be written down exactly. What cannot be described exactly in such simulations,

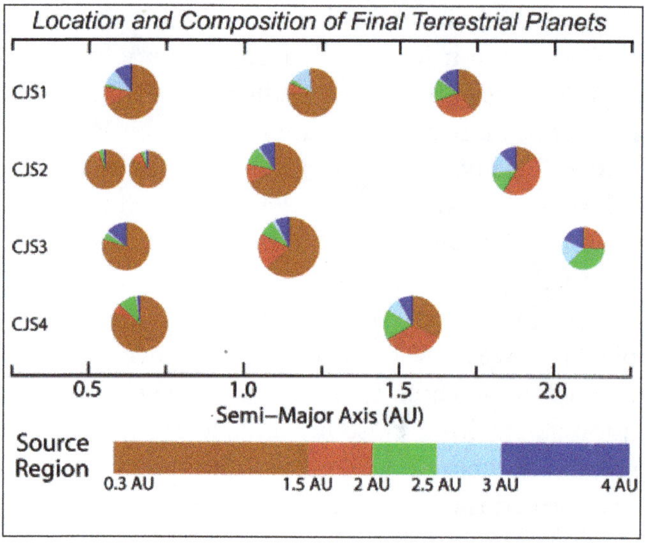

Figure 10.7. Terrestrial planet growth via accretion. The color bar at the bottom of the image indicates the initial positions of the seed particles. The results from four simulations are shown, revealing different planet mass and orbit outcomes. The pie-chart interior of each end planet indicates the percentage distribution of its source material, and the size of each planet's disk indicates its relative size. Image adapted from O'Brian *et al.* [10].

however, is the accretion process. Assumptions will always need to be made concerning the physical interaction of two objects when they are deemed to have collided — there is no specific theory to describe how much material is actually accreted, or ejected back into the disc, during a collision, and, indeed, how this varies with the size, speed, mass and the chemical composition of the bodies concerned. Be all this as it may, the simulations, in broad-brush form do provide insight into how planets form, and how planetary orbits evolve over time.

The numerical simulations all tend to start in the same manner. First, a number of test particles are placed on circular orbits, over a range of distances from the Sun. This is an attempt to simulate the dynamical organization of the first, smaller mass, planetesimals to form within the disk. These particles are then allowed to interact with each other gravitationally, and typically, some additional perturbing factors are introduced. In the case of modeling the growth of terrestrial planets, for example, it is generally assumed that Jupiter has already formed, and will accordingly add gravitationally perturbations to those test particles that are initially located close to it. All this is entirely practical and reasonable, and many research groups have studied the details of terrestrial planet growth. Figure 10.7 shows the results from an early study by David O'Brien (Planetary Science Institute, Tucson) and co-workers from 2006 [10]. In this case, the 1000 initial planetesimals (each assumed to be of the mass of Mars) are set within a source region extending from 0.3 AU to 4 AU, and Jupiter and Saturn are assumed to have already formed and move along circular orbits at heliocentric distances of 5.45 AU and 8.18 AU respectively (see the Nice Model discussion). The motion and accretion events are then followed and modeled until a set of dominant masses (the final planets) emerge. In this manner, a range of possible accretion outcomes are produced. Some runs produce a distribution (CJS1) that looks similar to what we see for the actual planets Venus, Earth, and Mars. Other runs produce two small-mass planets in the location where we find Venus (CJS2), others produce two large-mass planets (CJS4). The simulations also indicate that of order 50% of the initial planetesimals actually end up in a planet, with the remainder (about 5%) being absorbed by the Sun, and ejected from the system. The characteristic growth times can also be estimated, and this turns out to be of order 70 million years.

Simulations, such as those illustrated in Figure 10.7, provide some considerable confidence that at least some of the key formation processes have been identified. Even though the dynamics has been followed exactly, however, the planetary systems that result are determined by the random starting conditions and subsequent random wanderings of embryos. Run the simulation 100 times and 100 different planetary arrangements will emerge, but, importantly, the arrangements will tend to have similarities. The innermost planet will tend to be predominantly composed of those planetesimals initially located in the inner zone of the source region, while the outermost planet will tend to be composed of material that was initially situated in the middle and outer zone of the source region — but significant radial mixing of components is certainly predicted to have occurred. Recall, the significance of this relates to the basic temperature profile and chemistry of the protostellar disk out of which the initial planetesimals formed. Indeed, the material located beyond the ice line, at about 3 AU, will be ice-rich, and this will have a significant effect on the internal structure of a planet.

A question that must follow any series of numerical simulations is as follows: Do the end results look anything like reality? In the case of the terrestrial planets, the question concerns the location, mass, and essential composition of the planets Mercury to Mars. Again, in broad-brush form simulations such as those illustrated in Figure 10.7 do account for what is seen, but there are (as ever) issues. One of the issues that emerged from studies relating to terrestrial planet growth was that concerning Mars. The problem was not, so much, explaining the existence of a planet in the region between 1.5 and 2 AU from the Sun; rather it was that the simulations always seemed to predict the growth of a relatively high-mass planet. Coming in at just 1/10th the mass of the Earth, however, Mars is effectively the runt of the terrestrial planet litter.[8] This, so-called, stunted-Mars problem needed explaining: Was it a model quirk (or failing), or was it an indication that different initial conditions were required?

One way to limit the growth of Mars is to truncate the extent of the source region from which it can accrete material, the problem, however, is how to justify such a truncation. A possible answer to this is provided for by the, so-called, Grand Tack Model (GTM) [11]. In this scenario, Jupiter undergoes a significant inward migration, lasting for about 100,000 years, during which time it approaches as close as some 1.5 AU to the Sun (Figure 10.8). Jupiter, as discussed earlier, still forms beyond the ice-line, but now the

[8] This being said, Mercury comes in at half the mass of Mars.

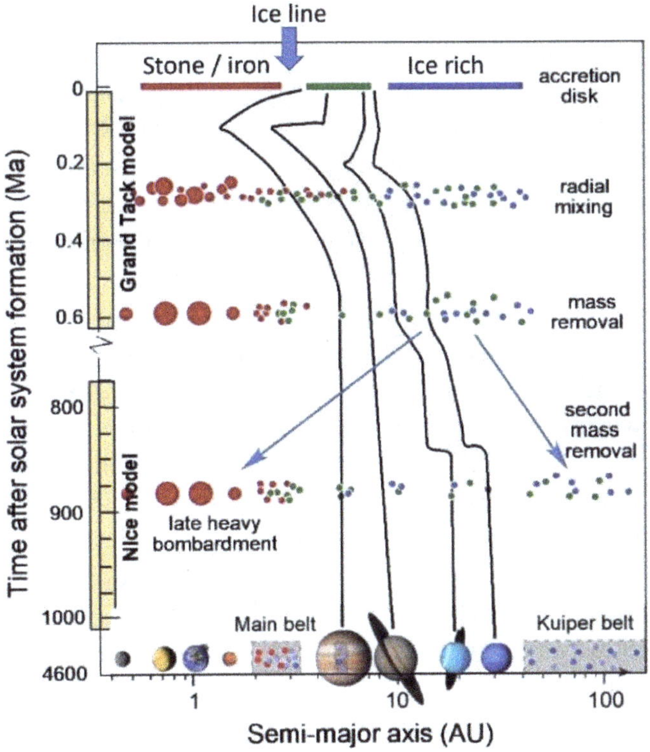

Figure 10.8. The grand tack model. The vertical axis indicates the time since planet formation in units of millions of years. The original disk distribution is shown as the red-green-blue line at the top of the diagram — the ice line is indicated at about 3 AU from the Sun. The basic chemistry is indicated as (red) stone/iron interior to the ice line and ice-rich (blue) beyond the ice line. Jupiter is taken to have formed at about 4 AU from the Sun. See text for further details.

details of its growth, specifically the manner in which it interacts with and captures gas from the disk, are taken into account.

Take another look at Figure 10.6. This image shows a series of concentric darker bands within the disk — these are the regions in which planets are forming. By robbing the disk of gas and dust, the planets effectively carve out low-density, depleted-matter zones, and the consequence of this, it turns out, is orbital migration. Two main types of migration are generally described, so-called, Type I and Type II, and the key physical principle at play is the exchange and transport of angular momentum between the planet and the disk material. As seen before, orbital angular momentum L is a quantity relating to the mass M, velocity V, and orbital radius R, with $L = MVR$. The important point here is that if the momentum changes, then it can be accounted for by concomitant changes in the mass, velocity, and/or radius terms. If a planet is being robbed of orbital angular momentum, then this can be accommodated by a reduction in its orbital radius — that is, the planet migrates inwards, moving closer to the Sun. In the case of low-mass stars, detailed numerical simulations indicate that the gas accretion results in the formation of a spiral-like, density-enhanced, structure within the disk — the arms of which proceed inwards and outwards into the disk, centered upon the planet. This extended feature results in the planet losing orbital angular momentum, and accordingly, it adjusts its orbit by moving inwards. Under such, Type I, conditions, detailed numerical studies indicate that the planet moves inward at a faster rate than the disk material. If nothing stops this type of migration then the planet will continue to migrate inward all the way to the inner edge of the disk, and potentially it could be absorbed by the parent star. Fortunately, in the latter case, as the mass of the planet increases, a threshold is reached beyond which the planet actually clears a distinct gap within the disk. This gap-opening process results in Type II migration, and the planet's inward migration rate is slowed significantly. It is thought that it is the process of gap-opening that enables the formation (that is survival) of a hot Jupiter. A single planet interacting with its natal disk will always end up losing orbital angular momentum, and will always migrate inwards. In the case of two or more interacting planets it transpires, however, disk interactions can result in outward migration, that is movement away from the parent star. It is this two-planet interaction mechanism that is proposed to operate in the GTM and which saved Jupiter in our solar system from becoming a hot Jupiter.

With reference to Figure 10.8, the GTM starts with the planetesimals distributed in a flattened disk, with the basic composition pivoting about the ice line at 3 AU from the Sun. Jupiter is taken to form very rapidly, within a few tens of thousands of years, at a distance of about 4 AU from the Sun. Under Type I migration, Jupiter begins to move inwards for a period of about 100,000 years. Over this same time interval, Saturn has also migrated inwards, moved closer to, and entered into a 3:2 resonance with Jupiter. By this time, Jupiter has migrated some 3 AU inwards from its initial starting point and has accordingly radially mixed, swept-up, and/or scattered out of the disk many of the rocky/iron planetesimals that might have otherwise gone into the formation of the terrestrial planets (as indicated in Figure 10.7). Once Saturn has entered into the 3:2 resonance with Jupiter, however, the pair of planets remain locked together, and disk interactions now cause the pair to move outward, away from the Sun.[9] This is the grand tack, or dog-leg path shown in Figure 10.8. After several 100,000 years, Jupiter and Saturn have regained their initial orbital radii, and after about half a million years in total, they attain orbital radii close to their presently observed values. At this stage, Jupiter and Saturn have passed through the region between 4 and 1 AU from the Sun twice. It is this clearing (mass loss and radial mixing) effect of the GTM that accounts for the Stunted Mars Problem and also accounts for the low total mass of the main belt asteroid region. Indeed, 1/3rd of the mass of the entire main belt asteroid region is contained in one object, the dwarf planet Ceres, and its total mass is only about 3% of that of Earth's Moon. At this stage, we are only halfway along the time history displayed in Figure 10.8. What follows the GTM, is even more planetary orbital rearrangement.

The problem that now needs addressing, is the growth times for Uranus and Neptune, and an explanation for what is called the Late Heavy Bombardment. The former problem has already been touched upon and relates to the accretion growth time of Jovian planets within the outer regions of solar nebular. If these planets formed at their observed locations, then their estimated formation times are well in excess of several tens of millions of years. Such a timescale, however, is far too long, given that the solar nebular only survived

[9] It is this same migration mechanism that results in the formation of systems such as TRAPPIST-1, in which a whole series of resonant orbits have built up.

for no more than about 10 million years. The way to solve this growth time mismatch is to have them form much closer toward the Sun, where the disk is denser and the accretion times are shorter, and then find a way of moving them outward to their presently observed locations. At this stage, however, with the solar nebula all but dispersed, there is no disk interaction mechanism to enable outward migration (as in the case of the GTM). What can happen, however, is the development of an orbital instability; the result of which is the rapid, and dramatic re-orbiting of the close-in formed Uranus and Neptune. This instability-driven, scenario is accounted for by what is called the Nice Model (NM) and is deemed to have occurred when the solar system was some 800 million years old. The initial Nice Model was developed at the Observatoire de la Cote d'Azur, in Nice (hence the name), France, in 2005, and it follows the development of an orbital instability brought about by Jupiter and Saturn entering into a 2:1 orbital resonance [12]. The relevance of the 800-million-year timescale relates to what is called the Late Heavy Bombardment (LHB), an epoch during which the Moon (and greater solar system) apparently underwent a brief period of intense cratering. Still controversial, and some researchers argue that it never even happened, the LHB requires a triggering mechanism that rapidly scatters and re-orbits the asteroids and icy planetesimals within the solar system. Under the Nice model, the trigger mechanism is the sudden re-orbiting of Uranus and Neptune (see Figure 10.9). Once again, the NM is driven by extensive numerical simulations, fined-tuned so as to produce the specified instability at the time that the LHB occurred. In order to achieve this, the NM sets the initial orbital radii of the four Jovian planets as 5.45, 8.18, 11.5, and 14.2 AU from the Sun. Surrounding the planets is an annulus, extending from 15.5 to 35 AU from the Sun, of icy planetesimals. The total mass of planetesimals is taken to be 35 Earth masses. Given these initial conditions, the numerical simulations reveal that gravitational scattering of planetesimals located on the inner edge of the disk results in the slow, but steady, outward migration of Saturn, Uranus, and Neptune. After several 100 million years, Saturn drifts into a 2:1 resonance with Jupiter, and it is this interaction that cocks the orbital re-distribution

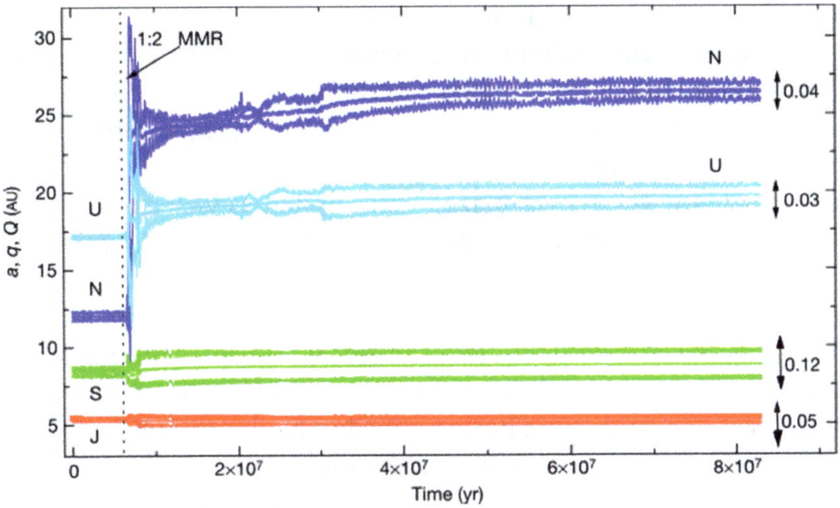

Figure 10.9. Nice model simulation of the orbital evolution of Jupiter, Saturn, Uranus, and Neptune. The vertical axis indicates the orbital semi-major axis (a), perihelion (q), and aphelion (Q) distances of each planet. The simulation timescale is given along the horizontal axis. The rapid re-orbiting of Uranus and Neptune comes about after Jupiter and Saturn lock into a 2:1 mean motion resonance (vertical dotted line). Note in this simulation an orbital ordering switch takes place, with Neptune (N) forming the interior to the orbit of Uranus (U). Image courtesy of Tsiganis *et al.*, Origin of the orbital architecture of the giant planets of the Solar System, *Nature* 435, 2005.

trigger. With Jupiter and Saturn locked in a 2:1 resonance, the eccentricity of Saturn's orbit begins to increase, resulting in it having closer interactions with Uranus. This process builds and builds until the orbit of Uranus is destabilized and rapidly re-orbited further away from the Sun. In this re-orbiting of Uranus, the orbit of Neptune is additionally changed and pushed deep into the annulus of icy planetesimals (see Figure 10.9). It is the re-orbiting of Neptune that triggers a large influx of icy planetesimals into the inner solar system, resulting in the LHB.

The NM provides a scenario that solves a number of problem issues. It solves the growth time problem for Uranus and Neptune by allowing them to form much closer-in toward the Sun, where the disk density is higher. It additionally provides for a trigger mechanism that allows for the re-orbiting of Uranus and Neptune, launching them into their currently observed orbits, and it offers an explanation for the LHB. It further allows (although not described here) for an explanation of the Plutinos that have been captured into a stable 3:2 orbital resonance with Neptune; the capture of the Trojan asteroids, which are in a 1:1 resonance with Jupiter, as well as the formation of the Hilda family of asteroids, which are in a stable 3:2 resonance with Jupiter. For all this, the NM generates more questions than it seems to answer. Problematically, however, the starting conditions have to be finely adjusted so as to provide the required time delay before orbital destabilization, and there is no theory or observations to guide us in determining initial planetary orbits or the extent and mass of the primordial annulus of icy planetesimals. In many ways, that the NM can be made to work at all in its best-selling feature. This, of course, is an uncomfortable position to be in, since there is no clear understanding of how unique the model results are. Certainly, between them, the GTM and the NM provide for a good story — they can account for the observations. Plato (recall Chapter 3) would be well-pleased with what astronomers have been able to read into nature. But, of course, the big question is, is it all, or any of it, true — is this really how the planets within our solar system found their presently observed orbital radii? Or, are we just finding engineered results within the vast landscape of a numerical fantasy land?

Numerical simulation is not the same as physical experimentation — the latter determines, measures, and codifies what is actually going on in nature, and the former mimics mathematically what might be going on. However, the GTM and the NM offer the possibility of understanding, at least in principle, some of the important processes that resulted in the formation of the planets about the Sun, and determined their final orbital distribution. The Titius–Bode rule bears no witness nor offers any understanding of the complexity of solar system architecture, as revealed in this chapter, and once again we are driven to dismiss it — the Titus–Bode rule is nothing more than a numerical coincidence and chimera. As pointed out by Raymond Pierrehumbert (University of Oxford), the word planet is derived from the Greek expression for "wanderer", and this now applies not only to their celestial motion, with respect to the background stars but also to their formation history [13].

References

[1] Batygin, K., *et al.* The planet 9 hypothesis. *Physics Reports*, **805**, 1–53, 2019.
[2] Shankman, C., *et al.* OSSOS VI: Striking biases in the detection of large semimajor axis Trans-Neptunian Objects. *The Astronomical Journal*, **154**, article 50, 2017.
[3] Aschwanden, M. Self-organizing systems in planetary physics: Harmonic resonances of planets and moon orbits. *New Astronomy*, **58**, 107–123, 2018.
[4] Roy, A. and Ovenden, M. On the occurrence of commensurable mean motions in the solar system. *Monthly Notices of the Royal Astronomical Society*, **114**, 232–241, 1954.
[5] Molchanov, A. M. The resonant structure of the solar system: The law of planetary distances. *Icarus*, **8**, issue 1–3, 203–215, 1968.
[6] Weiss, L., *et al.* The California-Kepler Survey. V. Peas in a Pod: Planets in a Kepler multi-planet system are similar in size and regularly spaced. *The Astronomical Journal*, **155**, article 48, 2018.
[7] Deleuil, M., *et al.* Observational constraints on the formation and evolution of Neptune-class exoplanets. *Space Science Reviews*, **216**, article 105, 2020.
[8] Evans, N. and Tabachnik, S. Possible long-lived asteroid orbits in the inner solar system. *Nature*, **399**, 41–43, 1999.
[9] Beech, M. *The Pillars of Creation — Giant Molecular Clouds, Star Formation, and Cosmic Recycling*. Springer, New York, 2017.
[10] O'Brien, D., Morbidelli, A., and Levison, H. Terrestrial planet formation with strong dynamical friction. *Icarus*, **184**, 39–58, 2006.
[11] Walsh, K., *et al.* A low mass for Mars from Jupiter's early gas-driven migration. *Nature*, **475**, 206–209, 2011.
[12] Tsiganis, K., *et al.* Origin of the orbital architecture of the giant planets of the solar system. *Nature*, **435**, 459–461, 2005.
[13] Pierrehumbert, R. *Planetary Systems — A Very Short Introduction.* Oxford University Press, Oxford, 2021.

Chapter 11

Remarkable Maths and the Maths of Remarkable

Mathematics is a place where you can do things which you can't do in the real world.

— Marcus du Sautoy

Definition. Remarkable — adj.: worth noticing, unusual.

In this final chapter, I want to look more closely at the properties of numbers and the ways in which mathematics, as a language of nature, is presented and used in developing models of the observable universe. Indeed, to the human mind, the universe offers an apparently endless supply of notable and remarkable objects and vistas. The world of mathematics is also replete with discoveries of unexpected wonder, and there are many, many circumstances of remarkable results with apparent (and real) connections occurring. Some of these latter connections are profound, others are nothing more than random parings — mathematical equivalents of the Titius–Bode rule — that seem to hint at the marvelous, but in actuality are empty and barren.

In terms of number play, the two classic constants are π and e. The first is known to almost everyone, the second perhaps less so. Likewise, π is ancient, while e is relatively new in terms of human usage. Both numbers are ubiquitous, being found literally everywhere in mathematics, physics, engineering, and indeed, the human-derived structure of nature. They often appear when least expected, and they hint at relationships and structures, between what are otherwise contrasting quantities, in mysterious ways. They are both irrational numbers, with infinitely extending decimal expansions, and yet, remarkably, one can manipulate approximations of them with a standard pocket calculator. Indeed, when it comes to numbers people can do obsessive things.[1] Indeed, calculator-button pressing can become an adventure full of surprising results. With no specific notion in mind, for example, I chanced to find, in a matter of minutes, with my calculator the unexpected result that $(\pi^4 + \pi^5)/e^6 = 1.000000042...$ this is a number so close to being an integer that one is almost convinced that there must be something interesting going on. But no, there really is nothing mathematically interesting taking place — it is none-the-less a notable and unexpected result.[2] For all this, it is the sort of result that has inspired (or, more correctly, corrupted and misguided) some investigators to promote the strangest of ideas. For example, let us make up a planetary distance rule, in which the order of the planetary distances $D(n)$ is given by the representation

$$D(n) = \frac{1}{4}\left(\frac{\pi^4 + \pi^5}{e^6}\right) \times S_n \tag{11.1}$$

where, for $n = 1$–9, $S_n = 1, 3, 4, 6, 10, 20, 40, 80, 120$, respectively. In this sequence, $n = 5$ corresponds to the mid-point location, $D(5) = 2.5$, of the main belt asteroids. There is no physical rhyme or reason to equation (11.1), it is entirely made up, but it has a form not wholly dissimilar to the planetary spacing law presented by Pliny

[1]My favorite such story is that concerning Mrs. Marva Drew of Waterloo, Iowa, who between 1968 and 1974 used a manual typewrite to type out every digit between 1 and 1 million. When asked why, she replied that she liked typing. Using three columns per page, a total of 2473 sheets of paper were required to complete the project.

[2]Even more unexpectedly, I subsequently discovered that the relationship has been discovered by other button-pressing explorers.

Chase[3] in 1873. It is the number sequence S_n that makes (11.1) appear to work, and it, of course, is entirely contrived to do exactly that. For all this, S_n does have some seemingly intriguing properties — it is a little like a Fibonacci sequence[4] with occasional "jumps" added in. There is, however, no specific formula for generating S_n. We have no idea what number (if it were required) would follow $S_9 = 120$ — it could be anything, and cannot be predicted on the basis of what has gone before in the sequence. We are, with equation (11.1) in the realm of Wittgenstein's rule-following paradox. In terms of describing planetary orbital spacings $D(n)$, equation (11.1) is a complete and utter sham, it contains no physical meaning, and the mathematical terms are just a flimsy façade covering over an ugly foundation.

What about other expressions that provide unexpected numerical results? The so-called Ramanujan number provides the following relationship between, e, π and, of all numbers, the square root of 163, with

$$e^{\pi\sqrt{163}} = 262537412640768743.99999999999925007259719820 \quad (11.2)$$

Incredibly, for so it appears, this exponential power is just a hairs-breadth away from being an integer. Indeed, this number was the focus of an April Fools joke conducted by mathematician Martin Gardner in *Scientific American*. Gardner's column for April 1975 indicated that the expression had, in fact, been shown to be a true integer, "as conjectured by Ramanujan". Actually, Ramanujan made no such conjecture, but he did notice the remarkably, near-integer properties of $e^{\pi\sqrt{43}}$ and $e^{\pi\sqrt{67}}$. Now, interestingly, in this case, the remarkable near-integer properties of the numbers studied by Ramanujan, and of $e^{\pi\sqrt{163}}$, are quite profound. Indeed, these

[3] In developing a 'vibrating' nebular theory for planetary spacings, Chase (*Proceedings of the American Philosophical Society*, **13**, 143–5, 1873) argued that planets would form at nodal points distributed according to a distance rule $R(N) = (\pi/32)(1 + \pi N)$, where $N = 1, 2, 3, 5, 9, 17, 33, 65, 97$. The N values were specifically chosen (not derived by any theory) in order to match the observed radial distribution of the planets within the solar system.

[4] This well-known sequence runs 1, 1, 2, 3, 5, 8, 13, 21, 34, 55, ... where each new term — after the second one — is the sum of the previous two terms.

numbers relate to a conjecture proposed by the German mathematician Carl Friedrich Gauss[5] in the early 19th century, and a proof (later requiring a few minor corrections) of the conjecture by an amateur mathematician, Kurt Heegner, which appeared in the mid-20th century. The conjecture at play has tendrils that stretch deep into the vast ocean of number theory, and specifically the factorization of whole numbers into prime number products.

The study of prime numbers is ancient, vast, subtle, ugly, and complex, and we cannot hope to fully illustrate its wonders here. However, we can (hopefully) provide some flavor of the luscious pickings that have emerged from its depths. To begin, prime numbers are defined as those natural numbers greater than 1 that cannot be expressed as the product of two smaller natural numbers. The first ten primes are accordingly 2, 3, 5, 7, 11, 13, 17, 19, 23, 29, and it was a classic proof by Euclid which established the fact that there must be an infinite number of primes.

Proof: There is no greatest prime number: Assume that there is a last prime number N. Take all the prime numbers between 2 and N and generate the product $B = 2 \times 3 \times 5 \times \cdots \times N$. Next, add 1 to B in order to produce the number $C = B + 1$. The number C is not divisible by any of the primes less than N, because N itself is divisible by all the primes. Now, since no two consecutive numbers can have a prime factor in common, so either C is a prime — which contradicts the assumption that N is the last prime — or, it is the product of smaller primes not included in the generation of the product B — which contradicts the construction methodology for B.

Proof by contradiction, accordingly, provides us with the knowledge that the primes are scattered amongst the natural numbers all the way to infinity. But there is more — much more — to the prime number story. Indeed, the mystery of the primes begins with its very first member, the number 2. This prime is strange since

---

[5] Often described as the *Prince of Mathematicians*, and *the Archimedes of the 19th century*, in accord with his many significant contributions to mathematics, physics and astronomy.

it is the only even prime number — the infinity of all other prime numbers being odd. Picking the primes out of the natural numbers is easy in the first few cases, but as the numbers increase in size so it becomes more time-consuming to test for primality. Unfortunately, there is no specific formula that generates all the primes smaller than some given value. For all this, however, there are ways of identifying all the primes smaller than some given limit. The oldest such method, the Sieve of Eratosthenes, is named after the Greek philosopher Eratosthenes and works by list deletion through successive divisions. If, for example, we take the first 25 integers

1, 2, 3, 4, 5, 6, 7, 8, 9, 10, 11, 12, 13, 14, 15, 16, 17, 18, 19, 20, 21, 22, 23, 24, 25

and underline all those numbers divisible by 2, we have

1, 2, 3, 4, 5, 6, 7, 8, 9, 10, 11, 12, 13, 14, 15, 16, 17, 18, 19, 20, 21, 22, 23, 24, 25

now find (underline) all those numbers divisible by 3 so that

1, 2, 3, 4, 5, 6, 7, 8, 9, 10, 11, 12, 13, 14, 15, 16, 17, 18, 19, 20, 21, 22, 23, 24, 25

next, we find (and underline) all those numbers divisible by 5 so that

1, 2, 3, 4, 5, 6, 7, 8, 9, 10, 11, 12, 13, 14, 15, 16, 17, 18, 19, 20, 21, 22, 23, 24, 25

By removing all the underlined numbers (i.e., all those terms that are divisible by numbers smaller than 25), we are left with just the prime numbers less than 25 — namely: 1, 2, 3, 5, 7, 11, 13, 17, 19, 23. This procedure is easy to implement on a numerical computer, and, for example, it can be stated with certainty that there are exactly 50,847,478 primes below 1,000,000,000.

Acting as the building blocks of the natural numbers, every integer can be factored into a unique product of primes, so, for example, $2020 = 2 \times 2 \times 5 \times 101$. This property of each integer having a unique prime factor sequence is encapsulated in what is grandly known as the *Fundamental Theory of Arithmetic.* It is additionally conjectured that every even integer greater than 2 can be expressed as the sum of two primes — with, e.g., $2020 = 2017 + 3$. This latter conjecture, known as the Goldbach conjecture, was first described in 1742, and it still requires a proof of statement.[6] Additionally, it appears that every integer greater than 5 can be written as the sum of three prime numbers — with, $2020 = 2011 + 7 + 2$, for example. The list goes on and on, and the prime numbers continue to surprise with their simplicity, their complexity, and their reach. While many mathematicians have tried to find straightforward (even complicated) formulae for generating the primes, none have ever succeeded, and it seems a safe bet to say that none ever will. There are some remarkable sequences, however, that add to the allure of the prime numbers. The 17th-century mathematician Marin Mersenne, for example, constructed what are now called Mersenne numbers M_N such that

$$M_N = 2^N - 1 \tag{11.3}$$

This formula produces a prime number when $N = 2, 3, 5, 7, 13, 17, 19, 31, \ldots$, giving the Mersenne primes 3, 7, 31, 127, 8191, 131,071, 524, 287, and 2,147,483,647, respectively. When introducing his numbers, Mersenne was building upon a theorem in Euclid's ninth book of the *Elements*, which states that: if $2^N - 1$ is a prime, then $2^{N-1}(2^N - 1)$ is equal to the sum of its proper divisors.[7] When $N = 5$, for example, $2^5 - 1 = 31$, which is a prime number, and 2^{N-1}

[6] Indeed, providing a proof to the Goldbach conjecture would make one both famous and rich, with substantial prize money being offered for an acceptable proof. That the conjecture is true seems certain, and numerically it has been shown to be true for all numbers smaller than 4×10^{18}.

[7] A proper divisor of the number M is a positive divisor that leaves no remainder.

$(2^N - 1) = 16 \times 31 = 496$. The proper divisors of 496 are 1, 2, 4, 8, 16, 31, 62, 124, 248, and this list of divisors does indeed sum to 496.

Clearly, many primes are missed in Mersenne's formulation, and it certainly does not work for all values of N. When $N = 4$, for example, $M_4 = 2^4 - 1 = 15 = 3 \times 5$ which is not a prime number. Pierre Fermat looked at a slightly more complicated arrangement and considered the function F_N

$$F_N = 2^{2^N} + 1 \qquad (11.4)$$

and found that it produced prime numbers for $N = 0, 1, 2, 3$, and 4, corresponding to 3, 5, 17, 257, and 65,537, respectively. It is thought, but not yet proven, that F_4 is the last of the Fermat primes. Once again, however, Fermat's formula produces only a limited selection of primes. Notably, however, to continue our story, a young Gauss, just shy of his 19th birthday, discovered a remarkable result concerning the geometrical construction of the heptadecagon (the 17-sided polygon). Indeed, Gauss was able to show that the ruler[8] and compass construction (Figure 11.1) of an n-sided regular polygon, where n is a prime, is only possible if n is a Fermat prime. Here was a completely unexpected and exciting result linking prime numbers to Euclidean geometry. Prior to the discovery by the adept Gauss, Euclid (2000 years earlier) had outlined construction techniques for 3 (equilateral triangle), 4 (square), 5 (pentagon), and 15-sided regular polygons. Technically, of course, there are a great number of ruler and compass constructions. Taking the technique for constructing the square (the 4-sided regular polygon), then one can simply keep bisecting the angles to produce the 8, 16, 32, 64, 128, ... sided regular polygons. The limiting factor in practice is the physical process of drawing arcs and lines — but remember, the discoveries of Euclid and Gauss relate to perfect mathematical rulers and compasses.

[8] Mathematicians generally refer to a ruler as being a straight edge. The idea here is that a ruler is specifically a measuring device, while a straight edge is simply a means of drawing a straight line.

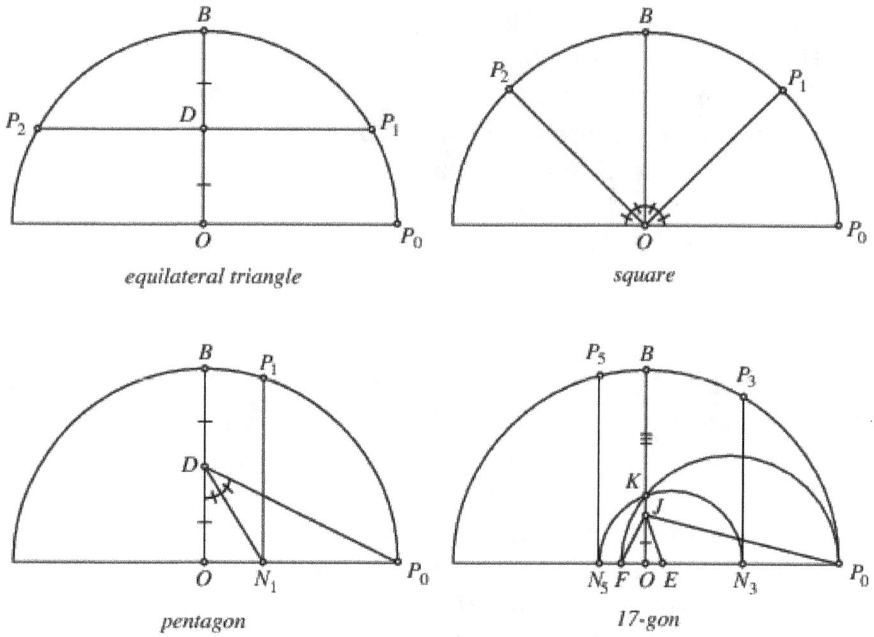

Figure 11.1. Compass and ruler methods for constructing the sides of the equilateral triangle (upper left), the square, the pentagon, and the heptadecagon (lower right). Image courtesy of Wolfram MathWorld.

It is the idea, rather than the reality, that is important to a mathematician. For all this, it is remarkable to think that there are ruler and compass construction techniques for the 257 and 65,537-sided polygons, albeit that the end results look, to the human eye, just like a circle. There may yet be even more extreme n-gon constructions to decipher — if, that is, there are any Fermat primes beyond F_4. Indeed, it is one thing to know that a construction is theoretically possible, but a very different think to find and annotate the practical methodology. Magnus Paucker and Friedrich Richelot independently found and published the solution methodology for the 257-gon in 1832, while that for the 65,537-gon was described by Johann Hermes in 1894. The construction outlined by Hermes was the result of a Herculean effort taking some 10 years to complete. For Hermes, the task was clearly an obsessive one, and he recorded his progress, from 1879 to 1889, in a *diarium* containing, when finished, 221 large manuscript pages. The text is dense and interspersed with extensive tables and spidery diagrams — to the unbeknownst reader it is as unreadable as the Voynich Manuscript. Simply entitled, "Diary about the division of the circle", the *diarium* contains page after page of technical reading, and it has never been published in its full form. Indeed, the manuscript is now kept in a specially constructed case, dull-brown on the outside, with a blue and white stripped interior lining, held in the archive of the Mathematics Institute at Gottingen University. The case for the Hermes manuscript is seemingly symbolic: it contains and preserves the results of an obsessive mind at play, but like a coffin (a final casing) it isolates the internal madness from the human world.

A particularly productive formula for generating primes was found by Leonard Euler in 1742, at which time he noted that the function $P(N)$, such that

$$P(N) = N^2 + N + 41 \tag{11.5}$$

generates prime numbers for all $0 \leq N \leq 39$. When $N = 40$, however, $P(40) = 1681 = 41 \times 41$ indicating the appearance of a non-prime number. Euler's formula provides for a marvelous run, but it does not capture every prime. Likewise, the formula $F(N) = N^2 - 79N + 1601$ generates primes for all $0 \leq N \leq 80$. In the same way that the Titius–Bode rule has formed an army of enthusiasts and a veritable cottage industry of tinkerers, so too has the computational game of finding the largest known prime number attracted a host of participants. This is a non-trivial game, even with modern-day computers, since the numbers being considered are now very large. The formula developed by Mersenne is a favorite to study, and as this page is written the record holder is the prime number generated for $2^N - 1$ when $N = 77{,}232{,}917$ — providing for a prime number made-up of 23,249,425 digits. This result required 6 days of non-stop CPU time[9] to verify. This record prime will no-doubt be superseded soon, and the game will play-on. For all the intrigue of prime numbers, however, such results seem hollow, and one cannot help but think that such knowledge is entirely useless and carries no mathematical or scientific profundity. The result is cold and mechanical, and there is no real beauty in it — such milestones are but a consequence of winning, albeit briefly, a hard-played mind game. This is not, in any way, a criticism: useless mind games can be fun, and entertaining, but they are only occasionally enlightening. Apart from applications in encryption and cyber security, both areas predicated on paranoia and self-interest, humanity has found no practical use for the primes. Likewise, nature has found little utility in the application of prime numbers. There are, however, species of cicada that only emerge to mate at intervals corresponding to prime numbers (typically 13 or 17 years), thereby, it is argued, to off-set any locking-in of a predator mating cycle. This, at first glance, seems like a good strategy for survival, but, of course, it makes no difference when the cicadas really emerge, whether at an odd, even, or prime number interval, since as soon as they do emerge the predator species that are then present will simply gorge themselves on the newly appeared bounty. There would be a survival advantage to the prime number mating-emergence interval if just one or two specialist predators ate cicadas and cicadas only, but this is not the case — cicadas are eaten by virtually any animal that can get hold of them. Beyond Earth, and the existential woes of cicadae, the generating sequence for the Titius–Bode rule (recall Chapter 9), out to planet Neptune, is 4, 7, 10, 16, 28, 52, 100, 196, 388, and of these, only the 2nd term is a prime number. The prime orbit is that corresponding to the semi-major axis of Venus. Prime numbers may be the building blocks of mathematics, but they are not foundational structures of the

[9] This computational result was found by Jonathan Pace in late 2017. The discovery, and record prime number with 23,249,425 digits, are the culmination of 14 years of effort by Pace — for which kudos is due. The search for so-called Mersenne primes is part hobby, part quest, and partly the search for the glory of holding an unusual record.

Titius–Bode rule, nor the cosmos. Indeed, one cannot help but feel that the physical impotency of the prime numbers reveals something profound about the relationship of mathematics to the universe.

With all the above in place, let us now return to Ramanujan's number, and the narrative concerning $e^{\pi\sqrt{163}}$. In forming his conjecture, Gauss was interested in identifying Pythagorean triplets (recall Chapter 2) and this led him to the consideration of how the sum $a^2 + b^2$ might be factorized, and specifically how the sum might be factorized in terms of a unique product of primes. These considerations eventually led Gauss to consider imaginary numbers — that is, numbers constructed around the square root of –1. In contrast to other numbers so-far described, where one can say, or visualize the expression, I'll take three of those, or 1/7th of that, or subtract –4 from the total of this, there is no such corresponding entity as $i = \sqrt{-1}$, it is an entirely imagined number having no tangible realization — it simply poses a mathematical action. For all this, imaginary (or complex) numbers of the form $z = a + ib$, where a and b are real numbers, are profoundly useful in mathematics, mechanics, and physical theory. Specifically, however, Gauss noted that by introducing this imaginary extension of the number field he could construct the factorization: $a^2 + b^2 = (a - ib)(a + ib)$ — this follows since when expanding the brackets, $i \times i = -1$. The question then was as follows: Could this factorization be cast in terms of a unique product of primes? Before answering this, a little additional notation is required. This involves the introduction of the, so-called, quadratic field $Z[i] = a + ib$, where $i = \sqrt{-1}$ and a and b are considered to be whole numbers. With this notion in place, Gauss found that unique prime factorization[10] could only take place for a limited range of quadratic fields $Z[d]$, and specifically so when $d = \sqrt{-1}$, $\sqrt{-2}$, $\sqrt{-3}$, $\sqrt{-7}$, $\sqrt{-11}$, $\sqrt{-19}$, $\sqrt{-43}$, $\sqrt{-67}$, or $\sqrt{-163}$. Any other values of d produced non-unique factorizations. If $d = \sqrt{-6}$ is applied, for example, then $15 = 3 \times 5 = (3 + \sqrt{-6})(3 - \sqrt{-6})$, providing for two different prime

---

[10] In this discussion, I am skipping over many important considerations and deep technical details.

factorizations: 3 and 5, and $a = 3$ and $b = 1$ in $Z[\sqrt{-6}]$. With this remarkable result in hand, Gauss made a wild (but intuition-based) conjecture. The range of values for which $Z[d]$ provides for the unique prime factorization condition was finite and indeed, ended at $d = \sqrt{-163}$. There are just nine numbers (1, 2, 3, 7, 11, 19, 43, 67, 163), out of the infinitely many whole numbers available, for which the unique prime factorization condition was possible. This was a conjecture since, at the time, Gauss could not be certain that there weren't additional numbers, indeed an infinite number of numbers, beyond 163 for which unique prime factorizations might be found. It was later shown through the work of Kurt Heegner, in 1952, that Gauss's conjecture was true — the sequence truly does end at 163. Indeed, the entire nine-number sequence from 1 through to 163 is now known as the Heegner numbers.

Backtracking in time slightly, in 1913, mathematician/physicist Georg Rainich found a remarkable connection between the (eventually named) Heegner numbers and Euler's prime formula — recall equation (11.5). Noting specifically, that $P(N) = N^2 + N + p$, where p is a prime number, generates prime numbers for $N = 0, 1, 2, 3, \ldots (p - 2)$ if and only if $1 - 4p$ is the negative of a Heegner number. Using, for example, the 8th Heegner number, the Rainich condition indicates $1 - 4p = -67$, giving $p = 17$ and $P(N) = N^2 + N + 17$. This particular formula produces primes for $N = 0, 1, 2, 3, \ldots, 15$.

So, where are we? Given the very brief introduction to the origin of the Heegner numbers, we can immediately notice that the exponential powers studied by Ramanujan involving 43 and 67, and the so-called Ramanujan number involving 163, are the last three numbers in the Heegner sequence. And, the question is, what is this telling us? The answer, in fact, takes us into exotic, difficult, and deep realms. Indeed, we are in the world of elliptic curves and modular functions. The elliptic curves of interest are those of the form $y^2 = x^3 + ax + b$, and the modular function of interest $f(j)$ is one such that $f(j + 1) = f(j)$ and $f(-1/j) = 2^k f(j)$, where $k \geq 0$. Elliptic curves, it turns out, are of immense importance in number theory, and they are at the very heart of Andrew Wiles' remarkable proof of Fermat's Last Theorem (recall Chapter 2). Of interest to us here, however, is the so-called *j*-function (or *j*-invariant). Two elliptic curves E and E′ are said to be isomorphic, that is, there is a structure-preserving mapping from one curve to the other if they have the same *j*-function: $j(E) = j(E')$. Now, since modular functions are periodic, they can be expanded in terms of a Fourier series,[11] and it transpires that the *j*-function can be expanded as follows:

$$j(\tau) = \frac{1}{q} + 744 + \text{Smaller terms} \tag{11.6}$$

where $q = e^{2\pi i\tau}$. None of this is immediately obvious, and I am being incredibly vague when I say *smaller terms* in the expansion, but to break the golden rule of mathematics, please take my word for it at the moment.[12] One of the important and thoroughly remarkable properties of the *j*-function is that when $\tau = (1+\sqrt{-d})/2$, so $j(\tau)$ will be an integer if and only if d is a Heegner number. Indeed, when $d =$ 163, so $j(\tau) = -(640320)^3$. Substituting this term into equation (11.6) then yields the result that

$$e^{\pi\sqrt{163}} = (640,320)^3 + 744 + \text{Smaller terms} \tag{11.7}$$

and the smaller terms in this case (expressed vaguely and without proof here) are all very close to being zero. This expansion was actually discovered by French mathematician Charles Hermite in 1859, and indeed, Hermite showed that if there was such an entity as a tenth Heegner number α beyond 163, then $e^{\pi\sqrt{\alpha}}$ would evaluate to a number even closer to an integer than $e^{\pi\sqrt{163}}$. Since there is no tenth Heegner number, however, with (11.7) we have reached the limits of possibility with the *j*-function, but we have discovered why the Ramanujan number is something altogether special.

The near integer property of $e^{\pi\sqrt{163}}$ has led us on a merry dance through the rudiments of geometry, complex numbers, and prime number theory, and perhaps the moral of the story is that every

---

[11] This process developed by Joseph Fourier decomposes a periodic curve into an infinite series of sine and cosine terms of appropriately varying amplitude.

[12] Better still, of course, investigate the *j*-function and its expansion yourself, there is a vast literature on this topic.

number has a deep story to tell — with some numbers having perhaps a deeper story to tell than others. There are many similar such oddities that have been noted, but whether they have any deep significance remains unclear. Gelfond's constant, investigated by Soviet mathematician Aleksander Gelfond, for example, reveals that $e^{\pi} - \pi =$ 19.999099979.... a number remarkably close to the integer 20. While Ramanujan's number has deep roots that stretch into the exotic realm of number theory, it is generally assumed that Gelfond's constant, just like our scam-result in equation (11.1), is nothing more than a numerical coincidence, and an amusement at best.

The mathematical world is rich in numbers, and coincidences are inevitably going to occur. Indeed, we would be extremely surprised if there were no coincidences or unexpected (innocent) results. It is surprising, and yet intriguing, for example, that in the vast infinite realm of the integers that there are just two Munchausen[13] numbers: 1 and 3435. These are the only two numbers, out of all the numbers one could possibly construct, that satisfy the self-raising condition of their digits. These numbers specifically satisfy the condition that $1 = 1^1$ and $3435 = 3^3 + 4^4 + 3^3 + 5^5$. Is this telling us something profound, or is it just another remarkable (amusing, unexpected, inevitable) numerical coincidence? It is almost certainly the latter, of course, rather than the former, but if the history of mathematics has taught us anything, such statements should be tempered with the rider: so far as is presently known.

11.1. Mathematics and Beauty

Mathematicians often talk about beauty. But it is a strange beauty that is hard to describe and very often difficult to see. Like all things that are considered beautify, it is a beauty expressed purely in the eye of the beholder, and more specifically it is something expressed in

[13] The great Baron Munchausen was a larger-than-life character, described in a series of stories by Rudolf Raspe in the late 18th century. One of the Baron's exploits saw him extract himself, and his horse, from a muddy pond by pulling upwards on his own hair. In effect pulling himself up by his hair-roots (rather than bootstraps).

the chemistry of the brain of the beholder. There is no such physical entity as beauty, it is undefinable, it is intangible, it is not mathematical, and it is purely a human construct. The fact that a mathematical result is called beautiful does not make it any more real, or better, than a mathematical result that looks ugly, and it does not make it a better description of reality. Physicist Paul Dirac is known to have insisted that given a choice between a good theory with ugly equations, and a bad theory with beautiful equations, he would choose the latter. Indeed, Dirac was critical of the renormalization procedures that had been introduced in quantum electrodynamics (QED), arguing (somewhat audaciously) that, "just because the results happen to be in agreement with observations does not prove that one's theory is correct". In reality, of course, what Dirac really meant by beauty (but apparently wasn't prepared to admit) was that it allowed for the development of a set of equations that he (and a relative few others) could actually solve in clever ways. Nature is not described by beautiful equations — indeed, it is almost invariably described by sets of rather ugly-looking ones. The physical universe, when all is done and dusted, is described by equations with all manner of in-built assumptions, arbitrary approximations, re-scaling terms, and dubious fudge-factors. Beautiful mathematics may get close to describing reality, but neither the beauty nor the mathematics is the essence of reality. Mathematician G. H. Hardy, famously wrote in his *A Mathematicians Apology* (first published in 1940) that there is no room in the world for ugly mathematics — he was as wrong in saying this as anyone can possible be. Indeed, Hardy was writing from a highly privileged, elitest, and decidedly snobbish position, his *Apology* revealing that the essential, school-boy, reason he became a mathematician in the first place was simply because it was one way that he could do better than, and indeed *best* other boys. His mathematical abilities, which were indeed formidable, were essentially being deployed as a means to bolster his own ego — and therefore, of course, they must be beautiful. Human nature allows us to sympathize with Dirac and Hardy, but it does not allow their opinions of what is beautiful and/or ugly to be either definitive or dogmatically authoritative.

In numerous poles and surveys, over many years, the equation that has consistently been ranked number one in terms of its beauty is Euler's equation $e^{i\pi} - 1 = 0$. Indeed, this is a remarkable equation and one that makes the mind boggle given that it is composed of two irrational numbers that can never be fully written down and a number $i = \sqrt{-1}$ that has no tangible counterpart in reality. That these three entities can be so arranged to equal 1 is certainly impressive, but why is it beautiful? My guess with respect to the last question is that Euler's equation is deemed beautiful entirely because of a human herding instinct. Since the equation is always number one on the beauty list, it must, therefore, be beautiful. This, of course, is a tautological argument, and it provides no basis upon which to gauge or define beauty. Again, my guess is that mathematicians who repeatedly vote for Euler's equation as being beautiful are really saying that they find the analysis behind its derivation remarkable and pleasing — such attributes, however, do not automatically make the end equation beautiful. Indeed, an equation on its own is just a collection of arbitrary symbols, and since mathematical symbols are entirely arbitrary in their origin, one could legitimately have a representation of Euler's formula that runs as, say, E▲👁☺▼*1♋● — which is hardly beautiful, yet is says the same thing.

It is sometimes argued that simplicity and compactness are the main attributes of beauty, but again, this is just individual bias. Simplicity and compactness are certainly attributions that can be applied to Euler's formula, but these same attributes carry a certain sense of dullness — even fixedness — that lacks real conviction. They instead imply a certain stifling, even blindness, of the imagination. Rather than inherent simplicity and compactness implying that something must be considered beautiful, they really box things in and sully the imagination and one's creativity. To the author's eye (brain and neurology), for example, Ramanujan's constant. Equation (11.2) provides for a much more beautiful, vibrant, exiting, and expansive result than that expressed by, say, Euler's equation. Euler's equation, once seen, implies a dull finality,[14] whereas Ramanujan's constant simply begs for further exploration.

[14] Technically, it simply indicates that a rotation of π radians is equivalent to a reflection (or multiplication) by –1.

Both Bertrand Russell and Herman Weyl associated beauty with truth, but there is no law to guarantee that beauty (however defined) must lead to truth, or indeed, lead to any useful understanding of the observable universe. In his James Scott Prize lecture, delivered to the Royal Society of Edinburgh in 1939, Paul Dirac argued that, "the mathematician plays a game in which he himself invents the rules while the physicist plays a game in which the rules are provided by Nature, but as time goes on it becomes increasingly evident that the rules which the mathematician finds interesting are the same as those which Nature has chosen". Such reasoning, of course, leads to an inevitable tautological self-supporting structure — and it accordingly carries very little conviction. Since physicists have chosen to play (indeed, can only play) the same game as the mathematician, that is they can only use mathematical tools in their explanations and theories, then the paths of their theories and mathematics will automatically converge. There are many physical theories that are described as being mathematically elegant, simple, and compact, but invariably, the simplicity and compaction are entirely within the mathematical notation, not the actual equations behind the compaction. The compaction and simplification are a mathematical sham and indeed, a sleight of hand. Einstein's general relativistic field equations, for example, can be written down in a compact mathematical form as $(4\pi G/c^4)T_{\mu\nu} = R_{\mu\nu} - \frac{1}{2}Rg_{\mu\nu} + \wedge g_{\mu\nu}$, but to use these equations in any practical fashion requires a tremendously complicated unpacking.[15] Furthermore, Einstein's equation admits to an infinite number of possible universes — so it is hardly a beautiful expression of reality. None of this discussion is to downplay the incredible results and intellectual brilliance behind the derivation of Einstein's field equations, and/or Maxwell's equations — but to say that the end equations are beautiful is entirely subjective. At best, to say such equations are beautiful is to define beauty entirely in terms of the idiom of the *emperor's new cloths*.

[15] In this formula, T is the energy–momentum tensor, R is the curvature scalar, and g is the metric tensor. The Λ terms is the so-called cosmological constant, G is the gravitational constant and c is the speed of light.

If an equation cannot be beautiful, what about a graph? One of the most often quoted examples of a mathematical beautiful object is that of the Mandelbrot set (Figure 11.2) [1]. Indeed, it has pleasing qualities to the eye, but its derivation is again entirely hidden from the observer's view, and indeed, it is an idealized highly complex object that does not (cannot) physically exist in reality. The Mandelbrot set is derived according to a very simple formula, and involves the iteration: $z_{n+1} = z_n^2 + c$, where z and c are complex numbers, with c being an arbitrary constant. The Mandelbrot set is therefore positioned with the Argand plane,[16] and constructed on the requirement that the iterations for z_n starting at $z = 0$ do not explode to infinity as n tends to infinity. If the iteration sequence does not diverge to infinity and remains finite, then it is part of the Mandelbrot set (and colored black in Figure 11.2). If the iteration sequence diverges to infinity, then it is not part of the Mandelbrot set (and colored blue in Figure 11.2). Analysis of the boundary of the Mandelbrot set reveals a fractal structure showing ever-deeper regressive features and patterns. The subtle color shading in the diagram of the Mandelbrot set, Figure 11.2, is entirely arbitrary, and based upon made-up coding rules relating to how rapidly the iterations on z_n diverge to infinity [2]. I find no great beauty in the Mandelbrot set — it is complex, obscure, contrived, remote, and far-removed from any reality. Visually, the Mandelbrot Set is too symmetric, too perfect, and the eye searches for some asymmetry to make it artistically appealing. It is a classical example of the dullness of perfection.[17]

---

[16] The Argand plane provides for a geometrical representation of a complex number, $z = a + ib$. The real number part of z, a, is read along the x-axis, while the imaginary part, b, is read along the y-axis (or more correctly now, the imaginary axis). The diagram was first described by the Swiss amateur mathematician Jean Robert Argand in 1806, although the Norwegian amateur mathematician Caspar Wessel made the exact same arguments earlier in 1797.

[17] Indeed, in terms of beauty, perfection is something that might be aimed for, but is never actually achieved. This idea runs parallel to the practice of ancient Persian rug weavers, who, recognizing their skills, deliberately introduced mistakes into their works according to the belief that only God could create something that was perfect.

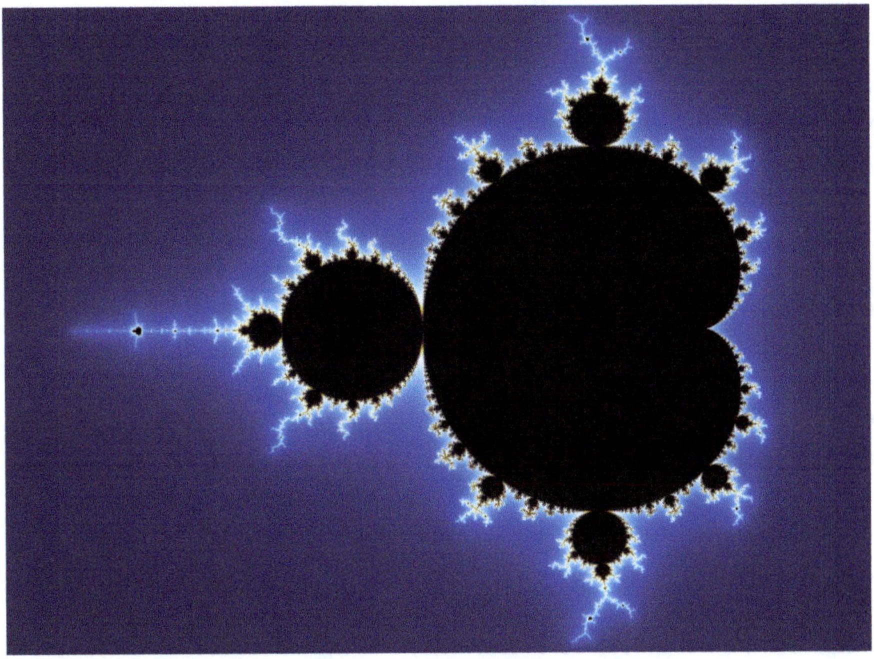

Figure 11.2. The Mandelbrot set. Image courtesy of Wikimedia Commons: https://en.wikipedia.org/wiki/File:Mandel_zoom_00_mandelbrot_set.jpg.

To rank as beautiful (to the human eye) one suspects that what is required is a subtle and harmonious blend of both chaotic and ordered elements. Such opposing characteristics are abundant in nature but are largely absent in images such as shown Figure 11.2. The very nature of their precise, ordered, numerical construction, via iterative computation, makes them sterile and artistically moribund — that is, not beautiful. Others will no doubt disagree with these statements, and to a certain extent, this simply underscores the fact that beauty is not intrinsic to the equations, or the image — but something else entirely. Mathematical beauty, so-called, in spite of the engaging arguments presented by Hardy, Dirac, and many others, is neither a guaranteed guide nor an infallible methodology by which physical reality can be found, and neither is it an infallible guide to what might be called a correct physical theory.

11.2. Discovered or Invented

Fundamental to the question of mathematical applicability is the deeper-still question which asks if mathematics is discovered or invented — this question was briefly discussed in Chapter 3. Indeed, the act and art of creation, that is the generation of new works and ideas, be it in the field of science, art, or literature, is an incredible human ability, and many of the best practitioners have expressed the notion of discovery rather than invention. J. R. R. Tolkien, for example, explained that he felt that the stories set in the various epochs of his Middle-earth were discovered rather than invented. Characters, he complained in various letters, just kept appearing unannounced. Likewise, J. K. Rowling indicated that the storied life of Harry Potter emerged as if it was discovered. Other stories, myths, and chronicles, however, are clearly invented for a very specific purpose by their authors. Are we to believe, in parallel with the Platonists, that there is some exterior universe to our own that contains all possible stories? Indeed, the invented-verses-discovered question is not unique to mathematics or philosophy, but it is rather an outcome of human consciousness. Certainly, for so it appears, mathematicians don't know where their results come from (other than by hard work), and indeed, mathematicians are probably the last people to ask what mathematics really is. To echo (tongue-in-check) the words of chemist Josiah Willard Gibbs, who once supposedly quipped during a lecture, "a mathematician may say anything he pleases, but a physicist must be at least partially sane". As a case in point, the most gifted, natural (innate?), mathematician of all time, Ramanujan attributed his mathematical results to the goddess Mahalakshmi,[18] who, he indicated, presented equations to him upon visionary scrolls. For Ramanujan, it would seem, therefore, that mathematics was a religious and gifted experience, rather than a process of discovery or invention. Philosopher and mathematician Alfred North Whitehead observed, in somewhat drier tones

[18] Mahalakshmi, also Lakshmi, is one of the principal Hindu goddesses and is attributed with wealth, fortune, beauty, joy, and prosperity.

than Ramanujan, "the pursuit of mathematics is a divine madness of the human spirit, a refuge from the goading of contingent happenings". Furthermore, Godfrey Hardy wrote in his *Apology* that he believed, "mathematical reality lies outside of us, that our function is to discover or observe it, and that the theorems which we prove, and which we describe grandiloquently as our creations are simply the notes of our observations". The revered peripatetic mathematician Paul Erdös once commented that "God has a transfinite Book, which contains all theorems and their best proofs". Fortunately, it would seem, and Erdös agreed, one did not need to believe in God to make progress in mathematics — but one did have to believe in the Book. There is little hope that any clear insight, or more specifically, consensus, on the discovery-versus-invented question will appear in the near future, but for all this, mathematics is still a marvelous enterprise. Indeed, the question is essentially ill-posed in the sense that it is clear that mathematics is both discovered and invented, but for all this, the discovery is wholly attached to the allowed mathematical rules working on the (human mind) invented parts.

In a now famously repeated essay entitled, "The unreasonable effectiveness of mathematics in the Natural sciences", published in the journal *Communications on Pure and Applied Mathematics*, in 1960, Eugene Wigner argued, "the miracle of the appropriateness of the language of mathematics for its formulation of the laws of physics is a wonderful gift which we neither understand nor deserver". This is essentially invoking the same circular argument given by Dirac in his 1939 address — although Wigner was particularly intrigued as to why complex numbers are so important in physical theory. This same fascination with complex numbers was echoed by Richard Hamming in his essay, with the same title as Wigner's, published in *The American Mathematical Monthly* in 1980. Indeed, Hamming went as far as to exclaim that, "God made the universe out of complex numbers". Why is it that Wigner and Hamming were so surprised by the effectiveness of (among other mathematical tools) complex numbers in physical theory? The answer to this question takes us right back to Plato.

Plato, recall Chapter 3, reasoned that there was a realm (a universe of ideals) separate from our reality (the observable universe) in which mathematical entities had a real and timeless existence. Accordingly, numbers like π and e and $\sqrt{2}$, with their infinite decimal tails, really do have a physical existence (unlike in the physical universe). This universe of ideal structures contains, ready assembled, all possible mathematical theorems and results, and accordingly, any mathematical result that applies to a physical theory is discovered — the answer, as it were, existed in the Platonic realm before it was needed or given meaning in some physical theory that mapped it into the observable universe. If one believes this, then it truly is unreasonable, and a profound mystery, as to why mathematical objects that exist in a non-experiential, idealized realm should have anything to do with the real universe in which we live. In contrast, all this mystery vanishes if one posits that mathematics is nothing more than an invented language and a product of the human mind. In the non-Platonic viewpoint, there is no mystery as to why mathematics is so effective in physical theory — there is, in fact, no miracle at all. Rather, mathematics is a mental abstraction, based upon observations, utility, and human inventiveness — it has meaning and real-world applicability because human abstraction put it there. Indeed, the history of mathematics shows a continued and gradual development and a concomitant application to physical theory. When mathematics fails to provide an answer to some physical problem, however, well, we can either abandon the problem or we can invent more and new mathematics to fill the knowledge gap. The Platonist might well argue at this stage that sometimes the needed mathematics had already been developed prior to its application to some physical theory — but this is largely an illusory, cherry-picked, answer that focusses on the few occasions where such circumstances have actually come about. The circular disk existed long before anyone thought of turning it into a wheel. Most of the time the ideas and proofs within mathematics have no contact with the physical world. Fermat's Last Theorem, for example, required a magnificent invention of tools and ideas before it could be proved true — and yet, not one physical theory hangs upon its proof or negation.[19]

Oxford mathematician and Nobel Laureate Roger Penrose, a voice always worth paying great attention to, as well as being a strong Platonist, has argued that the Mandelbrot set is an example of a mathematical discovery, rather than a human invention. But this is only superficially true. True, the set was unexpected and discovered by Benoit Mandelbrot (and other researchers) in the 1980s, but the discovery is a direct product of the prior invention of complex number theory — sooner or later it would have shown-up in someone's computer-based analysis. Rather than a discovery, it was an inevitability. The Mandelbrot set is a consequence of the human mind abstracting and inventing the non-existent, non-real number $\sqrt{-1}$ and then developing and playing with the mathematical language to go with that invention. The Mandelbrot set was not discovered, as such, but sat latent within the theory of complex numbers — this is no different to saying that Shakespeare's sonnets lay latent within the English language — the point being that the language and the written word, all human inventions, were needed to be put in place before the story, or poem, could emerge. Is, therefore, the real universe a mathematical construct? The answer to this question is a straight no. Can the working of the universe be described mathematically? The answer to this question is a straight yes. Mathematics, it is argued here, is very much invented, and decidedly not a pre-existing entity, and as such it is a pure human construct. Based upon the mind's ability to abstract details from the observable universe, mathematics is a language, composed just like any other spoken or written language, of symbols and rules. From such constructs apparent discoveries will follow, stories will be told,

---

[19] On a more human level, however, there is a story (possibly apocryphal) that German mathematician and physician Paul Wolfskehl became so distracted by a technical paper relating to a supposed proof of Fermat's Last Theorem that he missed his planed midnight deadline to commit suicide. Proving the point that Fermat's Last Theorem may well have sent many mathematicians mad, but it saved the life of at least one.

calculations will be made, and occasionally the end result will fall in step with reality, at other times it will be a fairy tale. Indeed, it would appear that the pillars and structures that underpin mathematics provide more props than are actually needed to explain reality, and accordingly, mathematics has become a tool for illusion as well as illustration.

It has been said, on many occasions before now, that scientists are often guilty of not taking the implications of their ideas and mathematical laws seriously enough, and to counter this some recent commentators have argued that we must embrace all the implied philosophical and mathematical oddities that several theories have thrown out.[20] In this manner, the observable universe is taken as being identical to the Platonic universe. But, again, this is no more than saying, "because we can't find anything better to describe what is happening, then what we have found (with all its oddities and infinities) must be true". Such an outlook reminds us of the poor philosophy (and muddled logic) nuanced by Sherlock Holmes[21] in *The Sign of Four*: "when you have eliminated the impossible, whatever remains, *however improbable*, must be the truth". Like many of Holmes's sound-bite statements, there is no real magic or truth in what he says. Rather, the statements are verbal conjuring tricks that seem fundamental but are really vague and inherently un-useful. To posit that all in the observable universe is mathematics is just pure philosophical escapism and an abdication of scientific responsibility.[22] Such dictatorial reasoning does not advance our understanding of the universe in any way, it simply allows us to pick and choose the mathematical theories that we like — the flashy, glittering bits,

---

[20] Some researchers, for example, argue for the many-worlds interpretation of quantum mechanics. A viewpoint that has us traveling through an infinitude of endlessly splitting universal histories.

[21] I am conscious here of invoking a fictional literary creation to expand upon a point concerning physical laws and the meaning of reality.

[22] Not, of course, that there is such a hard and fast thing as scientific responsibility — scientists cannot agree on many topics, and most scientists *do* science in the modern era not because they have any sense of responsibility toward it but simply because they have a *bent* for its practices and because they can make a living by doing it.

that we can actually work with. To say that all reality is embroiled within an infinity of infinities, of one mathematical form or another, is simply wishful thinking, and provides no foundation, not even a shaky one, upon which to base our understanding of reality. As H. G. Wells reminds us, "if anything is possible, then nothing is interesting". Mathematics provides a good, completely unsurprising, description of the physical world purely because it has been invented by the human mind to describe and explain the physical world. Or, as mathematician Sir Michael Atiyah has commented, "man has created mathematics by idealizing and abstracting elements of the physical world" — so, in summary, it is no surprise at all that mathematics describes the physical world so well, and the wonder expressed by Wigner and Hamming is entirely misplaced.

Mathematics can guide us in the search for order in, and understanding of the observable universe. Indeed, it is the very engine of scientific discovery, but its very fecundity, and an absolute acceptance of its fundamental reach, can equally send us in entirely wrong directions, and provide for entirely misguided physical theories — if you like, mathematics can produce any number of false-positives. What does it say of a physical theory, such as quantum mechanics, if it can be expressed in multiply different mathematical ways? If these different mathematical doctrines produce the same predictions, then what does this say about the mathematical description of reality? Planetary positions can be calculated just as accurately in the Ptolemaic system (recall Chapter 4) as they can in the Copernican one (recall Chapter 5) — indicating that a successful mathematical model (that is one which provides a commonality between what it predicts and what is actual observed) need not correspond to any physical reality. Likewise, the Dirac imperative that a physical theory should be mathematically elegant and/or beautiful does not always produce a correct physical outcome. At the British Association meeting in 1913 mathematician James Jean presciently commented upon Niels Bohr's quantum model of the atom that, "the only justification at present put forward for these assumptions is the very weighty one of success". Indeed, Jean was right, and while the mathematical model gave the right results, the mathematics did not proceed from a correct physical theory. Examples of additional coupling-failures (but in their time extremely popular theories) could include the steady state cosmological model of Hoyle, Bondi, and Gold; the Kaluza-Klein 5-dimensional extension of Einstein's general relativity; and the SU(5) grand unified field theory for combining the electroweak and strong forces. In more recent times, the many-dimensioned, mathematical constructs inherent in string-theory appear to be yet another example of physicists locked into a largely misguided mathematical game, that in reality sees them hopelessly trying to fit a square peg into a round hole.[23] It is entirely possible that string theory may be made to work, but it can never move into the predictive, experimental realm — at best, to echo Plato, it might only save the appearance of reality.

11.3. The Mathematics of Miracles

Writer and Christian apologetic, C. S. Lewis is known to have suggested that, "probably every age gets, within certain limits, the science it deserves", to which he added, "miracles do not, in fact, break the laws of nature" [3]. It is not obviously clear what Lewis intended by the expression, "the science it deserves", other than it being a pithy reflection upon the fact that scientific research tends to thrive under war-time conditions. The comment on miracles is additionally obscure (in spite of the claims by Lewis) since the very idea that something happens miraculously automatically places it outside of the realm of scientific study, and beyond the laws of nature. Lewis would have been on much safer ground if he had worried more about probabilities, where, what is most-often called miraculous is nothing more than something unexpectedly occurring (with a very low probability), but that something is still safely contained within the realm of natural law. The terms miraculous and highly improbable are often laid against such oddities as the Titius–Bode rule. "Surely", proponents, argue, "such a close description cannot come about by pure chance — that would be a miracle". Upon this basis, such proponents argue, that the

[23] Of course, it is the case that a square peg of side $L = \sqrt{2}R$, or smaller, can fit into a round hole of radius R.

Titius–Bode rule must be either a miracle or some natural law. And, since miracles are not something that can be addressed by science, the inferred implication is that the rule must be the result of some real phenomenon.

Cambridge mathematician John Littlewood attempted to debunk the idea of miracles by comparing such occurrences against the average human attention span. Being generous, Littlewood noted that the average human attention span is 10 seconds,[24] and to this he added that a miracle is something that occurs as a one-in-a-million event. From this simple starting point, Littlewood then went on to develop a law of the miraculous:

Littlewood's law of the miraculous: A person can expect to experience events with odds of one in a million, that is a miracle, at a rate of no less than once every 116 days (or about 3 times per year).

Proof: The average human attention span is about 10 seconds. Accordingly, the time to accumulate one million moments of attention requires a time span of 10^7 seconds, and this is equivalent to 115.7 days.

Corollary: Assuming the average human being sleeps for 8 hours a day, the typical time interval between experiencing a miracle can be no less than 173.6 days (or about 2 times per year).

Littlewood's law is problematic for several reasons. First, why is a miracle something that occurs with a frequency of one in a million — why not one-in-a-billion? And, second, there is no physical reason why the one-millionth event observed by a specific human-being should be a miraculous one. Indeed, given the combined population of even a relatively small city, Littlewood's law would suggest that miracles must be happening all the time — it just depends upon who is experiencing, at any given instant, their accumulated one-millionth moment of focused attention gathering. The point is, as physicist Freeman Dyson has noted, improbable (that is

[24]More recent studies suggest it is actually more like 8 seconds — a depressingly short duration.

miraculous) things happen around us all the time. Indeed, that such entities as stars exist is entirely predicated on very rare events happening all the time — and in vast numbers. At the Sun's core energy is generated via the so-called proton–proton (PP) chain. The first step in this reaction is the coming together of two protons P + P, and symbolically proceeds as $P + P \Rightarrow D + e^+ + \nu$, where the end products are a deuterium nucleus D, a positron e^+, and an electron neutrino ν. The deuterium nucleus soon interacts with another proton to produce a helium-3 nucleus and a gamma ray photon. The helium-3 nucleus then interacts with another helium-3 nucleus to produce the final end product of helium-4. In this series of steps, it transpires, a total of 4 protons are ultimately transformed into a single helium-4 nucleus (composed of two protons and two neutrons), along with the production of two positrons and two electron neutrinos. Energy is lost to the star through the escape of the neutrinos into space; the star acquires energy, that heats its interior, from the gamma-ray photon and the ultimate annihilation of the two positrons with two electrons. The total amount of energy generated per complete transformation of four protons is $E_{4P} = 4.4 \times 10^{-12}$ joules. Given the Sun's energy output at its surface (its luminosity) is $L = 3.85 \times 10^{26}$ joules per second so of order $N_{4P} \approx 10^{38}$ complete PP chain reactions must take place per second. This number of reactions amounts to a mass-to-energy conversion of $M_{4P} = N_{4P}(E_{4P}/c^2)$, where c is the speed of light,[25] giving $M_{4P} \approx 5 \times 10^9$ kg/s. The Sun has been converting of the order of five billion kilograms of matter into energy every second of its existence — so far, for some 4.56 billion years. Over the age of the solar system, the Sun has converted some 7×10^{26} kilograms of matter into energy due to the successive progression of the PP chain — the equivalent of about 117 Earths, but a mere 0.036% of its present mass. It is partly for this reason that stars must be massive conglomerations of matter — they are effectively feeding upon themselves. Indeed, the Sun is composed of about 10^{57} protons, but at any one instant, only a minuscule fraction, of order $N_{4P}/10^{57} = 10^{-19}$, of its protons are

[25] Here, I am applying Einstein's famous $E = mc^2$ formula.

actually involved in generating energy. Accordingly, the Sun and similar stars can happily generate energy in their interiors, remain stable for immense periods of time, and radiate energy into space at their surface — some of the Sun's energy, of course, heating the Earth and enabling life to exist upon its surface. There is more to just bulk, however, in the Sun's story. While straightforward calculations (as given above) allow us to estimate the number of atoms involved in powering the Sun, it turns out from more detailed calculations that the very first step in the PP chain, the generation of a deuterium nucleus, is a highly improbable event. Indeed, if we watch a given individual proton in the Sun, then the average time interval for it to approach and interact with another proton to produce a deuterium nucleus is about 10 billion years[26] — that's twice the present age of the solar system. Here, importantly, we have to remember what this average interaction time refers to and specifically remember that some interactions will occur much sooner and some much later. In the case of the Sun, and stars in general, it is the very rapid, far-from-average time of interaction events that drive the whole show — it is the tail that wags the dog. The $N_{4P} \approx 10^{38}$ complete PP chain reactions that must occur every second within the Sun's core are all dependent upon those protons that interact on timescales very, very, very much shorter than the main sequence lifetime. The only way that so many improbable interactions can come about is if the Sun is composed of a massive assembly of protons. Indeed, the lowest mass stars have a mass of about 1/10th that of the Sun, and accordingly to achieve stardom a minimum of some 10^{56} protons need to be brought together, in a small enough volume of space, so that enough very, very, very low probability interactions can take place per second in order to generate sufficient internal heat energy to halt the gravitational collapse.[27] That the Sun exists and remains stable, that the Earth exists, that

[26]This corresponds to what is called the main-sequence lifetime.

[27]If the nuclear reactions suddenly stopped within a star, and nothing opposed the collapse then it would form a black hole, in the case of the Sun, this collapse time would be about half an hour.

life on Earth exists, and that human beings evolved, are all very, very, very low probability events — but they assuredly came about, and they did so without the need for invoking any miracles in which the laws of physics, chemistry and biochemistry were broken. The coming true of very low-probability events is certainly something incredible, but this is not the same as requiring a miracle to occur.

The label *remarkable* or *miraculous* is purely an outcome of human perspective — it is not an attribute of reality. There is nothing remarkable, for example, in the Titius–Bode rule — it (and all the other rules found for planet ordering — recall Chapter 9) is nothing more than one of the many (indeed, infinite) non-significant coincidences that the corpus of numbers can generate. The Titius–Bode rule is not miraculous, and nor is it fundamental — indeed, one might argue, that such a rule is inevitable. The Titius–Bode rule stands amidst the delightful collection of numerical coincidences that permeate mathematics and physical theory. There are an infinite number of numbers, there are an infinite number of fractions, and an infinite number of ways in which numbers can be added, divided, subtracted, and manipulated — it is not at all surprising, therefore, that numbers and mathematics can describe nature and the universe. This being said, even numbers sometimes struggle to find enough traction to complete the tasks that have been asked of them.

11.4. The Law of Small Numbers (Again)

Srinivasa Ramanujan asserted that all numbers were interesting in one way or another — indeed, each one was akin to being a personal friend. The classical story that illuminates this digital passion stems from a visit by Cambridge mathematician G. H. Hardy to a bedridden and hospitalized Ramanujan. Presumably to make some conversation, Hardy commented that the license number for the taxicab that he had traveled in was 1729, which seemed like a rather dull number. Nonsense, Ramanujan immediately replied, it is an interesting number — it is the smallest number expressible as the sum of two cubes in two different ways.[28] Ramanujan's fascination with numbers, and there ever intriguing and multilayered relationships, can be cast in terms of a very general theorem:

Theorem: All numbers are interesting.[29]

Proof: This proof will proceed by contradiction. Assuming that the theorem is false there must be a set B (possibly infinite) of boring numbers. Within this set, there must, however, be a least boring number $B_0 \in B$. But the fact that there is the least boring number B_0 makes it, by default, an interesting number. We have accordingly a contradiction, and the theorem must therefore be true.

Mathematics abounds — it is a very fecund human creation. For all this, however, are we any closer now, at the end of this chapter and book, to understanding why the Titius–Bode rule (and its erstwhile companions) come about. In many ways, the rule has been set up as a *straw man* — a hapless target erected with the sole purpose of knocking it down. But there is more to the debate than this. The fundamental question still remains: Is the Titius–Bode rule a mathematical coincidence, or is it a fundamental rule about planetary system formation?

The essence of the Titius–Bode rule, and the fact that it is nothing more than a numerical coincidence, is fully encapsulated within the *Strong Law of Small Numbers*. This law was introduced, with multiple examples of its application, by mathematician Richard Guy (University of Calgary) in several articles published in the *Mathematics Magazine*, in the late 1980s [4,5]. While not a strict mathematical theorem, thereby requiring a formal proof, the Strong Law of Small Numbers is effectively a statement about arithmetic proved by exhaustion — the exhaustion, in this case, however, is one of being purely overwhelmed with case numbers.

---

[28] The two cases are $1729 = 12^3 + 1^3 = 10^3 + 9^3$. David Wells in his, *The Penguin Dictionary of Curious and Interesting Numbers* (Penguin Books, 1986), also informs us that 1729 is a Harshad (from the Sanskrit for *joy-giver*) number, meaning that it is divisible by the sum of its own digits, with $1729 = 19 \times 91$.

[29] This theorem and its proof are taken from *The Grapes of Math* (Doubleday, Canada, 2014) by Alex Bellos.

The Strong Law of Small Numbers: There aren't enough small numbers to meet the many demands made of them.

We have already seen a selection of examples that satisfy the Strong Law of Small Numbers — all the formulae for generating prime numbers (by Mersenne, Fermat, Euler, and Rainich) described above, for example, illustrate how insidious the Strong Law is. Indeed, the Strong Law applies to all those cases of mathematical deception, where, at first glance, it looks like some fundamental rule has been discovered, only for the rule to slowly (but assuredly) fall apart multiple steps into its application. Guy has summarized this aspect of the Second Law as follows: "you can't tell by looking". One nice example given by Guy in his second paper (example 41) concerning the Strong Law is that of the sequence generated by the formula $[e^{(n-1)/2}]$, $n = 0, 1, 2, 3, \ldots$, where $[x]$ is the so-called ceiling function that gives the smallest integer $\leq x$. The first 10 numbers in the sequence (for $n = 0, 1, 2, \ldots, 9$) are 1, 1, 2, 3, 5, 8, 13, 21, 34, 55 which corresponds exactly with the Fibonacci sequence of numbers. For $n = 10$ and 11, however, the numbers generated are 91 and 149, which are not members of the Fibonacci sequence — the appropriate numbers being 89 and 144. Indeed, for $n > 9$, the formula provides numbers that are increasingly divergent from the Fibonacci sequence. The formula fails, as a generating formula for the Fibonacci sequence, but only after the tenth term. After a highly impressive start, the generating formula eventually fails to deliver, providing results that increasingly differ from the Fibonacci sequence. The Titius–Bode rule shows exactly the same behavior with respect to accounting for planetary orbits. Indeed, the TB rule provides seemingly impressive results for the inner solar system (although recall the discussion in Chapter 9), but for Jupiter and planets beyond the predictions are increasingly different from those required. Being generous one could say that the TB rule works up to the 6th term (planet Saturn), which is a far less successful result than that provided by, for example, the generating function for the Fibonacci numbers just discussed.

Guy followed up his Strong Law of Small Numbers with a second observation concerning perceived numerical coincidences:

The Second Strong Law of Small Numbers: When two numbers look to be equal it isn't necessarily so.

This law applies not only to numerical sequences but also to numerical coincidences where it might seem that some underlying physical theory has been unearthed. A purely numerical example of the Second Strong Law is Ramanujan's number — equation (11.2). This is especially so in the spoof offered by Martin Gardner when he suggested that $e^{\pi\sqrt{163}}$ truly was an integer — indeed, with no other aid than a hand calculator one cannot tell otherwise. The second strong law might reasonably have been better called the Law of False Comparisons since it lies at the heart of many a misguided idea — such as the Titius–Bode rule. Just because two numbers, derived from two totally different fields of study, are approximately the same does not remotely mean that there has to be any connection between the two fields concerned. This rule (the Second Strong Law) is something that physicists and astronomers have broken on many occasions, and especially so when dealing with dimensionless constants. Indeed, much ink has been consumed, over many years, in trying to explain the fundamental constants of physics in terms of simple rational powers of integers and fundamental mathematical constants such as π and e. To some extent such constructions are predicated on the idea that the fundamental laws of nature should somehow be derivable by rational thought — in other words, the constants of nature cannot be just any set of random numbers — rather, they must be numbers that have some mathematical underpinning. This philosophical viewpoint essentially builds upon the Platonic ideal, but more likely than not, the vast majority of such derivations are nothing more than the Second Strong Law of Small Numbers at play. One well-known example of an apparent relationship between a physical constant and a pure number is that of Rydberg's constant R_∞. Named after Johannes Rydberg this constant appears as a fundamental unit in atomic spectroscopy and is one of the most accurately known (that is experimentally derived) of the physical constants. Expressed as a frequency, we have the following:

$$\frac{cR_\infty}{10^{15}} = 3.2898419602508 \approx \frac{\pi^2}{3} = 3.2898681336964\ldots$$

the accuracy of the pure numerical approximation is remarkable, but it has no known meaning and/or theoretical justification. It is a coincidence — a very good one, but a coincidence none the less. It is interesting to reflect, however, upon what observable effects might result if the Rydberg frequency truly did equal the $\pi^2/3$ term — to which the answer is effectively none. Yes, spectral lines would systematically fall at slightly different wavelengths, but the universe would look and behave in the same manner as it does now. Another well-known coincidence between a fundamental physical constant and simple mathematical powers is that of the mass ratio of the proton to the electron, with $M_{\text{proton}}/M_{\text{electron}} = 1836.15267343 \approx 6\pi^5 = 1836.11810871$. Again, the match between the numbers is intriguing, but there is no theoretical reason to explain why the proton has to be $6\pi^5$ times more massive than the electron.[30] Sticking with fundamental atomic constants, many researchers have drawn attention to the apparent rational approximations of the fine structure constant α. Also known as Sommerfield's constant, α quantifies the strength of the electromagnetic interaction between two elementary charged particles, and it is generally expressed in terms of its inverse as follows:

$$\alpha^{-1} = \frac{e^2}{4\pi\varepsilon_0\hbar c} = 137.035999084\ldots \tag{11.8}$$

where e is the elemental charge, ε_0 is the permittivity of free space, c is the speed of light and $\hbar$ is Planck's constant divided by 2π. The numerical expression on the righthand side of (11.8) is determined according to the standard values for the constants involved, but according to the theoretical considerations of Swiss mathematician

[30] It appears that this coincidence was first noted in a Letter to the Editor in the *Physical Review Journal*, **82**, 554 (1951). Perhaps ranking as the shortest physics letter ever published in any science journal (it amounts to 3 lines of text, contains 27 words, 4 numbers and 1 reference), the author (Friedrich Lenz) simply states that the equality of the numbers "may be of interest".

Armand Wyler[31] the fine structure constant can also be expressed in terms of the rational powers of integers and π. Wyler's constant (now so-called) is

$$\alpha^{-1} = 2^{19/4}3^{-7/4}5^{1/4}\pi^{11/4} = 137.03608245... \tag{11.9}$$

Indeed, Wyler's constant does provide a remarkably good approximation to the fine structure constant, but the following question remains: Is it meaningful? This latter question occurs since technically Wyler's paper did not actually consider interactions between particles, but rather he was interested in the rotational properties of a seven-dimensional conformal group (representing five spatial dimensions and two imaginary dimensions). The experimentally measured value for the fine structure constant is α^{-1} = 137.035999206, which technically makes (11.8) the better approximation. Additionally, and problematically, many other numerical approximations, that are just as good in terms of approximating the fine structure constant, have been derived. Ralph Roskies (Yale University), in fact, set about the systematic search of all combinations of powers of 2, 3, 5, and π with exponents between –5 and +5, and he found 5 other polynomial approximations that worked just as well as Wyler's — e.g., $\alpha^{-1} = 2^{-19/4}3^{10/3}5^{17/4}\pi^{-2} = 137.035938....$ The point of Roskies' analysis is that Wyler's constant is just a numerical coincidence — in other words, an example of Guy's Second Strong Law of Small Numbers. Other good numerical approximations of the fine structure constant are known, and these include the following:

[31] Wyler's story is ultimately tragic and reflects how challenging it can be to work within a vibrant scientific field with many self-aggrandised commentators. While Wyler's papers were published in 1969 and 1971 the work went essentially unnoticed except for his derivation of the fine structure constant. Various commentators were, however, highly critical of his approach, and after failing to adequately explain his reasoning and methodology to other researchers, Wyler suffered a carrier ending breakdown.

$$\alpha^{-1} = 4\pi^3 + \pi^2 + \pi$$

$$\alpha^{-1} = \frac{29}{\pi}\cos\left(\frac{\pi}{137}\right)\tan\left(\frac{\pi}{29 \times 137}\right)$$

$$\alpha^{-1} = \sqrt{\pi^2 + (137)^2}$$

None of the above approximations have anything in common with each other (nor the fine structure constant — probably) other than a chance numerical closeness — there is no hidden physics here, just a human desire to place meaning into a set of numbers that, purely by chance, seem to be close to each other. They are all progeny of the Second Strong Law of Small Numbers.

The above considerations suggest an important question: Can the fundamental constants of physics (in all their measured glory) be determined or anticipated by pure mathematical reasoning? While philosopher Immanuel Kant argued that reason alone is not enough to determine the true nature of reality, the renowned astronomer Arthur Eddington argued that all the constants of nature could be anticipated through pure reasoning — no observations necessary. Indeed, in his later career, Eddington developed his, so-called, Fundamental Theory, in which he argued *a priori*, or epistemological knowledge can be used to derive or intuit knowledge about the workings of the observable universe [6]. He qualified this near heretical idea, however, with the constraint that, "observation is the supreme Court of Appeal". Indeed, Eddington argued that the fine structure constant was related to the number N_{Edd} of protons in the universe, with

$$N_{\mathrm{Edd}} = 2^{256}/\alpha \tag{11.10}$$

The derivation of (11.10) need not directly concern us here, but it is generally taken today that Eddington's reasoning is not just obscure, but more specifically faulty. None the less, in a series of Lectures delivered at Trinity College, Cambridge, in 1938, Eddington pronounced: "I believe there are 15,747,724,136,275,002,577,605, 653,961,181,555,468,044,717,914,527,116,709,366,231, 425,076,185,631,031,296 protons in the universe and the same number of electrons". This vast 80-digit number ($\sim 1.57 \times 10^{79}$) is the evaluation of equation (11.10) — a number Eddington claimed to have evaluated longhand, with pen and paper, while crossing the Atlantic on a steamship.[32] While the theory behind the derivation of N_{Edd} is no longer taken to be valid, what is remarkable is that a straightforward calculation (recall Chapter 2) based upon the approximate number of galaxies in the observed universe (about 35 billion), each containing order 10^{12} stars, with a typical mass of ½ that of the Sun (which is composed of $\sim 10^{57}$ protons), yields a proton count of some 2×10^{79} — a number remarkably close to N_{Edd}. Not only did Eddington argue that the fine structure constant depended upon the number of particles in the universe, he also believed that the ratio of the electrostatic force between a proton and an electron to the gravitational force between them was set according to the N_{Edd}. Specifically,

$$F_{\text{pe}} = \frac{e^2}{Gm_p m_e} \approx 2 \times 10^{39} \approx \sqrt{N_{\text{Edd}}} \tag{11.11}$$

Once again, it was the apparent coincidence between the two numbers that drove the reasoning — the idea being that since there was no specific reason that the two numbers should be of the same order of magnitude, the fact that they were must mean something. Indeed, Dirac built an entire cosmological model, the so-called Large Number Hypothesis (LNH), around F_{pe}, N_{Edd}, and another dimensionless ratio R_{ue}, with the latter being the ratio of the size of the visible universe to the classical radius of the electron [8]. Specifically, Dirac noted that

$$R_{\text{ue}} = \frac{c / H_0}{e^2 / m_e c^2} \approx 6 \times 10^{39} \approx F_{\text{pe}} \approx \sqrt{N_{\text{Edd}}} \tag{11.12}$$

[32] When he made his calculation, Eddington used the then believed value of $\alpha = 1/136$. The author still remembers (just), as a student, in the mid-1980s, writing a BASIC program for an Apple IIc computer, to check Eddington's number — it is indeed correct. One suspects that very few people have either the aptitude or the confidence to multiply out such a large number by hand and then get the answer right [7].

where H_0 is the Hubble constant.[33] To Dirac, the coincidence between the numbers in (11.12), derived according to fundamental constants from the quantum world and cosmology, was far too good to be none-meaningful. There is a high price to pay, however, if one admits to the reality of (11.12). Indeed, if (11.12) is true, then, as Dirac asserts in the LNH, the gravitational constant G must, in fact, be a time-varying term. This result comes about since the size of the universe is related to its age, with the age of the universe being of order $1/H_0$. In other words, given that the universe is getting bigger over time, so R_{ue} increases as the universe ages. Imposing the constraint (somewhat arbitrarily) that all atomic constants must be truly constant, Dirac was left with no other option than to conclude that the gravitational constant itself must decrease as the universe ages: $G(t) = G_0(t_0/t)$, where t is time, t_0 is the age of the universe, and G_0 is the present experimentally measured value for the gravitational constant. Not only must the gravitational constant decrease as the universe ages, but the LNH, through relationship (11.11), also requires the continuous creation of matter throughout the universe — since N_{Edd} will have to keep pace with the time-increasing values of R_{ue} and F_{pe} — or *vice versa* depending upon one's perspective. In the case of the LNH, three numbers, having the same order of magnitude, have brought us a long way, and it has introduced several new and controversial ideas. The requirement of a decreasing gravitational constant with time is controversial since Einstein's theory of general relativity requires that it must, in fact, be constant, and additionally, there was (and still is) no experimental evidence to indicate that the gravitational constant has changed over the age of the universe. Additionally, the idea

[33] Over the past decade, a problematic tension has developed between the two primary methods for determining the value of Hubble's constant. The cosmic distance ladder approach yielding a value consistently larger than that based upon the method using the cosmic microwave background. The two methods are deemed to be appropriate and sound and should give the same value. That the values so derived cannot be brought into agreement has led some researchers to suggest that new physics will be required to explain the discrepancy.

of matter being continuously generated is controversial since, again, there is no observational evidence for this taking place, and, indeed, no known mechanism by which it might occur. Of course, no doubt, all the requirements of the new physics could, if one really wanted to, be accommodated by some new mathematical twist or another. For all this, however, neither Eddington's ideas nor Dirac's LNH are considered main-stream topics of research in the present era — the popularity (or unpopularity) of any given scientific idea, it should further be said, being no reasonable criteria upon which to judge of its correctness. While the great wisdom of Eddington and Dirac cannot be denied, it would appear that they both fell foul of the long shadow cast by the Second Strong Law of Small Numbers (although the numbers involved in their case are decidedly large). Indeed, as Eddington was later to write, "science is one thing, wisdom is another. Science is an edged tool, with which men play like children, and cut their own fingers".

11.5. Complete Mathematics

Given the real, observable universe is not some strange mathematical construct, and given that mathematics is a human invention, how is it that mathematics can provide such a good description of reality? At its heart, it would seem that mathematics is about patterns and regularities, and again, at its heart, the physical universe appears to be built around symmetries and patterns. Paul Lockhart has suggested in his book, *A Mathematicians Lament*, that, "mathematics is the art of abstract pattern-making — even more than story-telling, painting or music — is our most quintessentially human art form. This is what our brains do, whether we like it or not, we are biochemical pattern-recognition machines, and mathematics is nothing less than the distilled essence of *who we are*".

Under the auspices of Noether's theorem, introduced by Emmy Noether in the early 20th century, it is argued that any symmetry inherent to the action[34] of a physical system is equivalent to a conservation law. Symmetry here is taken to mean that some system characteristic does not change under a specific translation. So, for example, a translational symmetry in space corresponds to the law of the conservation of momentum, a rotational symmetry corresponds to the law of the conservation of angular momentum, and a time translational symmetry corresponds to the law of the conservation of energy. It is the ability of mathematics to describe the relationships, translations, and connections between patterns in a detailed and logical fashion that makes it so good at generating useful models of nature — conservation laws being especially important. This remarkable ability holds, however, only provided we concentrate on just that subset of physical problems for which a mathematical approach can be usefully applied. Derek Abbott has nicely written that "physical models are idealizations with the rough edges removed [and] our successful models are merely selected out from the many more failed ones" [9]. In other words, the only constraint that actually applies to the mathematics that has been invented, constructed, developed, and then applied within a model of some physical phenomenon, is that it provides predictions and output that agree with what is observed. We are back, inherently, to Plato's dictate that mathematics is there only to save the phenomenon.

Mathematics is constructed and founded upon an arbitrary, but well-thought-out and consistent, set of axioms — statements that are deemed to be fundamentally true, requiring no proof of correctness or indeed, real-world applicability. Like human rights and freedoms, axioms are self-evident truths. The axioms act to constrain what can happen in subsequent mathematical developments, and they allow for the consistent development of theorems. For all this, it turns out that no finite set of axioms can provide for the development of a

[34] Action here is an attribute of the physical system that can be minimized. Classical mechanics stipulates that of all the many possible trajectories that a physical system could follow, the one that it actually follows is the one in which the action attains its minimum value. This law is incapsulated in the Principle of Least Action.

complete[35] mathematical system in which every possible correct theorem can be proved true. This latter lack of completeness is encapsulated within Godel's incompleteness theorems — two theorems that shook the very core of the mathematical establishment when they were first published in 1931. Renowned mathematician David Hilbert called in 1920 for the development of a finite set of axioms from which all mathematics could be constructed. The system would be complete and consistent,[36] and based not upon arbitrary axioms, but axioms that could, "only be so and by no means otherwise". Hilbert's program was a grand notion, and it followed in the steps of *Principia Mathematica*, by Alfred Whitehead and Bertrand Russell (published in three volumes between 1910 and 1913), which inherently set out to reduce mathematics to symbolic logic and a minimum number of appropriate axioms.[37] The aims of Whitehead, Russell, and Hilbert were entirely laudable, and indeed, seem desirable, but Austrian mathematician Kurt Godel dashed all such schemes to ruin. Godel's first incompleteness theorem indicates that every possible formal (that is axiom-based) mathematical system must contain statements which are neither provable nor disprovable by the means allowed in the system. That is, any formal mathematical system must contain true statements for which no proof of its truth can be established. Godel's second incompleteness theorem indicates that if a formal mathematical system actually admits a proof of its consistency, then it must, in fact, be inconsistent. The upshot of Godel's theorems is that no finite list of axioms

[35] The idea behind completeness is that, within a given formal axiomatic system, the truth or fallacy of any legitimate mathematical statement can be established — there is no ambiguity, it is either one or the other.

[36] Consistency indicates the condition that no meaningful string of symbols can be constructed, within a mathematical system, that is both true and false at the same time — it must always be one or the other.

[37] One of its other aims was to rid set theory of any potential paradoxes. The classic paradox is that developed by Russell himself and proceeds as follows: Let R be the set of all sets that are not members of themselves. If R is not a member of itself, then its definition dictates that it must contain itself, and if it contains itself, then it contradicts its own definition as the set of all sets that are not members of themselves.

can guarantee that all true mathematical statements are provably true. While Godel's work undermined the whole idea of formalizing mathematics, it has had zero practical impact (so far) on how mathematicians actually work. No problem has ever been found for which Godel's incompleteness condition has surfaced. There are no inherently unsolvable problems, of course, because mathematicians and physicists largely work on those problems that can be solved through known mathematics. Philosopher Ludwig Wittgenstein dismissed the logician's game of building paradoxes along with the mathematicians' struggle for formalism, as meaningless exercises — they have no place in reality and there is no such thing as a physical problem (that is, a problem arising from observations made in the real universe) without a solution. The latter point essentially buys into the notion that mathematicians and physicists preselect those problems for which existing mathematics already allows solutions. And, if mathematics doesn't already exist to solve a problem, then new rules and new axioms are invented to fill in the gaps.

11.6. Wayward Mathematics?

Have physicists become beguiled by the beauty of mathematics? Is it the case that theoretical physicists have followed the enticing path of tractable (yet complicated) mathematical models too far — chasing not a physical reality, but rather some pure mathematical pipe-dream? The viewpoint and response to such questions are both charged and many-natured, and many cases can be made for and against the issue. Dirac, as described in Section 11.1, chased physical reality through a perceived (but not universally shared) chain of mathematical beauty, Eddington searched for physical reality in the abstract realm of pure reason. Neither found lasting success in all of their searches and/or in their methodologies.

The belief that mathematical analysis alone will lead to an understanding of physical reality has become nearly universal in the modern era. Certainly, the sciences are structured, and explored, and, perhaps more importantly, scientists are funded, on the basis of their immediate results, and the only accepted way of presenting such results is through the application and display of tables, diagrams, statistical regressions, equations, and mathematical models. This working methodology has proved highly successful in furthering the cause of science and scientists, but it is becoming increasingly questionable if it has actually advanced human understanding of the observable universe. Modern cosmology, as a case in point, is predicated on the ideas of inflation, dark matter, and dark energy, and yet none of these phenomena have any verified physical understanding. Yes, the present-day ΛCDM cosmological model is very good at explaining the observed properties of the universe, but it is none-the-less predicated on *ad hoc* and unproven (even unprovable) ideas (e.g., the inflation field), and a large dose of both wishful thinking and parameter fixing. The model, may, of course, turn out to be on the right track toward a good explanation, but in reality, it is just a stepping-stone (upon which we temporarily pause) toward the next stepping-stone approximation. Occam's razor, in the form of the simplest theory, is the best theory, is not a requirement or law of science, and indeed it is in many ways restrictive in the sense of developing and inventing new models and ideas to describe the workings of the cosmos. The universe is not simple, and nor will the theory and equations that describe it be simple in their interpretation and form. The modern mantra is that such august theories, all highly successful in their own way, as quantum mechanics, general relativity, and the standard cosmological model are beautiful because they can be described by beautiful mathematical equations — the beauty, however, is purely a self-aggrandizing prophecy propagated by, and entirely limited to, the physics community. A given theory is beautiful only because physicists (belonging to the accepted community) keep telling each other that it is beautiful. The beauty, so-called, is perceived only in the continued repetition of the word "beauty", and is not part of the actual message and/or meaning. Beauty is an artificial label simply added on once a theory has become successful — that is experimentally tested, with its predictions probed, and its precepts accepted by a majority of researchers. During the vital, imaginative, and daring development stage, the stage hidden away from public scrutiny, a theory is built upon the notion that anything that is not expressly forbidden mathematically or physically goes — it is an ugly and brutal affair.

String theory, as a theory of the fundamental structure of reality, is one area of research that once promised great returns, but, after nearly half a century of wrangling, has delivered little more than confusion and obscuration. Its many forms have been developed along purely mathematical grounds, and these have all set themselves up beyond any means of experimental verification. Indeed, current models of string theory place themselves firmly beyond the falsifiability criterion normally seen as a vital component of any body of scientific work. This latter condition, being introduced by Karl Popper in the mid-20th century, insists that a physical theory cannot, in fact, be proven correct — it is, at best, only provisionally correct even when it provides a working description of all the available data. The point is that the accumulation of more data, and deeper experimental probing at some future epoch, could reveal that a previously, thought to be sound, theory no longer provides a correct explanation for the new observations. Failing is not a bad thing — it means that a greater understanding is possible, indeed, necessary, and a specific theory can either be adjusted or refined to account for the new observations or else, it must be thrown out altogether — this latter situation being Eddington's supreme court of appeal at work. Popper accordingly argued that to be a valid scientific contribution a theory must contain one or more boilerplate conditions — points of failure if you will, that achievable experimental testing and observations can potentially find and reveal. Of course, it may be the case that our level of exploration of the physical world, accessible to actual experimentation, has reached a new level of complexity, and we will simply have to accept that some fundamental ideas are beyond any physical verification (ever). In this case, mathematical consistency perceived elegance and beauty (if you will), fashion, and popular opinion will have to carry the day. But if this is so, then it has also to be accepted that the emperor will need to change his non-existent clothes, on a repetitious basis, as one or another popular fad, fantasy, or fashion moves through the scientific community [10]. If this is so, then so be it, but it accordingly becomes increasingly important to remember that any mathematical representation is simply one of many possible ways of describing the properties of the observable universe. Equally, it is important to emphasize, then, that the mathematical representation cannot possibly be the thing itself — once again, we find that it is mathematics that is saving, but not being, the appearance.[38]

11.7. Alien Mathematics

If a lion could talk, reasoned philosopher Ludwig Wittgenstein, then we would not be able to understand him. His point is that language is based on culture, experience, and heritage. What is important, and relevant in the life, and mind of a lion, is something entirely different and alien to that which holds true for homo sapiens. Is mathematics, however, subject to the same cultural, and specific-being variation? Would an alien, landing on Earth, after traveling across the length of the galaxy recognize human-based mathematics? Although there is no written record that he actually made the suggestion, it is commonly said that Carl Friedrich Gauss proposed that one way to make contact with the intelligent life on Mars (then uniformly taken to be the case) would be to construct a giant triangular wheatfield in Siberia. Specifically, the argument goes, lay out a right-angled triangle for the wheat field and then plant square-formed forests along each edge — in this way, it would be clear to any Martian observer that the inhabitants on Earth knew about Pythagoras's theorem. While the Martians would not call such an arrangement Pythagorean, of course, they would recognize the mathematical properties of the structure, and thereby deduce the existence of intelligent beings on planet Earth. Later, Francis Galton in 1896 developed a three-term coding scheme[39] that might be used to communicate with the Martians, via a large helioscope arrangement. Once the attention of

[38] Indeed, it can be said that the power of mathematics resides in the very fact that it is not constrained by physical reality. Mathematics can move in many directions, and not all of them will have any contact with reality.

[39] The system was similar to that of Morse code but was composed of a dot, a dash, and a line, with each signal being defined according to a specified time duration.

the Martians had been obtained, by simple repeated light flashes, the first message to be transmitted was that indicating the numbers zero to nine. Nikola Tesla later still, in 1901, informed the world (without any proof being presented) that he had received radio signals from Mars, "brethren [he explained] we have a message from another world, unknown and remote. It reads ONE-TWO-THREE". Many decades later, in 1985, Carl Sagan in his novel *Contact* had the erstwhile aliens reveal their presence by transmitting a series of beeps that corresponded to a list of prime numbers. Mathematics, to Gauss, Galton, Tesla, Sagan, and many others before and since, is assumed to be universal. But, of course, given that mathematics is a human invention can it really be accessible and universally understood by other (alien) beings? Would a Martian or an alien from another planet in another star system even recognize human-made mathematics (not just its symbols, but its very methodology) as being meaningful? The answer to this is probably no, there is no specific reason why alien mathematics needs to look like, be similar to, or indeed, operate in the same way as human-derived mathematics. Francis Galton trivially recognized this fact in his 1896 sci-fi essay published in the *International Review*, and based his signaling plan on counting in modulo-8, rather than modulo-10. His reasoning for this is a little suspect, however, since it was based on the notion that the Martians were giant ant-like creatures (with six legs and two antennae). While there are physical phenomena in the observable universe that can be considered as universally acting — such as gravity — these phenomena can be described in any number of meaningful but totally different ways. Upon first contact, they, the aliens, will probably not recognize human-derived mathematical structures, and we will probably not recognize theirs, although there will likely be many cross-over similarities. Indeed, Stephen Wolfram has argued that humanity has navigated just one of the possible pathways that exist within the computational landscape and that other, equally as good, but entirely different pathways, could have been found and followed by alien mathematicians [11]. Once again, this identifies mathematics as being the specific tool that has been invented (by humans and aliens alike, and in their own way) to understand the universe. Underscoring the conclusion that when it comes to the human understanding of the universe, mathematics is the servant and not the master.

11.8. And Finally

For oft their quip and crank and fable is
Wondrous good and profitable.
And much deep subtle thought they hide
Neath veils torn easily aside.

Roman de la Rose, Guillarme de Lorris

It is important at this end-of-chapter stage to re-emphasize what I have been trying to explore in the above sections. First, there is no constitutional crisis, problem, or failure in present-day mathematics and/or theoretical physics. They are both absolutely phenomenal fields of study — vibrant, full of wonder, and awe. They are both truly inspiring fields of human endeavor, displaying incredible depth, exactitude, and breathtaking brilliance. Second, I am not advocating the abandonment of mathematics as the fundamental tool of scientific investigation. What is at issue, however, is the (seemingly growing) slavish attachment to mathematical expressions as being the only avenue for understanding the universe, and for establishing physical truths. In short, I am arguing that there is no Platonic universe of mathematical ideals in which everything that can possibly be imagined already exists in some perfect form — Plato's universe is a philosophical delusion and fantasy, albeit a comforting one. There is but one physical universe and we, as human beings, are privileged to be able to observe some of it, and in a few situations, it would appear that we actually understand some of its workings in the form of physical models described by, and inscribed within, the mathematical tools that human minds have abstracted and invented. For all this, however, when all is said and done, the real, observable world is complicated, twisted, brutal, fragmented, inelegant, and far removed from being a finely constructed, finely tuned, or just waiting to be discovered mathematical structure. Truth does not always hide within beauty, and physical reality is not always entrained within the lines of mathematical formulae.

References

[1] Mandelbrot, B. *The Fractal Geometry of Nature*. W. H. Freeman and Co., New York, 1977.
[2] Peitgen, H.-O. and Richter, P. *The Beauty of Fractals — Images of Complex Dynamical Systems*. Springer, New York, 1986.
[3] Lewis, C. *Miracles*. Geoffrey Bles, London, 1947.
[4] Guy, R. The strong law of small numbers. *American Mathematical Monthly*, **95**, 697–712, 1988.
[5] Guy, R. The second strong law of small numbers. *American Mathematical Monthly*, **63**, 3–20, 1990.
[6] Eddington, A. *The Philosophy of Physical Science*. Cambridge University Press, Cambridge, 1939.
[7] Beech, M. A lesson learnt from Eddington. *The Observatory*, **107**, 79–80, 1987.
[8] Dirac, P. Cosmological models and the large numbers hypothesis. *Proceedings of the Royal Society of London A*, **338**, 439–446, 1974.
[9] Abbott, D. The reasonable ineffectiveness of Mathematics. *Proceedings of the IEEE*, **101**, 2147–2153, 2013.
[10] Penrose, R. *Fashion, Faith and Fantasy in the New Physics of the Universe*. Princeton University Press, Princeton, 2016.
[11] Wolfram, S. *A New Kind of Science*. Wolfram Media, Champaign, 2002.

Afterword

All truly great thoughts are conceived while walking.

— Friedrich Nietzsche

When walking a labyrinth, it is easy to become lost — indeed, that is the point of an artificially made labyrinth. When traversing the labyrinth of history, it is easy to become overwhelmed, because that is the way history unfolds. In exploring the labyrinth that is the history of science, mathematics, and astronomy, it is important to remember that there are more wrong ways to move than there are correct ones, and projecting this forward, what we call "doing science" is a process made in a labyrinth with little illumination and with moving walls. In order, therefore, to traverse the labyrinth that the natural world presents to the senses, human beings have deployed the explorative tools of experimentation, analysis, hunches, guesswork, inference, and mathematics. Preeminent among these tools of exploration, however, is mathematics. Indeed, mathematics and the contemplation of the heavens have evolved and grown up together, the one drawing inspiration from the other. In this book we have mostly looked at the ways in which mathematics has been applied in an attempt to understand the origin of, and reasons behind the observed planetary orbits. The picture that has emerged from this study is that, while we have learned a great deal about the planetary system, there are aspects of its formation, operation, and evolution that are beyond the ability of mathematics to coral. Indeed, in contrast to ancient wisdom, mathematics is not the be-all and end-all of understanding, and there are indeed fundamental knowledge gaps that will never be bridged by any algebra. In part, these knowledge gaps exist because mathematics is purely a human construct, and partly because a stubborn, flawed, and misguided, development has come about in the modern era in which it is assumed that anything that is capable of being written down mathematically must have some physical counterpart. A very simple example of this disconnect between mathematics and the physical world has been pursued in this text, and this relates to the checkered history of the Titius–Bode rule. This rule, stubbornly still presented in astronomy textbooks to this day, has received far more attention and ink than it deserves, and when all is said and done, it has resulted in much time-wasting and unnecessary public (even professional) confusion. And, at the heart of the impotency behind the Titius–Bode rule is a belief in the idea that simple mathematical expressions belie the structure of reality. In contrast, we insist that the solar system is not just a complex structure, but that it is a structure that mathematics will never fully describe. Indeed, it is time to do away in general with the naive notion that the universe is some simple, beautiful, and entirely mathematical construct. Indeed, it is none of these things.

We live in remarkable times (as everyone throughout human history has expressed), but perhaps the most remarkable scientific result of modern times is the way in which string theory has ensnared the thoughts of its practitioners. The entire history of string theory has been one of squirming from one mathematical construct to another, with the result that it has produced nothing of any seeming value — indeed, it has only succeeded in producing an obscured and inherently pointless picture of what our universe might be. We do not live in, and nor is our universe a part of, the string field *Multiverse Landscape*. This *Landscape*, composed of a supposed 10^{500} (or even higher!) possible string field formulations, is really a fog, or miasma, evanesced from the ever-flexible wrappings of mathematical molding. The chaos of the *Landscape* exists purely because of the infinite twisting and undulations allowed and contrived within mathematics (which, recall is a human construct), not because theoretical physicists have discovered the way in which the universe actually works. This is not necessarily a bad thing; it is just the way things are if we want to have some framework on which to base our ignorance. Indeed, given the belligerent string field stance, we are forced to conclude that there are no specific laws of physics, just infinite variations depending upon where and when one resides in the infinite *Landscape*, and we are led to the inevitable conclusion that there truly is no theory of everything, only endless mathematical moldings, and endless gaps in our knowledge. If we are to accept the *Landscape* as being real then we must also accept that there are no fundamental particles and no fundamental forces. Again, this is not necessarily a bad thing, but the danger is that we actually fool ourselves into thinking that fundamental knowledge has been gained. What has really been found is an endlessly flexible mathematical description of our failure to fully understand the actual universe in which we live. Indeed, one questions if it is time to redefine physics as being nothing more than the study of a distinct patchwork of mathematical models (constructed by human beings) that provide useful descriptions of the way in which things work, in a particular field, to the level that they can be experimentally observed. Such an approach actually accounts for what most of the history of science has been, and there seems no reason to suppose that such a similar unfolding won't continue into the forever-future. What this further tells us, is that the supposed knowledge gap called the theory of everything is really a chimera, and nothing more than a surrealist dream.

Definition: *honest*: truthful, trustworthy; fairly earned.

There has been, in several sections of this book, a sense of the aged curmudgeon on a rant — the grumpy old professor emeritus venting his displeasures and frustrations concerning the ways in which the contemporary scientific discourse is developing. I am not the first, and certainly won't be the last such complainant. Indeed, I have, on occasion, knowingly overstated some arguments and taken an unpopular, that is non-standard, position to make a point. I leave it as an exercise to the reader to double-check these transgressions. Why, however, have I taken such a tack, when running with the wind and the fast-flowing tide is so much easier? The reasons are partly encapsulated in the famous observation by Arthur C. Clarke, who wisely noted in the first of what are now known as his four laws, "when a distinguished but elderly scientist states that something is possible, he is almost certainly right. When he states that something is impossible, he is very probably wrong". I make no affectation to distinction, but after many decades of being *in* and *doing* science I fear that it is moving in directions that no longer make sense. But the new generation will have to make of it what they will. Astronomy, cosmology, and elementary particle physics in particular, have become ensnared within the straight-jacket of mathematics and computation theory. Astronomers, for example, rarely (as human beings) observe the heavens anymore, but find themselves sitting in windowless offices running endless computer simulations, arguing over the finer points of Bayesian priors, and bringing forth ought of any real value. But I am old, and bow (just a little) to the wisdom outlined within the second of Clarke's laws, which states that, "the only way of discovering the limits of the possible is to venture a little way past them into the impossible". The only proviso I would add to this fine argument is to not spend all of one's time in the realm of the impossible. Sometimes the answer lies in the obvious, the ordinary, and the pedestrian, and to insist upon going beyond these is to merely tilt at imaginary (even self-made) windmills. Sometimes, as human beings, we can be too clever for our own good. Clarke's third law, "any sufficiently advanced technology is indistinguishable to magic", also works if one substitutes the word *idea* for *technology*, and again my fear (and it is a fear) in modern times is that far too much effort is being invested in the magic — magic which, after all is said and done, is all about the slight of hand, or more to the point, the trick of some mathematical equation. Having said this, I further bow to Clarke's fourth law, which is perhaps the most humanistic of his laws, "for every expert, there is an equal and opposite expert". For all this, we exist in a weird and wonderful universe. A universe, however, that contains structures that lie beyond the grasp of mathematical knowledge — there are gaps, gaps to wide, gaps to subtle, and gaps to chaotic, that they will never be bridge. There is nothing wrong in accepting the existence of such limitations, it is not a surrender of the imagination, but it is the recognition of who we are, what we can know, and what the wonderful universe really is.

Printed in the USA
CPSIA information can be obtained
at www.ICGtesting.com
LVHW010320280923
759473LV00001B/2